S

M

Sensors and
Control Systems
in Manufacturing

Sensors and Control Systems in Manufacturing

Sabrie Soloman

McGraw-Hill, Inc.

New York   San Francisco   Washington, D.C.   Auckland   Bogotá
Caracas   Lisbon   London   Madrid   Mexico City   Milan
Montreal   New Delhi   San Juan   Singapore
Sydney   Tokyo   Toronto

Library of Congress Cataloging-in-Publication Data

Soloman, Sabrie.
   Sensors and control systems in manufacturing / Sabrie Soloman.
      p.   cm.
      Includes bibliographical references and index.
      ISBN 0-07-059626-3
      1. Automatic data collection systems.   2. Process control—
Automation. 3. Detectors. I. Title.
TS158.6.S55   1994
670.42′7—dc20   93-46490
                                          CIP

1 2 3 4 5 6 7 8 9 0 DOC/DOC 9 0 9 8 7 6 5 4

ISBN 0-07-059626-3

The sponsoring editor for this book was Harold B. Crawford, the editing supervisor was Caroline Levine, and the production supervisor was Donald F. Schmidt. This book was set in Century Schoolbook by McGraw-Hill's Professional Book Group composition unit.

Printed and bound by R. R. Donnelley & Sons Company.

This book is printed on acid-free paper.

This book is dedicated to my two brothers Drs. Shoukry and Nasier Soloman who played a vital role in my career, and to my late parents who lovingly paved my way by their endless support— intellectually, emotionally, and spiritually. My gratitude also to my sister Mrs. Laurice Kelada and my brothers Messrs. Soloman and Boushra Soloman.

Contents

Chapter 3. Networking of Sensors and Control Systems in Manufacturing

Chapter 5. Advanced Sensor Technology in Production Manufacturing Applications

Foreword

The ongoing revolution in manufacturing, well into its second decade, has now (albeit somewhat belatedly) become recognized and joined by government, industry, and academe in this country. It is based on a number of concepts which have made their way into the professional jargon and have been brought to the public's awareness by technical and business writers: concurrent engineering, flexible manufacturing, just-in-time inventory, automation, and quality engineering. Each of these are ingredients that contribute to the ultimate goal, which, simply stated, is to achieve the highest quality products at the lowest possible cost, and to do so in a timely fashion. A tall order, this, but one on which depend the welfare of a host of individual companies and, even more importantly, the economic health of entire countries, with political and social implications beyond overstatement.

A principal ingredient in the process, perhaps the most important one, is the achievement and implementation of error-free production, at one and the same time a guarantor of quality and a minimizer of waste of materials and labor. At first impression, the term "error-free" will sound like a pious ideal, to be striven for but impossible to attain. A moment of reflection will persuade, however, that the aim need not be a philosophical abstraction. In the final analysis, it is the end product *alone* that must fall within the range of prescribed tolerances, not each of the many steps in the production process. That is to say, given within the context of computer-integrated manufacturing a sufficient array of monitors distributed throughout the workspace—i.e., sensors (and appropriate means to feed back and respond to, in real time, the information gathered by them) and control systems which can identify, rectify, or remove defects in the course of production—every item that reaches the end of the production line will be, ipso facto, acceptable.

The book before the reader contains not only an exposition of sensors and controls, but a host of invaluable asides, comments, and extended discourses on key topics of modern manufacturing. The author is an active practitioner of advanced manufacturing techniques and a highly regarded teacher of the subject. In this volume, he brings error-free production a step nearer to realization; the world of manufacturing engineering owes him a debt of gratitude.

H. Deresiewicz
Chairman
Department of Mechanical Engineering
Columbia University

Preface

Rising costs. Shorter lead time. Complex customer specifications. Competition from across the street—and around the world.

Business today faces an ever-increasing number of challenges. The manufacturers that develop more effective and efficient forms of production, development, and marketing, will be the ones who meet these challenges.

The use of advanced sensors and control technology makes a fundamental commitment to manufacturing solutions based on simple and affordable integration. With this technology, one can integrate manufacturing processes, react to rapidly changing production conditions, help personnel to react more effectively to complex qualitative decisions, and lower the cost of and improve product quality throughout the manufacturing enterprise.

The first step in achieving such flexibility is to establish an information system that can be reshaped whenever necessary, thus enabling it to respond to the changing requirements of the enterprise—and the environment. This reshaping must be accomplished with minimal cost and disruption to the ongoing operation.

Sensors and control technology will play a key role in achieving flexibility in the information system. However, this technology alone can not shorten lead time, reduce inventories, and minimize excess capacity to the extent required by today's manufacturing operation. This can be accomplished only by integrating various sensors with appropriate control means throughout the manufacturing operation within *computer integrated manufacturing* (CIM) strategy. The result is that individual manufacturing processes will be able to flow—communicate—and respond together as a unified cell, well structured for their functions.

In order to develop a sensory and control information system that will achieve these objectives, the enterprise must start with a specific long-range architectural strategy, one providing a foundation that accommodates today's needs as well as taking those of tomorrow's—including the support of new manufacturing processes, incorporating new data functions, and establishing new data bases and distributed channels—into account. The tools for this control and integration are available today.

The United States leads the world in inventing new products; however, many of these new products, ultimately, are manufactured by other countries. The inability of U.S. manufacturers to compete globally cannot be blamed on low-cost labor in other countries; more than one-half the trade deficit comes from foreign industries that pay higher wages.* The inability to apply affordable manufacturing systems for automation can be a contributing factor to this dilemma.

An *affordable manufacturing system* is, simply, a system that contains a variety of reliable parts, harmoniously joined together to generate a specific motion that will achieve a particular manufacturing operation, directed and controlled by simple and effective sensors and control systems. Modern manufacturing technology is prevalent in the design and control of engineering systems and also in applying the sensory and control technology in production systems to situations such as product fabrication and assembly.

When considering the design aspect of a system the following must be taken into account: dynamics, kinematics, statics, and even styling of parts. All of these play a vital role in forming optimum manufacturing parameters in system design. Motion generation and control through various sensors provides a review of manufacturing engineering concepts from a system's point of view, directed towards the manufacturing engineering problems.

Manufacturing engineering plays a key role in translating new product specifications from design engineering into process plans which are then used to manufacture the product. As the product is being designed, manufacturing technical evaluators work with design engineers to ascertain if sensors and control systems for process monitoring and control can be integrated into the design of a system, and at what cost. Tolerances, materials, clearances, appropriate handling of parts, acceptable types and positioning of sensors, and product assembly times are particularly important factors in this evaluation because they directly affect productivity, the guidelines of which are essential to this analysis.

Cost estimates are equally important. If a new process is needed because, for example, existing processes are deemed too expensive or incapable of producing the desired product, process engineers would

---

*"Back to Basics," *Business Week*, June 16, 1989.

be asked to either develop a new economical process or to change product design. Choosing the best alternative could be very difficult because such a decision is based on many conflicting objectives, e.g., customer specifications, cost, feasibility, timeliness, market share, parts availability, standard parts, availability of part tooling, sensors locations, and the like. A well-designed system incorporating sensors and control technology for the generation of motion for a manufacturing system plays a fundamental role in the new manufacturing thinking.

Advanced sensory and control technology is more than an implementation of new technologies. It is a long-range strategy that allows components of the entire manufacturing operation to work together to achieve the qualitative and quantitative goals of business. It must have top management commitment and may entail changing the mind-set of people in the organization and managing that change. However, the rewards are great since successful implementation of this technology is, in large measure, responsible for the success of computer-integrated manufacturing strategy today.

Sabrie Soloman

Acknowledgments

Mere thanks is insufficient to Ms. Sherrie L. Watkins for her immeasurable efforts in reviewing and preparing the graphics for this book. This book was also made possible by the efforts of my colleagues and friends in various universities and industries, and by the encouragement and wisdom of Dr. H. Deresiewicz of Columbia University.

1

Introduction

Integrated sensors and control systems are the way of the future. In times of disaster, even the most isolated outposts can be linked directly into the public telephone network by portable versions of satellite earth stations called *very small aperture terminals* (VSATs). They play a vital role in relief efforts such as those for the eruption of Mount Pinatubo in the Philippines, the massive oil spill in Valdez, Alaska, the 90,000-acre fire in the Idaho forest, and Hurricane Andrew's destruction in south Florida and the coast of Louisiana.

VSATs are unique types of sensors and control systems. They can be shipped and assembled quickly and facilitate communications by using more powerful antennas that are much smaller than conventional satellite dishes. These types of sensors and control systems provide excellent alternatives to complicated conventional communication systems, which in disasters often experience serious degradation because of damage or overload.

Multispectral sensors and control systems will play an expanding role to help offset the increasing congestion on America's roads by creating "smart" highways. At a moment's notice, they can gather data to help police, tow trucks, and ambulances respond to emergency crises. Understanding flow patterns and traffic composition would also help traffic engineers plan future traffic control strategies. The result of less congestion will be billions of vehicle hours saved each year.

The spacecraft Magellan, Fig. 1.1, is close to completing its third cycle of mapping the surface of planet Venus. The key to gathering data is the development of a synthetic aperture radar as a sensor and information-gathering control system, the sole scientific instrument aboard Magellan. Even before the first cycle ended, in mid-1991, Magellan had mapped 84 percent of Venus' surface, returning more digital data than all previous U.S. planetary missions combined, with

Figure 1.1 The Spacecraft Magellan. (*Courtesy Hughes Corp.*)

resolutions 10 times better than those provided by earlier missions. To optimize radar performance, a unique and simple computer software program was developed, capable of handling the nearly 950 commands per cycle. Each cycle takes one venusian day, the equivalent of 243 Earth days.

Manufacturing organizations in the United States are under intense competitive pressure. Major changes are being experienced with respect to resources, markets, manufacturing processes, and product strategies. As a result of international competition, only the most productive and cost-effective industries will survive.

Today's sensors and control systems have explosively expanded beyond their traditional production base into far-ranging commercial ventures. They will play an important role in the survival of innovative industries. Their role in information assimilation, and control of operations to maintain an error-free production environment, will help enterprises to stay effective on their competitive course.

1.1 Establishing an Automation Program

Manufacturers and vendors have learned the hard way that technology alone does not solve problems. A prime example is the gap between the information and the control worlds, which caused production planners to set their goals according to dubious assumptions concerning plant-floor activities, and plant supervisors then could not isolate production problems until well after they had arisen.

The problem of creating effective automation for an error-free production environment has drawn a long list of solutions. Some are as old as the term *computer-integrated manufacturing* (CIM) itself. However, in many cases, the problem turned out to be not technology, but the ability to integrate equipment, information, and people.

The debate over the value of computer-integrated manufacturing technology has been put to rest, although executives at every level in almost every industry are still questioning the cost of implementing CIM solutions. Recent economic belt tightening has forced industry to justify every capital expense, and CIM has drawn fire from budget-bound business people in all fields.

Too often, the implementations of CIM have created a compatibility nightmare in today's multivendor factory-floor environments. Too many end users have been forced to discard previous automation investments and/or spend huge sums on new equipment, hardware, software, and networks in order to effectively link together data from distinctly dissimilar sources. The expense of compatible equipment and the associated labor cost for elaborate networking are often prohibitive.

The claims of CIM open systems are often misleading. This is largely due to proprietary concerns, a limited-access database, and operating system compatibility restrictions. The systems fail to provide the transparent integration of process data and plant business information that makes CIM work.

In order to solve this problem, it is necessary to establish a clearly defined automation program. A common approach is to limit the problem description to a workable scope, eliminating the features that are not amenable to consideration. The problem is examined in terms of a simpler, workable model. A solution can then be based on model predictions.

The danger associated with this strategy is obvious: if the simplified model is not a good approximation of the actual problem, the solution will be inappropriate and may even worsen the problem.

Robust automation programs can be a valuable asset in deciding how to solve production problems. Advances in sensor technology have provided the means to make rapid, large-scale improvements in problem solving and have contributed in essential ways to today's manufacturing technology.

The infrastructure of an automation program must be closely linked with the use and implementation of sensors and control systems, within the framework of the organization. The problem becomes more difficult whenever it is extended to include the organizational setting. Organization theory is based on a fragmented and partially developed body of knowledge, and can provide only limited guidance in the formation of problem models. Managers commonly use their experience and instinct in dealing with complex production problems that include organizational aspects. As a result, creating a competitive manufacturing enterprise—one involving advanced automation technology utilizing sensors and control systems and organizational aspects—is a task that requires an understanding of both how to establish an automation program and how to integrate it with a dynamic organization.

In order to meet the goals of integrated sensory and control systems, an automated manufacturing system has to be built from compatible and intelligent subsystems. Ideally, a manufacturing system should be computer-controlled and should communicate with controllers and materials-handling systems at higher levels of the hierarchy as shown in Fig. 1.2.

1.2 Understanding Workstations, Work Cells, and Work Centers

Workstations, work cells, and work centers represent a coordinated cluster of a production system. A production machine with several processes is considered a workstation. A machine tool is also considered a workstation. Integrated workstations form a work cell. Several complementary workstations may be grouped together to construct a work cell. Similarly, integrated work cells may form a work center. This structure is the basic concept in modeling a flexible manufacturing system. The flexible manufacturing system is also the corner stone of the computer-integrated manufacturing strategy (Fig. 1.3).

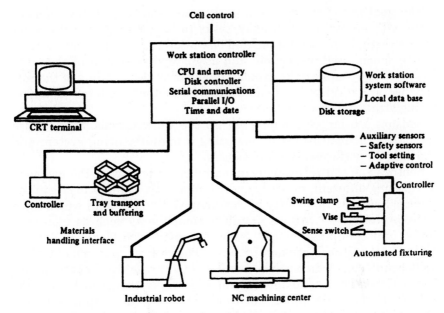

Figure 1.2   Computer-controlled manufacturing system.

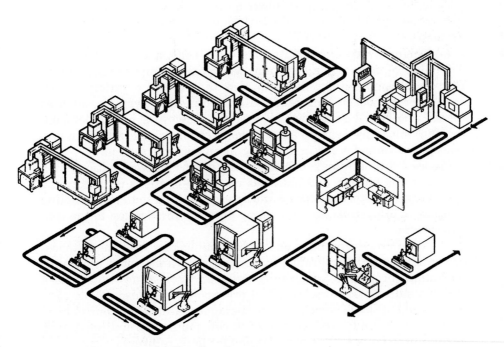

Figure 1.3   Workstation, work cell, and work center.

The goal is to provide the management and project development team with an overview of major tasks to be solved during the planning, design, implementation, and operation phases of computer-integrated machining, inspection, and assembly systems. Financial and technical disasters can be avoided if a clear understanding of the role of sensors and control systems in the computer-integrated manufacturing strategy is asserted.

Sensors are largely applied within the workstations. Sensors are the only practical means of operating a manufacturing system and tracking its performance continuously.

Sensors and control systems in manufacturing provide the means of integrating different, properly defined processes as input to create the expected output. Input may be raw material and/or data which have to be processed by various auxiliary components such as tools, fixtures, and clamping devices. Sensors provide the feedback data to describe the status of each process. The output may also be data and/or materials which can be processed by further cells of the manufacturing system. A flexible manufacturing system, which contains workstations, work cells, and work centers and is equipped with appropriate sensors and control systems, is a distributed management information system, linking together subsystems of machining, packaging, welding, painting, flame cutting, sheet-metal manufacturing, inspection, and assembly with material-handling and storage processes.

In designing various workstations, work cells, and work centers in a flexible manufacturing system, within the computer-integrated manufacturing strategy, the basic task is to create a variety of sensors interconnecting different material-handling systems, such as robots, automated guided-vehicle systems, conveyers, and pallet loading and unloading carts, to allow them to communicate with data processing networks for successful integration with the system.

Figure 1.4 illustrates a cell consisting of several workstations with its input and output, and indicates its basic functions in performing the conversion process, storing workpieces, linking material-handling systems to other cells, and providing data communication to the control system.

The data processing links enable communication with the databases containing part programs, inspection programs, robot programs, packaging programs, machining data, and real-time control data through suitable sensors. The data processing links also enable communication of the feedback data to the upper level of the control hierarchy. Accordingly, the entire work-cell facility is equipped with current data for real-time analysis and for fault recovery.

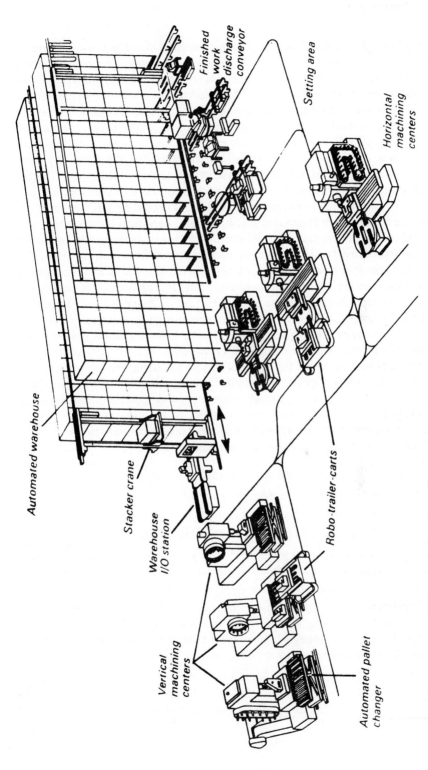

Figure 1.4 Conversion process in a manufacturing cell.

Automated warehouse

Finished work discharge conveyor

Setting area

Horizontal machining centers

Stacker crane

Warehouse I/O station

Robo-trailer-carts

Vertical machining centers

Automated pallet changer

7

A cluster of manufacturing cells grouped together for particular production operations is called a *work center*. Various work centers can be linked together via satellite communication links irrespective of the location of each center. Manufacturing centers can be located several hundred feet apart or several thousand miles apart. Adequate sensors and control systems together with effective communication links will provide practical real-time data analysis for further determination.

The output of the cell is the product of the module of the flexible manufacturing system. It consists of a finished or semifinished part as well as data in a computer-readable format that will instruct the next cell how to achieve its output requirement. The data are conveyed through the distributed communication networks. If, for example, a part is required to be surfaced to a specific datum in a particular cell, sensors will be adjusted to read the required acceptable datum during the surfacing process. Once the operation is successfully completed, the part must once again be transferred to another cell for further machining or inspection processes. The next cell is not necessarily physically adjacent; it may be the previous cell, as programmed for the required conversion process.

The primary reason for the emphasis on integrating sensors and control systems into every manufacturing operation is the worldwide exponentially increasing demand for error-free production operations. Sensors and control technology can achieve impressive results only if effectively integrated with corporate manufacturing strategy.

The following benefits can be achieved:

1. *Productivity.*   A greater output and a lower unit cost.

2. *Quality.*   Product is more uniform and consistent.

3. *Production reliability.*   The intelligent, self-correcting sensory and feedback system increases the overall reliability of production.

4. *Lead time.*   Parts can be randomly produced in batches of one or in reasonably high numbers, and the lead time can be reduced by 50 to 75 percent.

5. *Expenses.*   Overall capital expenses are 5 to 10 percent lower. The cost of integrating sensors and feedback control systems into the manufacturing source is less than that of stand-alone sensors and feedback system.

6. *Greater utilization.*   Integration is the only available technology with which a machine tool can be utilized as much as 85 percent of the time—and the time spent cutting can also be over 90 percent.

In contrast, a part, from stock to finished item, spends only 5 percent of its time on the machine tool, and actual productive work takes only 30 percent of this 5 percent. The time for useful work on stand-alone machines without integrated sensory and control systems is as little 1 to 1.5 percent of the time available (see Tables 1.1 and 1.2).

To achieve the impressive results indicated in Table 1.1, the integrated manufacturing system carrying the sensory and control feedback systems must maintain a high degree of flexibility. If any cell breaks down for any reason, the production planning and control system can reroute and reschedule the production or, in other words, reassign the system environment. This can be achieved only if both the processes and the routing of parts are programmable. The sensory and control systems will provide instantaneous descriptions of the status of parts to the production and planning system.

If different processes are rigidly integrated into a special-purpose, highly productive system such as a transfer line for large batch production, then neither modular development nor flexible operation is possible.

TABLE 1.1 Time Utilization of Integrated Manufacturing Center Carrying Sensory and Control Systems

	Active, %	Idle, %
Tool positioning and tool changing	25	
Machining process	5	
Loading and inspection	15	
Maintenance	20	
Setup	15	
Idle time		15
Total	85	15

TABLE 1.2 Productivity Losses of Stand-alone Manufacturing Center Excluding Sensory and Control Systems

	Active, %	Idle, %
Machine tool in wait mode		35
Labor control		35
Support services		15
Machining process	15	
Total	15	85

However, if the cells and their communication links to the outside world are programmable, much useful feedback data may be gained. Data on tool life, measured dimensions of machined surfaces by in-process gauging and production control, and fault recovery derived from sensors and control systems can enable the manufacturing system to increase its own productivity, learn its own limits, and inform the part programmers of them. The data may also be very useful to the flexible manufacturing system designers for further analysis. In non-real-time control systems, the data cannot usually be collected, except by manual methods, which are time-consuming and unreliable.

1.3 Classification of Control Processes

An *engineering integrated system* can be defined as a machine responsible for certain production output, a controller to execute certain commands, and sensors to determine the status of the production processes. The machine is expected to provide a certain product as an output, such as computer numerical control (CNC) machines, packaging machines, and high-speed press machine. The controller provides certain commands arranged in a specific sequence designed for a particular operation. The controller sends its commands in the form of signals, usually electric pulses. The machine is equipped with various devices, such as solenoid valves and step motors, that receive the signals and respond according to their functions. The sensors provide a clear description of the status of the machine performance. They give detailed accounts of every process in the production operation (Fig. 1.5).

Once a process is executed successfully, according to a specific sequence of operations, the controller can send additional commands for further processes until all processes are executed. This completes one cycle. At the end of each cycle a command is sent to begin a new loop until the production demand is met.

In an automatic process, the machine, the controller, and the sensors interact with one another to exchange information. Mainly, there are two types of interaction between the controller and the rest of the system: through either an open-loop control system or a closed-loop control system.

An *open-loop control system* (Fig. 1.6) can be defined as a system in which there is no feedback. Motor motion is expected to faithfully follow the input command. Stepping motors are an example of open-loop control.

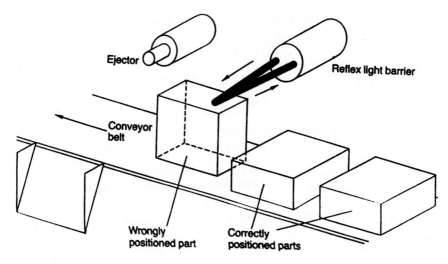

Figure 1.5 Sensors providing machine status.

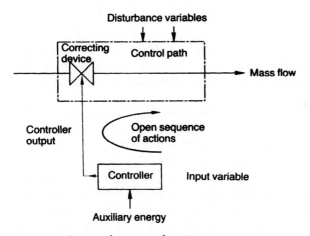

Figure 1.6 An open-loop control system.

A *closed-loop control system* (Fig. 1.7) can be defined as a system in which the output is compared to the command, with the result being used to force the output to follow the command. Servo systems are an example of closed-loop control.

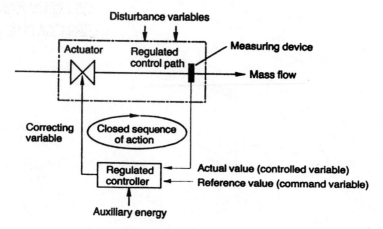

Figure 1.7   A closed-loop control system.

1.4   Open- and Closed-Loop Control Systems

In an open-loop control system, the actual value in Fig. 1.6 may differ from the reference value in the system. In a closed-loop system, the actual value is constantly monitored against the reference value described in Fig. 1.7.

The mass flow illustrated in Fig. 1.8 describes the amount of matter per unit time flowing through a pipeline that must be regulated. The current flow rate can be recorded by a measuring device, and a correcting device such as a valve may be set to a specific flow rate. The system, if left on its own, may suffer fluctuations and disturbances which will change the flow rate. In such an open-loop system, the reading of the current flow rate is the *actual value,* and the *reference value* is the desired value of the flow rate. The reference value may differ from the actual value, which then remains unaltered.

If the flow rate falls below the reference value because of a drop in pressure, as illustrated in Fig. 1.9, the valve must be opened further to maintain the desired actual value. Where disturbances occur, the course of the actual value must be continuously observed. When adjustment is made to continuously regulate the actual value, the loop of action governing measurement, comparison, adjustment, and reaction within the process is called a *closed loop.*

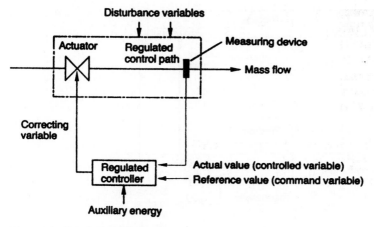

Figure 1.8   Regulation of mass flow.

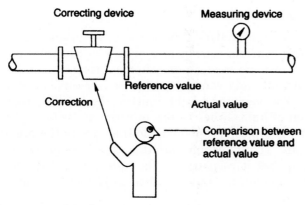

Figure 1.9   Reference value.

1.4   Understanding Photoelectric Sensors

In order to successfully automate a process, it is necessary to obtain information about its status. The sensors are the part of the control system which is responsible for collecting and preparing process status data and for passing it onto a processor (Fig. 1.10).

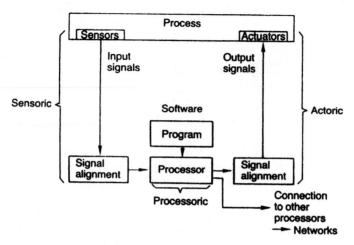

Figure 1.10   Components of controlled process.

1.4.1   Principles of operation

Photoelectric controls use light to detect the presence or absence of an object. All photoelectric controls consist of a sensor, a control unit, and an output device. A logic module or other accessories can be added to the basic control to add versatility. The sensor consists of a source and a detector. The source is a light-emitting diode (LED) that emits a powerful beam of light either in the infrared or visible light spectrum. The detector is typically a photodiode that senses the presence or absence of light. The detection amplifier in all photoelectric controls is designed so that it responds to the light emitted by the source; ambient light, including sunlight up to 3100 metercandles, does not affect operation.

The source and detector may be separated or may be mounted in the same sensor head, depending on the particular series and application (Fig. 1.11).

The control unit modulates and demodulates the light sent and

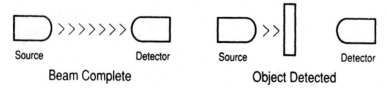

Figure 1.11   Components of photoelectric control.

received by the source and detector. This assures that the photoelectric control responds only to its light source. The control unit also controls the output device in *self-contained* photoelectric controls; the control unit and sensor are built into an integral unit.

Controls can be configured to operate as light-actuated devices. The output is triggered when the detector sees light. They can also be dark-actuated devices, where the output is triggered when the detector does not see light.

Output devices may include relays such as *double pole, double throw* (DPDT) and *single pole, double throw* (SPDT). Output devices may also include a triac or other high-current device and may be programmable-controller–compatible.

Logic modules are optional devices that allow addition of logic functions to a photoelectric control. For example, instead of providing a simple ON/OFF signal, a photoelectric control can (with a logic module) provide time-delay, one-shot, retriggerable one-shot, motion-detection, and counting functions.

1.4.2 Manufacturing applications of photodetectors

The following applications of photoelectric sensors are based on normal practices at home, at the workplace, and in various industries. The effective employment of photoelectric sensors can lead to successful integration of data in manufacturing operations to maintain an error-free environment and assist in obtaining instantaneous information for dynamic interaction.

A photoelectric sensor is a semiconductor component that reacts to light or emits light. The light may be either in the visible range or the invisible infrared range. These characteristics of photoelectric components have led to the development of a wide range of photoelectric sensors.

A photoelectric reflex sensor equipped with a time-delay module set for *delay dark* ignores momentary beam breaks. If the beam is blocked longer than the predetermined delay period, the output energizes to sound an alarm or stop the conveyer (Fig. 1.12).

A set of photoelectric through-beam sensors can determine the height of a scissor lift as illustrated in Fig. 1.13. For example, when the control is set for *dark-to-light* energizing, the lift rises after a layer has been removed and stops when the next layer breaks the beam again.

Cans on a conveyer are diverted to two other conveyers controlled by a polarized photoelectric reflex sensor with a divider module (Fig. 1.14). Items can be counted and diverted in groups of 2, 6, 12, or 24. A

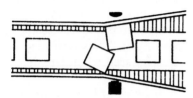

Figure 1.12 Jam detection with photoelectric sensor.

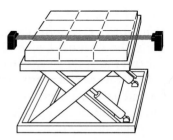

Figure 1.13 Stack height measurement with photoelectric sensor.

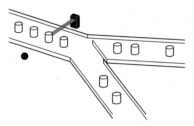

Figure 1.14 Batch counting and diverting with photoelectric sensor.

polarized sensor is used so that shiny surfaces may not falsely trigger the sensor control.

Two photoelectric control sensors can work together to inspect a fill level in cartons on a conveyer (Fig. 1.15). A reflex photoelectric sensor detects the position of the carton and energizes another synchronized photoelectric sensor located above the contents. If the photoelectric sensor located above the carton does not "see" the fill level, the carton does not pass inspection.

A single reflex photoelectric sensor detects boxes anywhere across the width of a conveyer. Interfacing the sensor with a programmable controller provides totals at specific time intervals (Fig. 1.16).

High-temperature environments are accommodated by the use of fiber optics. The conveyer motion in a 450°F cookie oven can be detected as shown in Fig. 1.17. If the motion stops, the one-shot logic module detects light or dark that lasts too long, and the output device shuts the oven down.

Placing the photoelectric sensor to detect a saw tooth (Fig. 1.18)

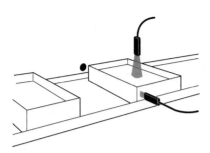

Figure 1.15 Measuring carton fill with photoelectric sensor.

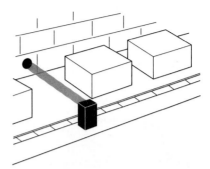

Figure 1.16 Box counting with photoelectric sensor.

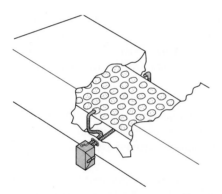

Figure 1.17 Detecting proper cookie arrangement.

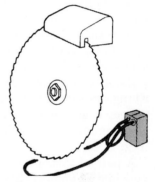

Figure 1.18 Sawtooth inspection.

enables the programmable controller to receive an input signal which rotates the blade into position for sharpening of the next tooth.

A through-beam photoelectric sensor is used to time the toll gate in Fig. 1.19. To eliminate toll cheating, the gate lowers the instant the rear of the paid car passes the control. The rugged sensor can handle harsh weather, abuse, and 24-h operation.

A safe and secure garage is achieved through the use of a through-beam photoelectric sensor interfaced to the door controller. The door shuts automatically after a car leaves, and if the beam is broken while the door is lowering, the motor reverses direction and raises the door again (Fig. 1.20).

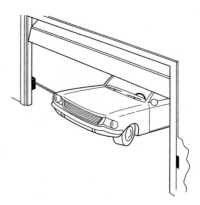

Figure 1.19 Toll-booth control with photoelectrical sensor.

Figure 1.20 Garage door control with photoelectric control.

A photoelectric sensor that generates a "curtain of light" detects the length of a loop on a web drive system by measuring the amount of light returned from an array of retroreflectors. With this information, the analog control unit instructs a motor controller to speed up or slow down the web drive (Fig. 1.21).

Small objects moving through a curtain of light are counted by a change in reflected light. A low contrast logic module inside the photoelectric sensor unit responds to slight but abrupt signal variations while ignoring slow changes such as those caused by dust buildup (Fig. 1.22).

A pair of through-beam photoelectric sensors scan over and under multiple strands of thread. If a thread breaks and passes through one of the beams, the low-contrast logic module detects the sudden changes in signal strength and energizes the output. As this logic module does not react to slow changes in signal strength, it can operate in a dusty environment with little maintenance (Fig. 1.23).

A remote photoelectric source and detector pair inspects for passage of light through a hypodermic needle (Fig. 1.24). The small, waterproof stainless-steel housing is appropriate for crowded machinery spaces and frequent wash-downs. High signal strength allows quality inspection of hole sizes down to 0.015 mm.

Index marks on the edge of a paper are detected by a fiber-optic photoelectric source/detector sensor to control a cutting shear down line (Fig. 1.25).

Liquids are monitored in a clear tank through beam sensors and an analog control. Because the control produces a voltage signal proportional to the amount of detected light, liquid mixtures and densities can be controlled (Fig. 1.26).

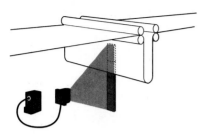

Figure 1.21   Web loop control.

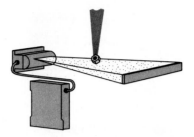

Figure 1.22   Small parts detection.

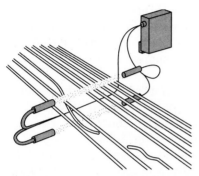

Figure 1.23   Broken-thread detection.

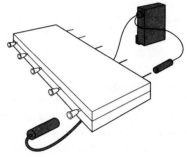

Figure 1.24   Hypodermic needle quality assurance.

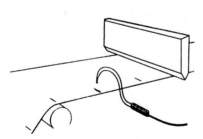

Figure 1.25   Indexing mark detection.

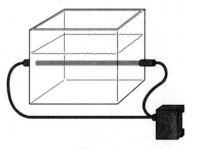

Figure 1.26   Liquid clarity control.

Remote photoelectric sensors inspect for the presence of holes in a metal casting (Fig. 1.27). Because each hole is inspected, accurate information is recorded. A rugged sensor housing and extremely high signal strength handle dirt and grease with minimum maintenance. The modular control unit allows for dense packaging in small enclosures.

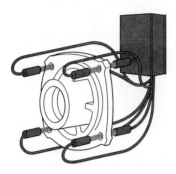

Figure 1.27 Multihole casting inspection.

In a web flaw detection application, a web passes over an array of retroreflectors (Fig. 1.28). When light is returned to the sensor head, the output is energized and the web shuts down. High web speeds can be maintained because of the superior response time of the control unit.

A reflex photoelectric sensor with a motion control module counts the revolutions of a wheel to monitor over/underspeed of a rotating object. Speed is controlled by a programmable controller. The rate ranges from 2.4 to 12,000 counts per minute (Fig. 1.29).

When the two through-beam photoelectric sensors in Fig. 1.30 observe the same signal strength, the output is zero. When the capacity of the web changes, as in a splice, the signal strengths are thrown out of balance and the output is energized. This system can be used on webs of different colors and opacities with no system reconfiguration.

Understanding the environment is important to effective implementation of an error-free environment. An awareness of the characteristics of photoelectric controls and the different ways in which they can be used will establish a strong foundation. This understanding

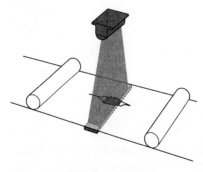

Figure 1.28 Web flaw detection.

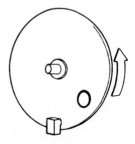

Figure 1.29 Over/under-speed of rotating disk.

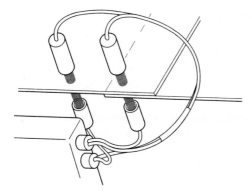

Figure 1.30 Web splice detection.

also will allow the user to obtain a descriptive picture of the condition of each manufacturing process in the production environment.

Table 1.3 highlights key questions the user must consider.

1.5   Detection Methods

The are three modes of detection used by photoelectric sensors:

1. Through-beam detection
2. Reflex detection
3. Proximity detection

TABLE 1.3   Key Characteristics of Sensors

Key point	Consideration
1. Range	How far is the object to be detected?
2. Environment	How dirty or dark is the environment?
3. Accessibility	What accessibility is there to both sides of the object to be detected?
4. Wiring	Is wiring possible to one or both sides of the object?
5. Size	What size is the object?
6. Consistency	Is object consistent in size, shape, and reflectivity?
7. Requirements	What are the mechanical and electrical requirements?
8. Output Signal	What kind of output is needed?
9. Logic functions	Are logic functions needed at the sensing point?
10. Integration	Is the system required to be integrated?

1.5.1   Through-beam detection method

The through-beam method requires that the source and detector are positioned opposite each other and the light beam is sent directly from source to detector (Fig. 1.31). When an object passes between the source and detector, the beam is broken, signaling detection of the object.

Through-beam detection generally provides the longest range of the three operating modes and provides high power at shorter range to penetrate steam, dirt, or other contaminants between the source and detector. Alignment of the source and detector must be accurate.

1.5.2   Reflex detection method

The reflex method requires that the source and detector are installed at the same side of the object to be detected (Fig. 1.32). The light beam is transmitted from the source to a retroreflector that returns the light to the detector. When an object breaks a reflected beam, the object is detected.

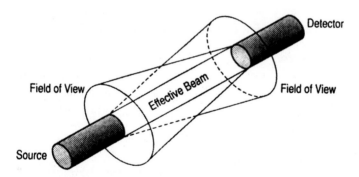

Figure 1.31   Through-beam-detection.

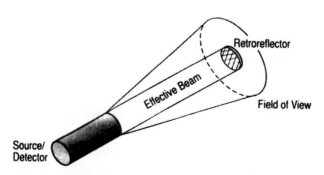

Figure 1.32   Reflex detection.

The reflex method is widely used because it is flexible and easy to install and provides the best cost-performance ratio of the three methods. The object to be detected must be less reflective than the retroreflector.

1.5.3   Proximity detection method

The proximity method requires that the source and detector are installed on the same side of the object to be detected and aimed at a point in front of the sensor (Fig. 1.33). When an object passes in front of the source and detector, light from the source is reflected from the object's surface back to the detector, and the object is detected.

Each sensor type has a specific operating range. In general, through-beam sensors offer the greatest range, followed by reflex sensors, then by proximity sensors.

The maximum range for through-beam sensors is of primary importance. At any distance less than the maximum range, the sensor has more than enough power to detect an object.

The optimum range for the proximity and reflex sensors is more significant than the maximum range. The optimum range is the range at which the sensor has the most power available to detect objects. The optimum range is best shown by an excess gain chart (Fig. 1.34).

Excess gain is a measure of sensing power available in excess of that required to detect an object. An excess gain of 1 means there is just enough power to detect an object, under the best conditions without obstacles placed in the light beam. The distance at which the excess gain equals 1 is the maximum range. An excess gain of 100 means there is 100 times the power required to detect an object. Generally, the more excess gain available at the required range, the more consistently the control will operate.

For each distance within the range of sensor, there is a specific

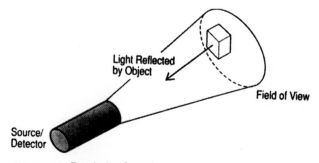

Figure 1.33 Proximity detection.

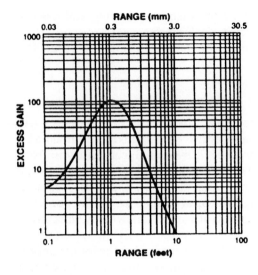

Figure 1.34 Photoelectric excess gain and range.

excess gain. Through-beam controls generally provide the most excess gain, followed by reflex and then proximity sensors.

General guidelines can be provided for the amount of excess gain required for the amount of contamination in an environment. Environments can be relatively clean, lightly dirty, dirty, very dirty, and extremely dirty. Table 1.4 illustrates the excess gain recommended for these types of environments for each sensing mode.

TABLE 1.4   Excess Gain Chart

Environment	Through beam	Reflex	Proximity
Relatively clean	1.25 per side	1.6 per side	
Office clean	1.6 total	2.6 total	2.6 total
Lightly dirty	1.8 per side	3.2 per side	
Warehouse, post office	3.2 total	10.5 total	3.2 total
Dirty	8 per side	64 per side	
Steel mill, saw mill	64 total		64 total
Very dirty	25 per side		
Steam tunnel, painting, rubber or grinding, cutting with coolant, paper plant	626 total		
Extremely dirty	100 per side		
Coal bins or areas where thick layers build quickly	10,000 total		

Example   If, in a through-beam setup, the source is in a lightly dirty environment where excess gain is 1.8, and the detector is in a very dirty environment where excess gain is 25, the recommended excess gain is $1.8 \times 25 = 45$, from Table 1.4.

1.6   Proximity Sensors

Proximity sensing is the technique of detecting the presence or absence of an object with an electronic noncontact sensor.

Mechanical limit switches were the first devices to detect objects in industrial applications. A mechanical arm touching the target object moves a plunger or rotates a shaft which causes an electrical contact to close or open. Subsequent signals will produce other control functions through the connecting system. The switch may be activating a simple control relay, or a sophisticated programmable logic control device, or a direct interface to a computer network. This simple activity, once done successfully, will enable varieties of manufacturing operations to direct a combination of production plans according to the computer-integrated manufacturing strategy.

Inductive proximity sensors are used in place of limit switches for noncontact sensing of metallic objects. Capacitive proximity switches are used on the same basis as inductive proximity sensors; however, capacitive sensors can also detect nonmetallic objects. Both inductive and capacitive sensors are limit switches with ranges up to 100 mm.

The distinct advantage of photoelectric sensors over inductive or capacitive sensors is their increased range. However, dirt, oil mist, and other environmental factors will hinder operation of photoelectric sensors during the vital operation of reporting the status of a manufacturing process. This may lead to significant waste and buildup of false data.

1.6.1   Typical applications of inductive proximity sensors

Motion position detection (Fig. 1.35)

1. Detection of rotating motion
2. Zero-speed indication
3. Speed regulation

Motion control (Fig. 1.36)

1. Shaft travel limiting
2. Movement indication
3. Valve open/closed

Figure 1.35 Motion /position detection with inductive proximity sensor.

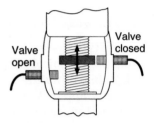

Figure 1.36 Motion control, inductive proximity sensor.

Conveyer system control (Fig. 1.37)

1. Transfer lines
2. Assembly line control
3. Packaging machine control

Process control (Fig 1.38)

1. Product complete
2. Automatic filling
3. Product selection

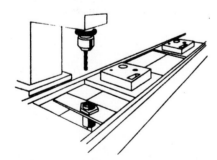

Figure 1.37 Conveyer system control, inductive proximity sensor.

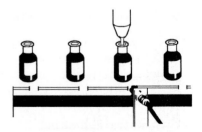

Figure 1.38 Process control, inductive proximity sensor.

Machine control (Fig. 1.39)

1. Fault condition indication
2. Broken tool indication
3. Sequence control

Verification and counting (Fig. 1.40)

1. Product selection
2. Return loop control
3. Product count

1.6.2 Typical applications of capacitive proximity sensors

Liquid level detection (Fig. 1.41)

1. Tube high/low liquid level
2. Overflow limit
3. Dry tank

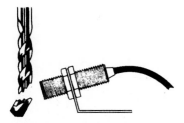

Figure 1.39 Machine control, inductive proximity sensor.

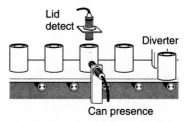

Figure 1.40 Verification and counting, inductive proximity sensor.

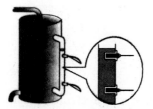

Figure 1.41 Liquid level detection, capacitive proximity sensor.

Figure 1.42 Bulk material level control, capacitive proximity sensor.

Figure 1.43 Process control, capacitive proximity sensor.

Bulk material level control (Fig. 1.42)

1. Low level limit
2. Overflow limit
3. Material present

Process control (Fig. 1.43)

1. Product present
2. Bottle fill level
3. Product count

1.7   Understanding Inductive Proximity Sensors

1.7.1   Principles of operation

An inductive proximity sensor consists of four basic elements (Fig. 1.44).

1. Sensor coil and ferrite core
2. Oscillator circuit
3. Detector circuit
4. Solid-state output circuit

The oscillator circuit generates a radio-frequency electromagnetic field that radiates from the ferrite core and coil assembly. The field is centered around the axis of the ferrite core, which shapes the field

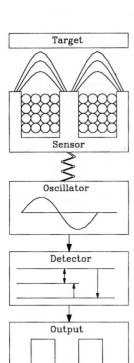

Figure 1.44 Operating principle of inductive proximity sensor.

and directs it at the sensor face. When a metal target approaches and enters the field, *eddy currents* are induced into the surfaces of the target. This results in a loading effect, or "damping," that causes a reduction in amplitude of the oscillator signal (Fig. 1.45).

The detector circuit detects the change in oscillator amplitude

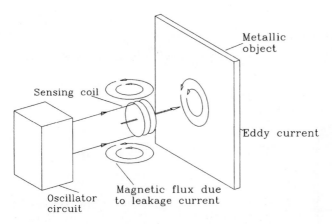

Figure 1.45 Induced eddy current.

(Fig. 1.46). The detector circuit will switch ON at a specific operate amplitude. This ON signal generates a signal to turn ON the solid-state output. This is often referred to as the *damped* condition. As the target leaves the sensing field, the oscillator responds with an increase in amplitude. As the amplitude increases above a specific value, it is detected by the detector circuit, which switches OFF, causing the output signal to return the normal or OFF (*undamped*) state.

The difference between the operate and the release amplitude in the oscillator and corresponding detector circuit is referred to as *hysteresis* (H) of the sensor. It corresponds to a difference in point of target detection and release distance between the sensor face and the target surface (Fig. 1.47).

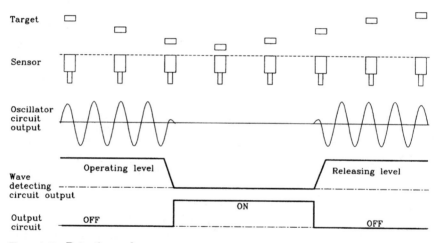

Figure 1.46 Detection cycle.

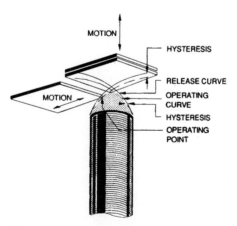

Figure 1.47 Core assembly.

1.7.2 Inductive proximity sensing range

The *sensing range* of an inductive proximity sensor refers to the distance between the sensor face and the target. It also includes the shape of the sensing field generated through the coil and core. There are several mechanical and environmental factors that affect the sensing range:

Mechanical factors	Environmental factors
Core size	Ambient temperature
Core shield	Surrounding electrical conditions
Target material	Surrounding mechanical conditions
Target size	Variation between devices
	Target shape

The geometry of the sensing field can be determined by the construction factor of the core and coil. An open coil with no core produces an omnidirectional field. The geometry of an air-core is a toroid. Such sensors could be actuated by a target approaching from any direction, making them undesirable for practical industrial applications (Fig. 1.48).

Ferrite material in the shape of a cup core is used to shape the sensing field. The ferrite material absorbs the magnetic field, but enhances the field intensity and directs the field out of the open end of the core (Fig. 1.49).

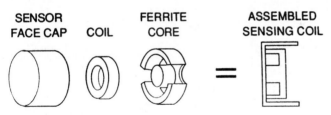

Figure 1.48   Open coil without core.

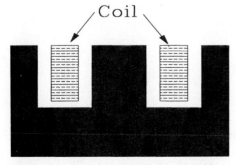

Figure 1.49   Cup-shaped coil/core assembly.

A standard field range sensor is illustrated in Fig. 1.50. It is often referred to as *shielded* sensing coil. The ferrite contains the field so that it emanates straight from the sensing face. Figure 1.51 shows the typical standard-range sensing-field plot.

An extended range coil and core assembly does not have the ferrite around the perimeter of the coil (Fig. 1.52). This unshielded device accordingly has an extended range. Figure 1.53 illustrates a typical extended sensing-field plot.

1.7.3   Sensing distance

The electromagnetic field emanates from the coil and core at the face of the sensor and is centered around the axis of the core. The nominal sensing range is a function of the coil diameter and the power that is available to operate the electromagnetic field.

The sensing range is subject to manufacturing tolerances and circuit variations. Typically it varies by 10 percent. Similarly, tempera-

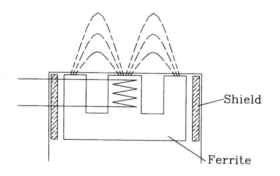

Shield

Figure 1.50   Standard range core coil.

Ferrite

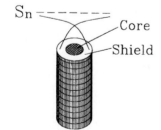

Core

Shield

Figure 1.51   Standard range field plot

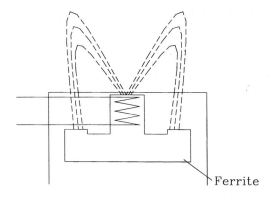

Figure 1.52 Extended range core and coil.

Ferrite

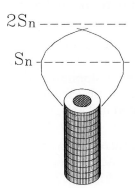

Figure 1.53 Extended range field plot.

ture drift can affect the sensing range by 10 percent. Applied to the nominal sensing switch, these variations mean that sensing range can be as much as 120 percent or as little as 81 percent of the nominal stated range (Fig. 1.54).

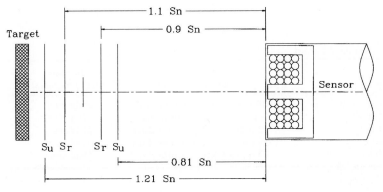

Figure 1.54 Sensing distance tolerances.

$$S_r = 0.9 < S_n < 1.1$$

$$S_n = 0.9 < S_r < 1.1$$

where S_n = nominal sensing range
  S_r = effective sensing range
  S = usable sensing range

1.7.4 Target material and size

As a target approaches the sensing field, eddy currents are induced in the target. In order to ensure that the target has the desired damping affect on the sensor, the target has to be of appropriate size and material. Metallic targets can be defined as:

- *Ferrous.* Containing iron, nickel, or cobalt
- *Nonferrous.* All other metallic materials, such as aluminum, copper, and brass. Eddy currents induced in ferrous targets are stronger than in nonferrous targets, as illustrated in Table 1.5.

An increase in target size will not produce an increase in sensing range. However, a decrease in target size will produce a decrease in sensing range, and may also increase response time. Figure 1.55 illustrates the relationship of target size and target material to the sensing range of a limit-switch-type sensor with nominal 13-mm sensing range.

Table 1.6, shows correction factors by which the rated nominal sensing range of most inductive proximity sensors can be multiplied. This will determine the effective sensing range for a device sensing a stated target material of standard target size.

TABLE 1.5   Standard Target

Device	Standard target dimensions	Standard target material
Modular limit switch type	45 mm square × 1 mm thick	Mild steel
8 mm tubular	8 mm square × 1 mm thick	Mild steel
12 mm tubular	12 mm square × 1 mm thick	Mild steel
18 mm tubular	18 mm square × 1 mm thick	Mild steel
30 mm tubular	30 mm square × 1 mm thick	Mild steel

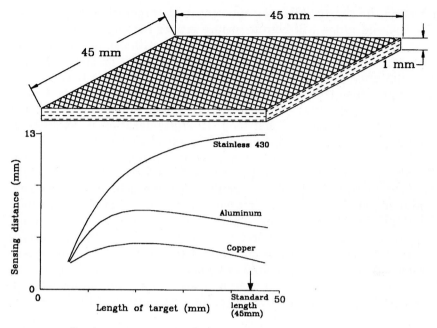

Figure 1.55 Sensing range correction factor.

TABLE 1.6   Target Material Correction

Target material	Limit-switch type	Pancake type	Tubular, mm			
			8	12	18	30
Steel (1020)	1.0	1.0	1.0	1.0	1.0	1.0
Stainless steel (400)	1.03	0.90	0.90	0.90	1.0	1.0
Stainless steel (300)	0.85	0.70	0.60	0.70	0.70	0.65
Brass	0.50	0.54	0.35	0.45	0.45	0.45
Aluminum	0.47	0.50	0.35	0.40	0.45	0.40
Copper	0.40	0.46	0.30	0.25	0.35	0.30

Example   18-mm tubular extended range   8 mm

Copper target correction factor $\times$ 0.35

Sensing range detecting copper standard target   2.80 mm

1.7.5 Target shape

Standard targets are assumed to be of a flat, square shape with the stated dimensions. Targets of round shape or with a pocketed surface have to be of adequate dimensions to cause the necessary dampening effect on the sensor. Allowing the sensor–to-target distance less than the nominal range will help to assure proper sensing. Also using the next larger size or an extended-range sensor will also minimize problems with other than standard target dimensions or shape. Figure 1.56 illustrates the axial (head-on) approach, indicating that the target approaches the face of the sensor on the axis of the coil core. When the target approaches axially, the sensor should not be located such that it becomes an end stop. If axial operation is considered, good application practice is to allow for 25 percent overtravel.

   Lateral (side-by) approach means the target approaches the face of the sensor perpendicular to the axis of the coil core (Fig. 1.57). Good application practice (GAP), a terminology often used in "world-class" manufacturing strategies, dictates that the tip of the sensing field envelope should not be used. That is the point where sensing range variations start to occur. Therefore, it is recommended that the target pass not more than 75 percent of the sensing distance D from the sensor face. Also, the target should not pass any closer than the basic tolerance incorporated in the machine design, to prevent damage to the sensor. Hysteresis can be greater for an axial approach (Fig. 1.58).

1.7.6   Variation between devices

Variations of sensing range between sensors of the same type often occur. With modern manufacturing technologies and techniques, variations are held to a minimum. The variations can be attributed to collective tolerance variations of the electrical components in the sensor

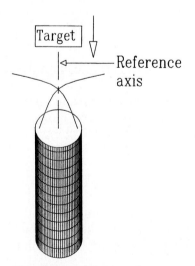

Figure 1.56   Axial approach.

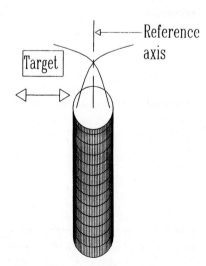

Figure 1.57   Lateral approach.

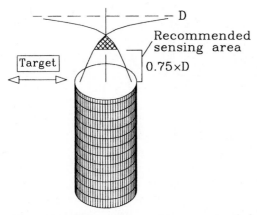

Figure 1.58 Lateral approach—recommended sensing distance.

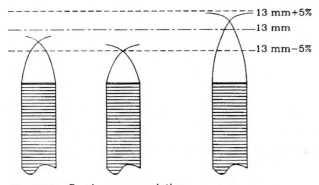

Figure 1.59 Sensing range variation.

circuit and to subtle differences in the manufacturing process from one device to the next; 5 percent variation is typical (Fig. 1.59).

Sensing distance also will vary from one temperature extreme to the other because of the effect of temperature change on the components of the sensor. Typical temperature ranges are –25°C (–3°F) to +70°C (+180°F). Figure 1.60 illustrates sensing range variation with temperature.

1.7.7 Surrounding conditions

There are several environmental factors that must also be considered in order to obtain reliable information from inductive proximity sensors. These surrounding factors are

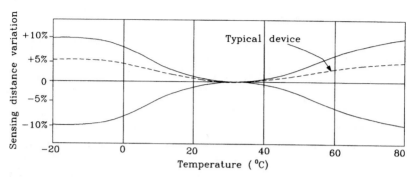

Figure 1.60   Sensing range variation—with temperature.

1. *Embeddable mounting.*   The shielded sensor in Fig. 1.61 is often referred to as a *flush-mounted* sensor. Shielded sensors are not affected by the surrounding metal.

2. *Flying metal chips.*   A chip removed from metal during milling and drilling operations may affect the sensor performance depending on the size of the chip, its location on the sensing face, and type of material. In these applications, the sensor face should be oriented so that gravity will prevent chips from accumulating on the sensor face. If this is not possible, then coolant fluid should wash the sensor face to remove the chips. Generally, a chip does not have sufficient surface area to cause a sensor turn on. If a chip lands on the center of the sensor face, it will have a negligible effect, but elsewhere on the sensor face, it will extend the range of the sensor.

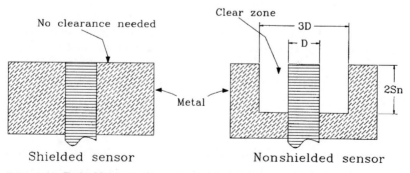

Figure 1.61   Embeddable and nonembeddable sensors.

3. *Adjacent sensors.*   When two similar sensors are located adjacent to or opposite each other, the interaction of their fields can affect operation. Figure 1.62 provides the guidelines for placing two similar sensors adjacent to each other. *Alternate-frequency* heads will allow adjacent mounting of sensors without interaction of their sensing fields.

4. *Magnetic fields.*   Electrical wiring in the vicinity of the sensor face may affect sensor operation. If the magnetic field around the electrical wiring reaches an intensity that would saturate the sensor ferrite or coil, the sensor will not function properly. Use of inductive sensors in the presence of high-frequency radiation can also unexpectedly affect their operation. Sensors specially designed for welding application can be used with programmable logic control (PLC). The PLC can be programmed to ignore the signal from the sensor for the period that the high-frequency welder is operated. A slight OFF-time delay assures proper operation of the sensor.

5. *Radio-frequency interference (RFI).*   Radio transceivers, often called *walkie-talkie* devices, can produce a signal that can cause an inductive proximity sensor to operate falsely. The radio transceiver produces a radio-frequency signal similar to the signal produced by the oscillator circuit of the sensor. The effect that RFI has on an

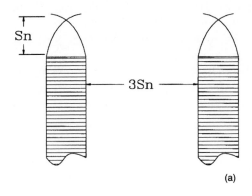

(a)

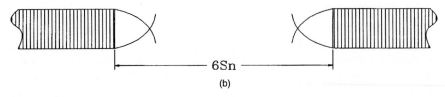

(b)

Figure 1.62 Adjacent sensors.

inductive proximity switch can vary. The factors that determine this variation are as follows:

a. *Distance between RFI source and the sensor.*   Typically, inductive proximity switches are not affected by RFI when a transceiver is 1 ft away from the inductive switch. However, if closer than 1 ft, the switch may operate without a target present.

b. *Signal frequency.*   The signal frequency may be the determining factor that will cause a particular device to false-operate.

c. *Signal intensity.*   Radio-frequency transceivers usually are portable devices with power rating of 5 W maximum.

d. *Inductive proximity package.*   The sensor package construction may determine how well the device resists RFI.

e. *Approach to the sensor.*   A transceiver approaching the connecting cable of a switch may affect it at a greater distance than if it was brought closer to the sensing face. As RFI protection varies from device to device and manufacturer to manufacturer, most manufacturers have taken steps to provide the maximum protection against false operation due to RFI.

6. *Showering arc.*   Showering arc is the term applied to induced line current/voltage spikes. The spike is produced by the electrical arc on an electromechanical switch or contactor closure. The current spike is induced from lines connected to the electromechanical switch to the lines connected to the inductive proximity switch, if the lines are adjacent and parallel to one another. The result can be false operation of the inductive proximity switch. The spike intensity is determined by the level of induced voltage and the duration of the spike. Avoiding running cables for control devices in the same wiring channel as those for the contactor or similar leads may eliminate spikes. Most electrical code specifications require separation of control device leads from electromechanical switch and contractor leads.

1.8   Understanding Capacitive Proximity Sensors

1.8.1   Principles of operation

A capacitive proximity sensor operates much like an inductive proximity sensor. However, the means of sensing is considerably different. Capacitive sensing is based on dielectric capacitance. *Capacitance* is the property of insulators to store an electric charge. A capacitor consists of two plates separated by an insulator, usually called a *dielectric*. When the switch is closed (Fig. 1.63) a charge is stored on the two plates.

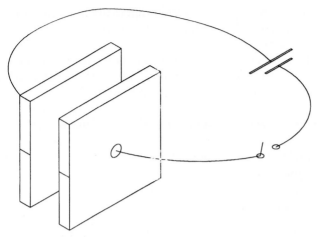

Figure 1.63 Capacitive principle.

The distance between the plates determines the ability of a capacitor to store a charge and can be calibrated as a function of stored charge to determine discrete ON and OFF switching status.

Figure 1.64 illustrates the principle as it applies to the capacitive sensor. One capacitive plate is part of the switch, the sensor face (the enclosure) is the insulator, and the target is the other plate. Ground is the common path.

The capacitive proximity sensor has the same four basic elements as an inductive proximity sensor:

1. Sensor (the dielectric plate)
2. Oscillator circuit
3. Detector circuit
4. Solid-state output circuit

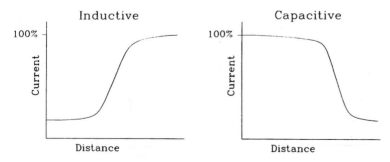

Figure 1.64 Capacitive sensor.

The oscillator circuit in a capacitive switch operates like one in an inductive proximity switch. The oscillator circuit includes feedback capacitance from the external target plate and the internal plate. In a capacitive switch, the oscillator starts oscillating when sufficient feedback capacitance is detected. In an inductive proximity switch, the oscillation is damped when the target is present (Fig. 1.65).

In both capacitive and inductive switch types, the difference between the operate and the release amplitude in the oscillator and corresponding detector circuit is referred to as *hysteresis* of the sensor. It corresponds to the difference between target detection and release distances from the sensor face.

1.8.2 Features of capacitive sensors

The major characteristics of capacitive proximity sensors are

1. They can detect nonmetallic targets.
2. They can detect lightweight or small objects that cannot be detected by mechanical limit switches.
3. They provide a high switching rate for rapid response in object counting applications.
4. They can detect liquid targets through nonmetallic barriers, (glass, plastic, etc.).
5. They have long operational life with a virtually unlimited number of operating cycles.
6. The solid-state output provides a bounce-free contact signal.

Capacitive proximity sensors have two major limitations:

1. They are affected by moisture and humidity.
2. They must have an extended range for effective sensing.

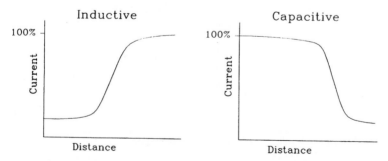

Figure 1.65 Oscillator damping of inductive and capacitive sensors.

1.8.3 Sensing range

Capacitive proximity sensors have a greater sensing range than inductive proximity sensors as illustrated below.

Tubular diameter, mm	Inductive extended range, mm	Capacitive extended range, mm
18	8	10
30	15	20
34	—	40

Sensing distance for capacitive proximity sensors is a matter of plate diameter, as coil size is for inductive proximity sensors. Capacitive sensors basically measure a dielectric gap. Accordingly, it is desirable to be able to compensate for target and application conditions with a sensitivity adjustment for the sensing range. Most capacitive proximity switches are equipped with a sensitivity adjustment potentiometer (Fig. 1.66).

1.8.4 Target material and size

The sensing range of capacitive sensors, like that of inductive proximity sensors, is determined by the type of material. Table 1.7 lists the

Figure 1.66 Sensitivity adjustment.

TABLE 1.7 Target Material Correction

Material	Factor
Mild steel	1.0
Cast iron	1.0
Aluminum and copper	1.0
Stainless steel	1.0
Brass	1.0
Water	1.0
Polyvinylchloride (PVC)	0.5
Glass	0.5
Ceramic	0.4
Wood	≥0.2
Lubrication oil	0.1

sensing-range derating factors which apply to capacitive proximity sensors. Capacitive sensors can be used to detect a target material through a nonmetallic interposing material like glass or plastic. This is beneficial in detecting a liquid through the wall of a plastic tank or through a glass sight tube. The transparent interposing material has no effect on sensing. For all practical purposes, the target size can be determined in the same manner as for inductive proximity sensors.

1.8.5 Surrounding conditions

Capacitive proximity devices are affected by component tolerances and temperature variations. As with inductive devices, capacitive proximity devices are affected by the following surrounding conditions:

1. *Embeddable mounting.* Capacitive sensors are generally treated as nonshielded, nonembeddable devices.

2. *Flying chips.* Capacitive devices are more sensitive to metallic and nonmetallic chips.

3. *Adjacent sensors.* Allow more space than with inductive proximity devices because of the greater sensing range of capacitive devices.

4. *Target background.* Relative humidity may cause a capacitive device to operate even when a target is not present. Also, the greater sensing range and ability to sense nonmetallic target materials dictate greater care in applying capacitive devices with target background conditions

5. *Magnetic fields.* Capacitive devices are not usually applied in welding environment.

6. *Radio-frequency interference.* Capacitive sensor circuitry can be affected by RFI in the same way that an inductive device can.

7. *Showering arc.* An induced electrical noise will affect the circuitry of a capacitive device in the same way that it does an inductive device.

1.9 Understanding Limit Switches

A limit switch is constructed much like the ordinary light switch used in home and office. It has the same ON/OFF characteristics. The limit switch usually has a pressure-sensitive mechanical arm. When an object applies pressure on the mechanical arm, the switch circuit is

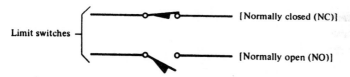

Figure 1.67 Normally open—normally closed microswitches.

energized. An object might have a magnet attached that causes a contact to rise and close when the object passes over the arm.

Limit switches can be either *normally open* (NO) or *normally closed* (NC) and may have multiple poles (Fig. 1.67). A normally open switch has continuity when pressure is applied and a contact is made, while a normally closed switch opens when pressure is applied and a contact is separated. A single-pole switch allows one circuit to be opened or closed upon switch contact, whereas a multiple-pole switch allows multiple circuits to be opened or closed.

Limit switches are mechanical devices. They have three potential problems:

1. They are subject to mechanical failure.

2. Their mean time between failures (MTBF) is low compared to noncontact sensors.

3. Their speed of operation is relatively slow; the switching speed of photoelectric microsensors is up to 3000 times faster.

1.10 Inductive and Capacitive Sensors in Manufacturing

Inductive and capacitive proximity sensors interface to control circuits through an output circuit, for manufacturing applications. Also, the control circuit type is a determining factor in choosing an output circuit. Control circuits, whether powered by ac, dc, or ac/dc, can be categorized as either *load powered* or *line powered*.

The load-powered devices are similar to limit switches. They are connected in series with the controlled load. These devices have two connection points and are often referred to as *two-wire switches*. Operating current is drawn through the load. When the switch is not operated, the switch must draw a minimum operating current referred to as *residual current*. When the switch is operated or damped (i.e., a target is present), the current required to keep the

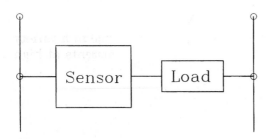

Figure 1.68 Load-powered residual current.

sensor operating is the minimum holding current (Fig. 1.68). The residual current is not a consideration for low-impedance loads such as relays and motor starters. However, high-impedance loads, most commonly programmable logic controllers, require residual current of less than 2 mA. Most sensors offer 1.7 mA or less.

In some manufacturing applications, a particular type of PLC will require less than 1.7 mA residual current. In such applications a loading resistor is added in parallel to the input to the PLC load. Then minimum holding current may range up to 20 mA, depending on the sensor specification. If the load impedance is too high, there will not be enough load current level to sustain the switch state.

Inductive proximity sensors with holding current of 4 mA or less can be considered low-holding-current sensors. These devices can be used with PLCs without concern for minimum holding current.

Line-powered devices derive current, usually called *burden current,* from the line and not through the controller load. These devices are called three-wire switches because they have three connections (Fig. 1.69).

The operating current for a three-wire sensor is burden current, and is typically 20 mA. Since the operating current does not pass through the load, it is not a major concern for the circuit design.

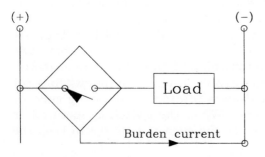

Figure 1.69 Line-powered burden current.

1.10.1   Relays

An output circuit relay is a mechanical switch available in a variety of contact configurations. Relays can handle load currents at high voltages, allowing the sensor to directly interface with motors, large solenoids, and other inductive loads. They can switch either ac or dc loads. Contact life depends on the load current and frequency of operation. Relays are subject to contact wear and resistance buildup. Because of contact bounce, they can produce erratic results with counters and programmable controllers unless the input is filtered. They can add 10 to 25 ms to an inductive or capacitive switch response time because of their mechanical nature (Fig. 1.70).

Relays are familiar to manufacturing personnel. They are often used with inductive or capacitive proximity sensors, as they provide multiple contacts. The good and bad features of a relay are summarized below.

Relay advantages	Relay disadvantages
Switches high currents/loads	Slow response time
Multiple contacts	Mechanical wear
Switches ac or dc voltages	Contact bounce
Tolerant of inrush current	Affected by shocks and vibration

1.10.2   Triac devices

A triac is a solid-state device designed to control ac current (Fig. 1.71). Triac switches turn ON in less than a microsecond when the gate (control leg) is energized, and shut OFF at the zero crossing of the ac power cycle.

Because a triac is a solid-state device, it is not subject to the mechanical limitations of a relay such as mechanical bounce, pitting, corrosion of contacts, and shock and vibration. Switching response

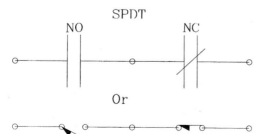

Figure 1.70   Relay output.

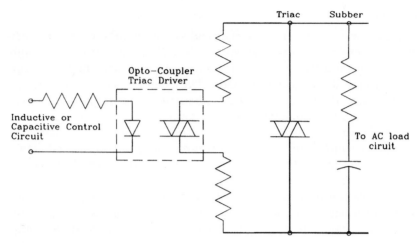

Figure 1.71   Triac circuit.

time is limited only by the time it takes the 60-Hz ac power to go through one-half cycle (8.33 ms) (Fig. 1.72).

As long as a triac is used within its rated maximum current and voltage specifications, life expectancy is virtually infinite. Triac devices used with inductive or capacitive sensors generally are rated at 2-A loads or less. Triac limitations can be summarized as follows: (1) shorting the load will destroy a triac and (2) directly connected inductive loads or large voltage spikes from other sources can false-trigger a triac.

To reduce the effect of these spikes, a snubber circuit composed of a resistor and capacitor in series is connected across the device. Depending on the maximum switching load, an appropriate snubber network for switch protection is used. The snubber network con-

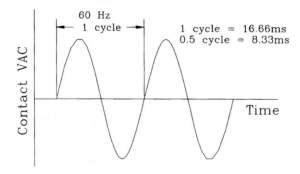

Figure 1.72   AC power cycle.

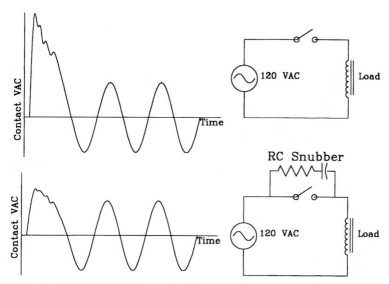

Figure 1.73   Snubber circuit.

tributes to the OFF state leakage to the load. The leakage must be considered when loads requiring little current, such as PLCs, are switched. In the ON state, a drop of about 1.7 V rms is common (Fig. 1.73). Good and bad features of triacs are listed below.

Triac advantages	Triac disadvantages
Fast response time (8.33 ms)	Can be falsely triggered by large inductive current
Tolerant of large inrush currents	Snubber contributes to OFF state leakage current
Can be directly interfaced with programmable controllers	Can be destroyed by short circuits
Infinite life when operated within rated voltage/current limits	

1.10.3   Transistor dc switches

Transistors are solid-state dc switching devices. They are most commonly used with low-voltage dc-powered inductive and capacitive sensors as the output switch. Two types are employed, depending on the function (Fig. 1.74).

In an NPN transistor, the current source provides a contact closure to the dc positive rail. The NPN current sink provides a contact to the

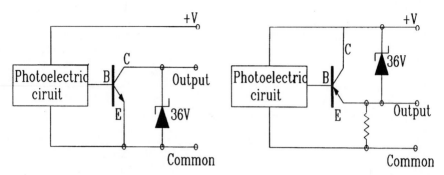

Figure 1.74 DC circuit logic.

dc common. The transistor can be thought of as a single-pole switch that must be operated within its voltage and maximum current ratings (Fig. 1.75).

Any short circuit on the load will immediately destroy a transistor that is not short-circuit protected. Switching inductive loads creates voltage spikes that exceed many times the maximum rating of the transistor. Peak voltage clamps such as zener diodes or transorbs are utilized to protect the output device. Transistor outputs are typically rated to switch loads of 250 mA at 30 V dc maximum (Fig. 1.76).

Transistor advantages	Transistor disadvantages
Virtually instantaneous response	Low current handling capacity
Low OFF state leakage and voltage drop	Cannot handle inrush current unless clamped
Infinite life when operated within rated current/voltage	Can be destroyed by short circuit unless protected
Not affected by shock and vibration	

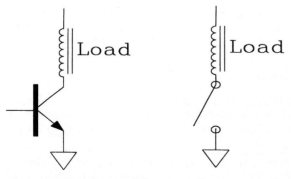

Figure 1.75 Transistor switch.

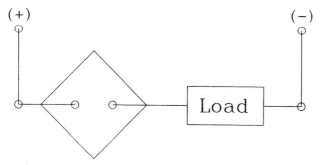

Figure 1.76 Voltage clamp.

1.10.3.1 Output configuration. Output configurations are categorized as follows:

1. Single output—normally open (NO)
2. Single output—normally closed (NC)
3. Programmable output—NO or NC
4. Complementary output—NO and NC

The functions of normally open and normally closed output are defined in Table 1.8.

1.10.4 Inductive and capacitive control/output circuits

A single output sensor has either an NO or an NC configuration and cannot be changed to the other configuration (Fig. 1.77).

A programmable output sensor has one output, NO or NC, depending on how the output is wired when installed. These sensors are exclusively two-wire ac or dc (Fig. 1.78).

A complementary output sensor has two outputs, one NO and one NC. Both outputs change state simultaneously when the target enters or leaves the sensing field. These sensors are exclusively three-wire ac or dc (Fig. 1.79).

TABLE 1.8 Output Logic

Output configuration	Target state	Oscillator state	Output
NO	Absent	Undamped	Nonconducting (OFF)
	Present	Damped	Conducting (ON)
NC	Absent	Undamped	Conducting (ON)
	Present	Damped	Nonconducting (OFF)

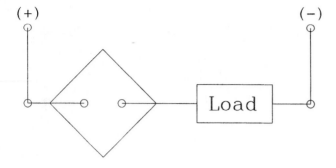

Figure 1.77 Single output.

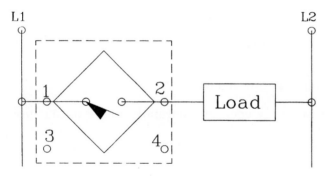

Figure 1.78 Programmable output.

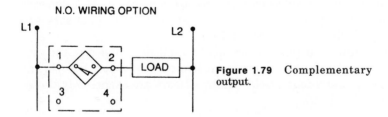

Figure 1.79 Complementary output.

The choice of control circuit and output logic plays an important part in determining the reliability of data collection. The choice of control circuit and output logic depends on the following parameters:

1. *AC or DC control voltage.* Use of ac control may seem to require

the use of an ac-configured sensor. However, interface circuitry can allow for dc sensors even if the main control voltage source is ac.

2. *Control circuit current requirements.*   Usually control circuits operating in the 200- to 300-mA range can use either ac or dc sensors. Circuits with 0.5-A and higher current will dictate the type of sensor to be used.

3. *Application output requirements.*   NO output is the most commonly used output type. Controlled circuit configurations may dictate use of NC or complementary-type configured sensors.

4. *Switching speed requirements.*   AC circuits are limited in their operations per second. DC circuits may be required for applications involving counting or high speed.

5. *Connecting logic device.*   The device to which the sensor is connected—such as programmable controller, relay, solenoid, or timer/counter—is usually the most important factor in sensor circuit and output configuration.

1.10.5   Accessories for sensor circuits

Sensor circuits and their output configurations must have various types of indicators and protection devices, such as:

1. Light-emitting diode (LED) indicators
2. Short-circuit protectors
3. Reverse-polarity protectors—dc three-wire
4. Wire terminators—color-coded wire
5. Pin connector type and pin-out designator

1.10.5.1   LED indicators.   LED indicators provide diagnostic information on the status of sensors, e.g., operated or not operated, that is vital in computer-integrated manufacturing. Two LEDs also indicate the status of complementary-type sensor switches and power ON/OFF status, and short-circuit condition.

1.10.5.2   Short-circuit protection.   Short-circuit protection is intended to protect the switch circuit from excessive current caused by wiring short circuits, line power spikes from high inrush sources, or lightning strikes. This option involves special circuitry which either limits the current through the output device or turns the switch OFF. The turn-off–type switch remains inoperative until the short circuit has been cleared—with power disconnected. Then power is reapplied to the sensor. A second LED is usually furnished with this type of device to indicate the shorted condition.

1.10.5.3 Reverse-polarity protection. Reverse-polarity protection is special circuitry that prevents damage in a three-wire dc sourcing (PNP) or sinking (NPN) device when it is connected to control circuitry incorrectly. Although reverse polarity is relatively common, not all switches are equipped with this option.

1.10.5.4 Wire termination. Wire terminals are common on limit-switch enclosure-type sensors. The terminal designations are numbered and correspond to the device wiring diagram (Fig. 1.80). Cable/wire stub terminations are most common on tubular sensors. Color-coded conductors are essential for correct wiring. Most sensor wires are color-coded to comply with industry wire color-code standards.

1.10.5.5 Pin connectors. Pin-connector-terminal sensors feature a male pin connector receptacle on the switch or at the end of the wire/cable stub. The female receptacle is at the end of the matching cable cord. Most industry-standard pin connectors are either the mini type—approximately 18 mm in diameter—or the micro type—approximately 12 mm in diameter (Fig. 1.81).

1.10.6 Inductive and capacitive switching logic

The outputs of two or more inductive or capacitive proximity sensors can be wired together in series or parallel to perform logic functions. The ON, or activated, switch function can be either a normally open or a normally closed output function, depending on the desired control logic.

Although sensors are the most effective means for the data acquisition role in manufacturing, care must be exercised when sensors are

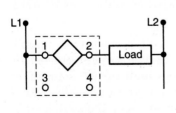

Figure 1.80 Wire terminal.

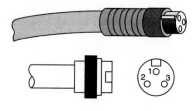

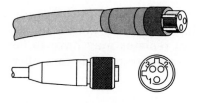

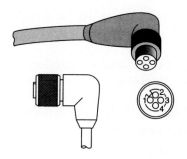

Figure 1.81 Pin connector.

integrated with various production operations. The following factors will affect the performance of the switch logic circuit:

1. Excessive leakage current in parallel-connected load-powered devices

2. Excessive voltage drop in series-connected devices

3. Inductive feedback with line-powered sensors with parallel connections

1.10.6.1 Parallel-connection logic—OR function. The binary OR logic in Table (1.9) indicates that the circuit output is ON (1) if one or more of the sensors in parallel connection is ON.

TABLE 1.9   Binary Logic Chart—Parallel
Connection

A	B	C	OUT
0	0	0	0
0	0	1	1
0	1	0	1
1	0	0	1

0 = OFF
1 = ON

It is important to note that, in two-wire devices, the OFF state residual current is additive (Fig. 1.82). If the circuit is affected by the total leakage applied, a shunt (loading) resistor may have to be applied. This is a problem in switching to a programmable controller or other high-impedance device.

Example $I_a + I_b + I_c = I_t$

$$1.7 + 1.7 + 1.7 = 5.1 \text{ mA}$$

Three-wire 10 to 30 V can also be connected in parallel for a logic OR circuit configuration. Figure 1.83 shows a current *sourcing* (PNP) parallel connection.

Figure 1.84 shows a current *sinking* (NPN) parallel connection. It may be necessary to utilize blocking diodes to prevent inductive feedback (or reverse polarity) when one of the sensors in parallel is damped while the other is undamped. Figure 1.85 demonstrates the use of blocking diodes in this type of parallel connection.

1.10.6.2   Series-connection logic—AND function. Figure 1.86 shows AND function logic indicating that the series-connected devices must be ON (1) in order for the series-connected circuit to be ON.

The voltage drop across each device in series will reduce the available voltage the load will receive. Sensors, as a general rule, have a

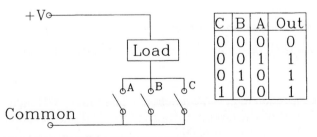

Figure 1.82   Parallel sensor arrangement.

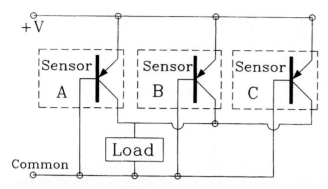

Figure 1.83 Sourcing (PNP) parallel sensor arrangement.

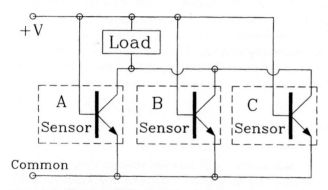

Figure 1.84 Sinking (NPN) parallel sensor arrangement.

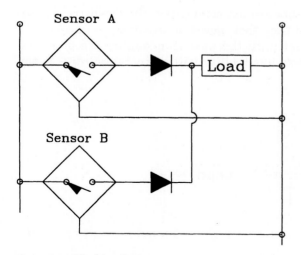

Figure 1.85 Blocking diodes.

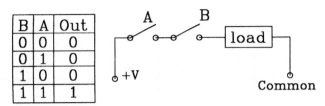

B	A	Out
0	0	0
0	1	0
1	0	0
1	1	1

Figure 1.86   Series AND logic.

7- to 9-V drop per device. The minimum operating voltage of the circuit and the sum of the voltage drop per sensor will determine the number of devices in a series-connected circuit. Figure 1.87 shows a typical two-wire ac series-connected circuit.

Series connection is generally applied to two-wire devices, most commonly two-wire ac. 10- to 30-V dc two-wire connections are not usually practical for series connection because of the voltage drop per device and minimum operating voltage. Three-wire devices are generally not used for series connection. However, the following characteristics should be considered for three-wire series-connected circuits (Fig. 1.88):

1. Each sensor must carry the load current and the burden current for all the downstream sensors (Fig. 1.88).

2. When conducting, each sensor will have a voltage drop in series with the load, reducing the available voltage to the load. As with two-wire devices, this and the minimum operating voltage will limit the number of devices wired in series.

3. When upstream sensors are not conducting, the downstream sensors are disconnected from their power source and are incapable of responding to a target until the upstream sensors are activated (damped). Time before availability will be increased due to the response in series.

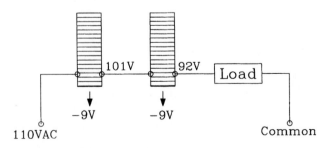

Figure 1.87   Series connected, load powered.

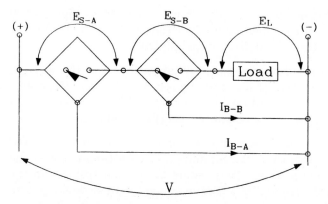

Figure 1.88   Series connected, line powered.

Series and parallel connections that perform logic functions with connection to a PLC are not common practice. Utilizing sensors this way involves the above considerations. It is usually easier to connect directly to the PLC inputs and perform the desired logic function through the PLC program.

1.10.7   Inductive and capacitive sensor response time—speed of operation

When a sensor receives initial power on system power-up, the sensor cannot operate. The sensor operates only after a delay called *time delay before availability* (Fig. 1.89).

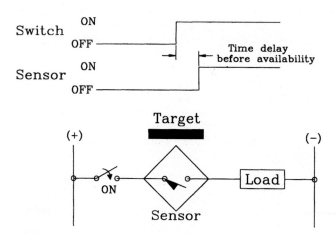

Figure 1.89   Time delay prior to availability.

In ac sensors, this delay is typically 35 ms. It can be as high as 100 ms in ac circuits with very low residual current and high noise immunity. In dc sensors, the time delay is typically 30 ms.

1.10.7.1 Response and release time. A target entering the sensing field of either an inductive or a capacitive sensor will cause the detector circuit to change state and initiate an output. This process takes a certain amount of time, called *response time* (Fig. 1.90).

Response time for an ac sensor is typically less than 10 ms. DC devices respond in microseconds. Similarly, when a target leaves the sensing field, there is a slight delay before the switch restores to the OFF state. This is the *release time*. Release time for an ac device is typically one cycle (16.66 ms). The dc device release time is typically 3 ms.

1.10.7.2 High-speed operation. Mechanical devices such as limit switches and relays do not operate at speeds suitable for high-speed counting or other fast-operating-circuit needs. Solid-state devices, however, can operate at speeds of 10, 15, or more operations per second. DC devices can operate at speeds of 500, 1000, or more operations per second.

In order to properly achieve high-speed operation, there are some basic principles that need be applied.

1.10.7.3 Maximum target length. There is a response delay when a sensor has a target entering the sensing field, as previously stated. There is a similar delay for the respective load to operate. The time from when the sensor conducts and the load operates is the *load*

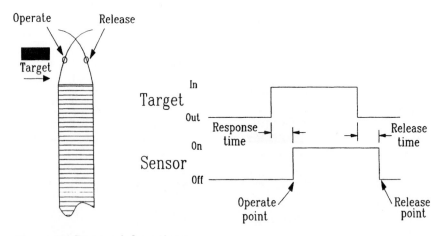

Figure 1.90 Response/release times.

response time. Together these delays make up the *system response time T_o*.

Similarly, there are delays when the target reaches the release point in the sensing field caused by the sensor release time and the corresponding *load release time*. In order to ensure that the sensor will operate reliably and repeatedly, the target must stay in the field long enough to allow the load to respond. This is called *dwell time*. Figure 1.91 illustrates the time functions for reliable, repeatable sensor operation. Figure 1.92 illustrates the dwell range.

1.10.7.4   Target duty cycle.   Response (turn-on) times for the sensor and the controlled load may be considerably different from the release (turn-off) times for the same devices. Conditions for the target duty cycles are illustrated by Fig. 1.93. Note that the target is not out of the sensing field long enough to allow the sensor to turn off the con-

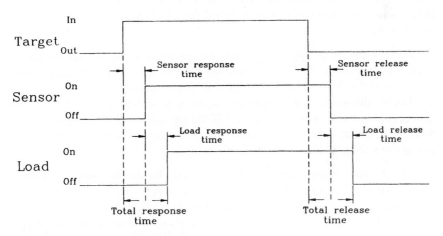

Figure 1.91   Maximum target length.

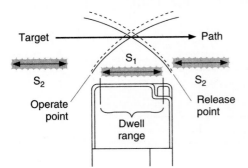

Figure 1.92   Dwell range.

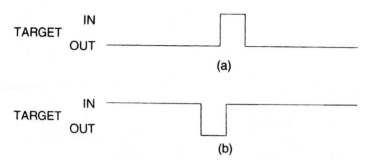

Figure 1.93 Target duty cycle. (a) Critical response time (turn-on). (b) critical release time (turn-off).

trolled load. The application must be arranged so that both sensor and load turn ON and OFF reliably and repeatedly.

1.10.7.5 Timing functions. When an inductive control is operating a logic function, an output is generated for the length of time an object is detected (Fig. 1.94).

1.10.7.6 ON delay logic. ON delay logic allows the output signal to turn on only after the object has been detected for a predetermined period of time. The output will turn off immediately after the object is no longer detected. This logic is useful if a sensor must avoid false interruption from a small object. ON delay is useful in bin fill or jam detection, since it will not false-trigger in the normal flow of objects going past (Fig. 1.95).

1.10.7.7 OFF delay logic. OFF delay logic holds the output on for a predetermined period of time after an object is no longer detected. The output is turned on as soon as the object is detected. OFF delay ensures that the output will not drop out despite a short period of signal loss. If an object is once again detected before the output times out, the signal will remain ON. OFF delay logic is useful in applications susceptible to periodic signal loss (Fig. 1.96).

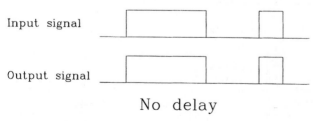

Figure 1.94 No delay.

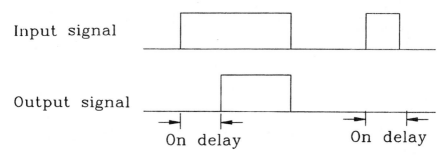

Input signal

Output signal

On delay   On delay

Figure 1.95 ON delay.

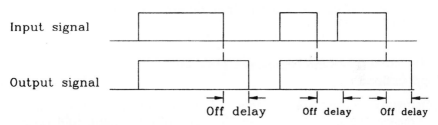

Input signal

Output signal

Off delay   Off delay   Off delay

Figure 1.96 OFF delay.

1.10.7.8 ON/OFF delay logic. ON/OFF delay logic combines ON and OFF delay so that the output will be generated only after the object has been detected for a predetermined period of time, and will drop out only after the object is no longer detected for a predetermined period of time. Combining ON and OFF delay smoothes the output of the inductive proximity control (Fig. 1.97).

1.10.7.9 One-shot delay. One-shot logic generates an output of predetermined length no matter how long an object is detected. A standard one-shot must time out before it can be retriggered. One-shot

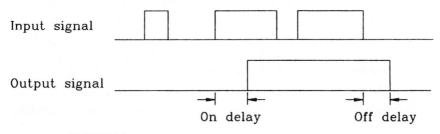

Input signal

Output signal

On delay   Off delay

Figure 1.97 ON/OFF delay.

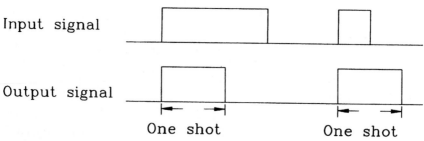

Input signal

Output signal

One shot   One shot

Figure 1.98   One-shot.

logic is useful in applications that require an output of specified length (Fig. 1.98).

1.11   Understanding Microwave Sensing Applications

Microwave sensors are valuable tools in the industrial environment for measuring motion, velocity, direction of movement, and range. They are rugged devices capable of operating in hostile environments. They are intrinsically safe, since they have no moving parts, and require low power. They will not cause any harm to operators and function effectively in explosive environments. They can successfully measure large military and industrial objects over a large distances and can provide a great deal of information about the objects, as observed during the Persian Gulf War in 1991.

Microwave technology has long been an effective method of measuring the parameters of motion and presence. Applications range from simple intrusion alarms which merely indicate that an object has entered its field of view to complex military radar systems which define the existence, location, and direction of motion.

Microwave sensing technology can be classified into five categories:

1. *Motion sensing.*   Sensing a moving object in a defined domain, for example, detecting an intruder in a prohibited area.

2. *Presence sensing.*   Sensing that an object exists in a defined domain at a given time. This concept is vital in industrial control systems where the arrival of an object may not be noticed.

3. *Velocity sensing.*   Sensing the linear speed of an object in a specified direction. This concept is used by police to detect speeding cars.

4. *Direction-of-motion sensing.*   Determining whether a target is moving away from or toward the microwave sensor device. This concept is particularly important for manufacturers of automated guided vehicle systems for obstacle avoidance. It is also used to detect whether objects or personnel are approaching or departing from automatic doors.

5. *Range sensing.*   Measuring the distance from the sensor to an object of interest. Applications include sensing the level of oil or chemical solutions in tanks and containers.

1.11.1 Characteristics of microwave sensors

Microwave sensor general characteristics important in industrial and commercial applications are

1. *No contact.*   Microwave sensors operate without actually contacting the object. This is particularly important if the object is in a hostile environment or sensitive to wear. They can monitor the speed of power-plant generator shafts, continuously monitoring acceleration and deceleration in order to maintain a constant rotational speed. Microwave sensors can effectively penetrate nonmetallic surfaces, such as fiberglass tanks, to detect liquid levels. They can also detect objects in packaged cartons.

2. *Rugged.*   Microwave sensors have no moving parts and have proven their reliability in extensive military use. They are packaged in sealed industrial enclosures to endure the rigors of the production environment.

3. *Environmental reliability.*   Microwave sensors operate reliably in harsh, inhospitable environments. They can operate from –55°C to +125°C in dusty, dirty, gusty, polluted, and poisonous areas.

4. *Intrinsically safe.*   Industrial microwave sensors can be operated in an explosive atmosphere because they do not generate sparks due to friction or electrostatic discharge. Microwave energy is so low that it presents no concern about hazard in industrial applications.

5. *Long range.*   Microwave sensors are capable of detecting objects at distances of 25 to 45,000 mm or greater, depending on the target size, microwave power available, and the antenna design.

6. *Size of microwave sensors.*   Microwave sensors are larger than inductive, capacitive, and limit switch sensors. However, use of

higher microwave frequencies and advances in microwave circuit development allow the overall package to be significantly smaller and less costly.

7. *Target size.*   Microwave sensors are better suited to detect large objects than smaller ones such as a single grain of sand.

1.11.2   Principles of operation

Microwave sensors consist of three major parts: (1) transmission source, (2) focusing antenna, and (3) signal processing receiver.

Usually the transmission and receiver are combined together in one module, which is called a *transceiver.* A typical module of this type is used by intrusion alarm manufacturers for an indoor alarm system. The transceiver contains a Gunn diode mounted in a small precession cavity which, upon application of power, oscillates at microwave frequencies. A special cavity design will cause this oscillation to occur at 10.525 GHz, which is one of the few frequencies that the U.S. Federal Communications Commission (FCC) has set aside for motion detectors. Some of this energy is coupled through an iris into an adjoining waveguide. Power output is in the 10- to 20-mW range. The dc input power for this stage (8 V at 150 mA) should be well-regulated, since the oscillator is voltage-sensitive. The sensitivity of the system can be significantly reduced by noise (interference).

At the end of the waveguide assembly, a flange is fastened to the antenna. The antenna focuses the microwave energy into a beam, the characteristics of which are determined by the application. Antennas are specified by beam width or gain. The higher the gain, the longer the range and the narrower the beam. An intrusion alarm protecting a certain domain would require a wide-beam antenna to cover the area, while a traffic control microwave sensor would require a narrow-beam, high-gain antenna to focus down the road.

Regardless of the antenna selection, when the beam of microwave energy strikes an object, some of the microwave energy is reflected back to the module. The amount of energy will depend on the composition and shape of the target. Metallic surfaces will reflect a great deal, while styrofoam and plastic will be virtually transparent. A large target area will also reflect more than a small one.

The reflected power measured at the receiver decreases by the fourth power of the distance to the target. This relationship must be taken into consideration when choosing the transmitted power, antenna gain, and signal processing circuitry for a specific application.

When the reflected energy returns to the transceiver, the mixer diode will combine it with a portion of the transmitted signal. If the target is moving toward or away from the module, the phase relationships of these two signals will change and the signal out of the mixer will be an audio frequency proportional to the speed of the target. This is called the *Doppler frequency*. This is of primary concern in measuring velocity and direction of motion. If the target is moving across in front of the module, there will not be a Doppler frequency, but there will be sufficient change in the mixer output to allow the signal processing circuitry to detect it as unqualified motion in the field.

The signal from the mixer will be in the microvolt to millivolt range so that amplification will be needed to provide a useful level. This amplification should also include 60-Hz and 120-Hz notch filters to eliminate interference from power lines and fluorescent light fixtures, respectively. The remaining bandwidth should be tailored to the application.

Besides amplification, a comparator and output circuitry relays are added to suit the application (Fig. 1.99).

1.11.3 Detecting motion with microwave sensors

The presence of an object in the microwave field disturbs the radiated field. There may be a Doppler frequency associated with the disturbance. The signal from the mixer to the signal processing circuitry may vary with a large amplitude and long duration so that it can be

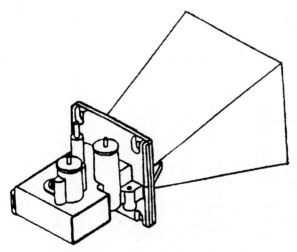

Figure 1.99 Typical microwave motion sensor module.

detected. The amplitude gain and the delay period are of specific importance in tailoring the device for particular application, such as motion detection. These sensors are primarily used in intrusion alarm applications where it is only necessary to detect the movement rather than derive further information about the intruder. The sensitivity would be set to the minimum necessary level to detect a person-sized object moving in the domain to be protected to prevent pets or other nonhostile moving objects from causing a false alarm. In addition, some response delay would be introduced for the same reason, requiring continuous movement for some short period of time.

Other applications include parts counting on conveyer belts; serial object counting in general; mold ejection monitoring, particularly in hostile environments; obstacle avoidance in automated guided vehicle systems; fill indication in tanks; and invisible protection screens. In general, this type of sensor is useful where the objects to be sensed are moving in the field of interest (Fig. 1.100).

Other devices which compete for the same applications are ultrasonic, photoelectric, and infrared sensors.

In the intrusion alarm manufacturing industry, microwave sensors have the advantages of longer range and insensitivity to certain environmental conditions. Ultrasonic sensors are sensitive to drafts and high-frequency ambient noise caused by bells and steam escaping from radiators. Infrared sensors are sensitive to thermal gradients caused by lights turning on and off. The effectiveness of infrared sensors is severely reduced at high ambient temperatures. However, utilizing dual technologies is recommended to minimize false alarms—

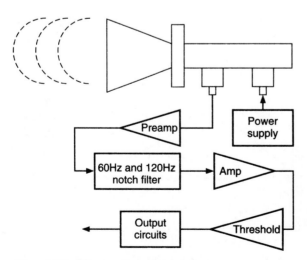

Figure 1.100 Microwave motion sensor.

combining microwave technology with infrared technology, for example. It is necessary for the intruder to be sensed by both technologies before an alarm is given. In other applications, microwave sensors can show advantages over photoelectric sensors in the areas of longer range, increased area of coverage, operation in hostile environments, and in applications where it is necessary to see through one medium (such as a cardboard box or the side of a nonmetallic tank) to sense the object on the other side.

If the target is moving toward or away from the transceiver there will be an audio-frequency (Doppler) signal out of the mixer diode which is proportional to the velocity of the target. The frequency of this signal is given by the formula:

$$F_d = 2V \, (F_t/c)$$

where F_d = Doppler frequency
V = velocity of the target
F_t = transmitted microwave frequency
c = speed of light

If the transmitted frequency is 10.525 GHz (the motion detector frequency), this equation simplifies to:

$$F_d = 31.366 \text{ Hz} \times V \qquad \text{in miles/hour}$$

or

$$F_d = 19.490 \text{ kHz} \times V \qquad \text{in kilometers/hour}$$

or

$$F_d = 84.313 \text{ kHz} \times V \qquad \text{in furlongs/fortnight}$$

This assumes that the target is traveling directly at or away from the transceiver. If there is an angle involved, then the equation becomes

$$F_d = 2V(F_t/c) \cos \Theta$$

where Θ is the angle between the transceiver and the line of movement of the target. Evidently, as the target is moving across the face of the transceiver, $\cos \Theta = 0$, and the frequency is 0. If the angle is kept below 18°, however, the measured frequency will be within 5 percent of the center frequency (Fig. 1.101).

Signal processing for this module must include amplification, a comparison network to shape the signal into logic levels, and a timing and counting circuit to either drive a display device or compare the frequency to certain limits. If more than one moving object is in the microwave field, it may be necessary to discriminate on the basis of amplitude or frequency bandwidth, limiting to exclude unwanted fre-

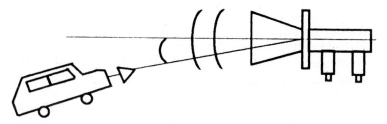

Figure 1.101   Angular velocity measurement.

quencies. Velocities near 3 km/h and 6 km/h are also difficult to measure with this system since the corresponding Doppler frequencies are 60 and 120 Hz, which are prime interference frequencies from power lines and fluorescent fixtures. Extra shielding or isolation will be necessary in this case. False alarm rate may also be reduced by counting a specific number of cycles before triggering an output. This will actually correspond to the target moving a defined distance.

Microwave sensors are well-suited for measuring velocity of objects, which most other sensors cannot do directly. Inductive, photoelectric, and other sensors can measure radial velocity. For example, inductive photoelectric sensors measure radial velocity when configured as a tachometer, and if the rotating element is configured as a trailing wheel, then linear velocity can be defined. Photoelectric sensors can also be set up with appropriate signal processing to measure the time that a moving object takes to break two consecutive beams. This restricts the measurement to a specific location. Multiple beams would be needed to measure velocity over a distance, whereas a single microwave sensor could accomplish the same result.

Aside from their use in police radars, microwave sensors can measure the speed of baseball pitches. There are many industrial applications for these sensors as well. Microwave sensors are an excellent means of closed-loop speed control of a relatively high-speed rotating shaft (3600 r/min). Other applications include autonomous-vehicle speed monitoring and independent safety monitoring equipment for heavy and high-speed machine tools. Also, a microwave sensor will detect an overvelocity condition (Fig. 1.102).

A microwave sensor, mounted on a tractor or other farm equipment to measure ground speed, will play an important role in reducing excessive distribution of seeds and fertilizer per acre. The ordinary wheel driven-speedometer is not sufficiently accurate because of wheel slippage. Accurate speed measurement is necessary in these vehicles, so that seeds and fertilizer are spread at a specific rate by

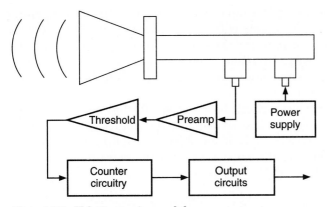

Figure 1.102   Velocity sensing module.

the accessory equipment; An over- or underestimate of the speed will result in the wrong density per acre.

1.11.4   Detecting presence with microwave sensors

A static object can be detected in the field of a microwave sensor. The purpose of this detection is to determine that the object is still in the field of interest and has not departed. This is particularly desirous in control systems where the controller is performing other tasks and then accesses the sensor to determine whether there is a sensed object at that particular time. In this situation, presence sensing is especially advantageous since the output can be verified by further interrogations to eliminate false sensing.

To detect the presence of an object, a microwave sensor with separate transmitter and receiver must be used. A transceiver in this application is not adequate, although the transmitter and the receiver can be mounted in the same enclosure. The receiver must not sense any energy unless the object is present in the field. A means to modulate the transmitter is needed, and the receiver should be narrowband to amplify and detect the modulated reflection. The sensitivity of the receiver must be adjustable to allow for ambient reflections.

Microwave sensors have been extensively and successfully tested at various fast-food drive-through vending locations. Other types of sensors such as ultrasonic and photoelectric sensors, were also tested, less successfully. They were sensitive to the environment. It was discovered that frost heaving of the ground would eventually cause their buried loop to fail, and the cost of underground excavation to replace the loop was exorbitant.

Another application of the microwave sensor is the door-opening market. The microwave sensor will check for safety reasons the area behind a swinging door to detect whether there is an individual or an object in the path way. Ultrasonic sensors may perform the same task, yet range and environmental conditions often make a microwave sensor more desirable.

A microwave sensor can check boxes to verify that objects actually have been packed therein. The sensor has the ability to see through the box itself and triggers only if an object is contained in the box. This technology relies on the sensed object being more reflective than the package, a condition that is often met.

1.11.5   Measuring velocity with microwave sensors

Microwave sensors are ideally suited to measuring linear velocity. Police radar is a simple example of a Doppler-frequency-based velocity sensor. This technology can be applied wherever it is necessary to determine velocity in a noncontact manner.

1.11.6   Detecting direction of motion with microwave sensors

Direction of motion—whether a target is moving toward or away from the microwave sensor—can be determined by the use of the Doppler-frequency concept, (Fig. 1.103) by adding an extra mixer diode to the

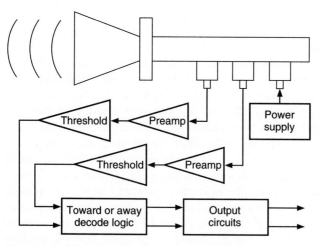

Figure 1.103   Direction of motion sensor schematic.

module. A discriminating effect is generated by the additional diode, which is located in the waveguide such that the Doppler outputs from the two mixers differ in phase by one-quarter wavelength, or 90°. These outputs will be separately amplified and converted into logic levels. The resulting signals can then be fed into a digital phase-discrimination circuit to determine the direction of motion. Such circuits are commonly found in motion control applications in conjunction with optical encoders. Figure 1.104 shows the phase relationships of the different directions.

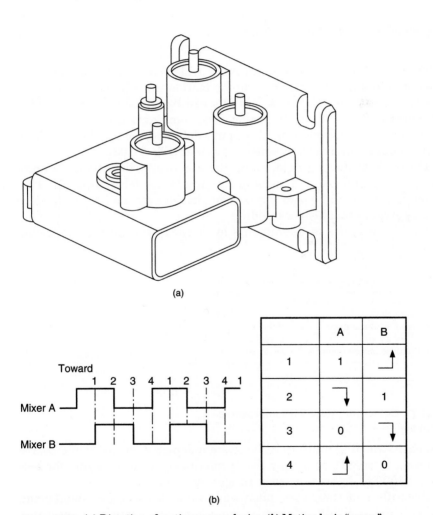

(a)

(b)

Figure 1.104   (a) Direction of motion sensor device. (b) Motion logic "away."

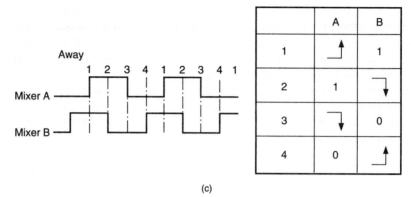

Figure 1.104   (c) Direction logic "toward."

Outputs from this module can vary widely to suit the application. The simplest is two outputs, one for motion and the other for direction (toward or away). These outputs can be added to a third, which provides the velocity of the target. The combination of signals could be analyzed to provide a final output when specific amplitude, direction, distance, and velocity criteria are met (Fig. 1.103).

In the door-opening field, using the amplitude, direction, distance, and velocity information reduces the number of false openings. This extends the life of the door mechanism, besides saving heat if the door is an entrance to a heated building.

In this case, the measurements by circuitry indicate the following characteristics:

Characteristic	Measurement
Person-sized object	Amplitude of return
Moving at walking pace	Velocity
Toward or away	Direction
Specific time before opening	Distance

1.11.7   Detecting range with microwave sensors

An early-warning military radar system depends on costly microwave sensors. A small yacht may use a microwave sensor selling for less than $1000 to detect targets at ranges up to 5 mi.

Regardless of their cost, microwave range sensors for commercial, industrial, and military applications employ essentially the same measuring technique. They transmit a narrow pulse of energy and measure the time required for the return from the target. Since microwave energy propagates at the speed of light, the time for the pulse to reach the target and return is 2 ns per foot of range. If the range to the target is 1 mi, the time required is 10.56 µs.

Although the microwave power needed is sufficient to raise the sensor temperature to 500°F, the design of the signal processing circuitry to measure the response is not difficult. However, if the target is very close to the transmitter, then the short response time may pose a real problem. At 3 ft, the time response is 6 ns. For 1-in resolution, the circuitry must be able to resolve 167 ps. This may pose a significant problem.

The alternative method to resolve a target at a short range involves changing the frequency of continuous oscillator. This method is better-suited to industrial applications. An oscillator starting at 10.525 GHz and sweeping at 50 MHz in 10 ms in the 6 ns mentioned above will have changed its frequency by:

$$(6 \text{ ns} \times 50 \text{ MHz}/0.01 \text{ s}) = 30 \text{ Hz}$$

The returning wave will be still at 10.525 GHz. The output from the mixer diode as the sweep continues will be the 30-Hz difference. If this frequency is averaged over time, it is not difficult to resolve a range to 0.001 in.

The above calculation indicates that the frequency is high for far-away objects and low for targets that are close. This leads to two conclusions:

1. The closer the object is, the lower the frequency, and therefore the longer the measurement will take.

2. The signal processing amplifier should have a gain which increases with frequency.

Problems can arise when the target is moving or there are multiple targets in the area of interest. Movement can be detected by comparing consecutive readings and can be used as a discrimination technique. Multiple targets can be defined by narrow-beam antennas to reduce the width of the area of interest. Gain adjustments are also required to eliminate all but the largest target. Audio-bandwidth filters may be used to divide the range into sectors for greater accuracy.

Other sensor types, such as photoelectric and inductive sensors, may be utilized to measure distances. Inductive sensors are used in tank level measurements. They must be coupled with a moving component which floats on the surface of the substance to be measured.

Photoelectric sensors measure position by focusing a beam on a point in space and measuring the reflection on a linear array. This can give very precise measurements over a limited range but is subject to adverse environmental conditions. Photoelectric sensors can focus a camera on a target for a better picture. Ultrasonic sensors may perform the same function, but their range is limited and they can be defeated by hostile environment.

Microwave sensors for measuring range have an impressive array of applications, including measurement of the level of liquid or solid in a tank, sophisticated intrusion alarms, autonomous guided vehicle industrial systems, and noncontact limit switching. In tank level sensing in the chemical industry (Fig. 1.105) the microwave sensor is mounted at the top of the tank and measures the distance from that position to the surface of the contents. Since the electronic circuitries can be isolated from the tank contents by a sealed window, it is intrinsically safe. It has the advantage of being a noncontact system, which means that there are no moving parts to break or be cleaned. This allows the microwave sensor to be used on aggressive chemicals, liquids, liquefied gases, highly viscous substances, and solids such as grain and coal.

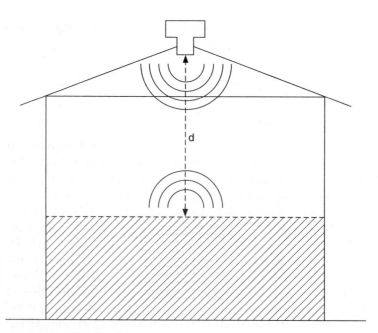

Figure 1.105 Tank level sensor.

1.11.8   Microwave technology advancement

Advances in technology have opened up important new applications for microwave sensors. The expected governmental permission to utilize higher frequencies and the decreasing size of signal processing circuitry will significantly reduce the cost of the sensors and will enable them to detect even smaller targets at a higher resolution. Microwave integrated circuit technology (MICT), presently developed for the Military Microwave Integrated Circuit (MIMIC) program will overflow into the industrial market, causing increases in performance and in analysis capabilities. Consequently, the utilization of computer-integrated manufacturing technology will be broadened.

1.12   Understanding Laser Sensors

Two theories about the nature of light have been recognized. The particle theory was the first presented to explain the phenomena that were observed concerning light. According to this theory, light is a particle with mass, producing reflected beams. It was believed that light sources actually generated large quantities of these particles. Through the years, however, many phenomena of light could not be explained by the particle theory, such as reflection of light as it passes through optically transparent materials.

The second theory considered that light was a wave, traveling with characteristics similar to those of water waves. Many, but not all, phenomena of light can be explained by this theory.

A dual theory of light has been proposed, and is presently considered to be the true explanation of light propagation. This theory suggests that light travels in small packets of wave energy called *photons*. Even though photons are *bundles* of wave energy, they have momentum like particles of mass. Thus, light is wave energy traveling with some of the characteristics of a moving particle. The total transmitted as light is the sum of energies of all the individual photons emitted.

Velocity, frequency, and wavelength are related by the equation:

$$c = f\lambda$$

where c = velocity of light, km/s
   f = frequency, Hz
   λ = wavelength, m

This equation shows that the frequency of a wave is inversely proportional to the wavelength; that is, higher-frequency waves have shorter wavelengths.

1.12.1   Properties of laser light

Laser stands for *light amplification by stimulated emission of radiation.* Laser light is monochromatic, whereas standard white light consists of all the colors in the spectrum and is broken into its component colors when it passes through a standard glass prism (Figs. 1.106, 1.107).

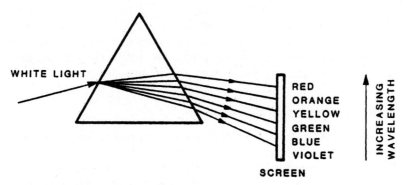

Figure 1.106   Standard white light.

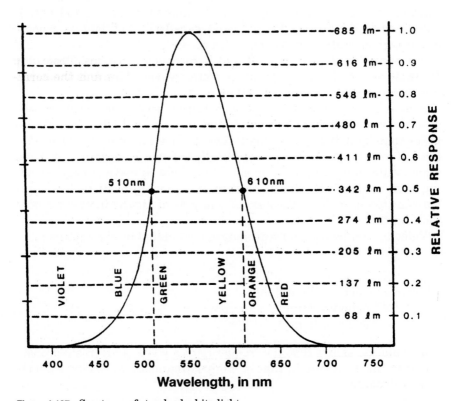

Figure 1.107   Spectrum of standard white light.

1.12.2   Essential laser components

Laser systems consist of four essential components:

1. The active medium
2. The excitation mechanism
3. The feedback mechanism
4. The output coupler

1.12.2.1   The active medium.   The active medium is the collection of atoms, ions, or molecules in which stimulated emission occurs. It is in this medium that laser light is produced. The active medium can be a solid, liquid, gas, or semiconductor material. Often the laser takes its name from that of the active medium. For example, the ruby laser has a crystal of ruby as its active medium while the CO_2 laser has carbon dioxide gas.

The wavelength emitted by a laser is a function of the active medium. This is because the atoms within the active medium have their own characteristic energy levels at which they release photons. It will be shown later that only certain energy levels within the atom can be used to enhance stimulated emission. Therefore a given active medium can produce a limited number of laser wavelengths, and two different active media cannot produce the same wavelengths. Table 1.10 contains a list of materials commonly used in lasers and the corresponding wavelengths that these materials produce.

The active medium is the substance that actually lases. In the helium-neon laser, only the helium lases.

TABLE 1.10   Wavelengths of Laser Materials

Type of active medium	Common material	Wavelength produced, nm
Solid	Ruby	694
	Nd:YAG	1,060
	Nd:glass	1,060
	Erbium	1,612
Liquid	Organic dyes	360–650
Gas	Argon (ionized)	488
	Helium-neon	632.8
	Krypton (ionized)	647
	CO_2	10,600
Semiconductor	Gallium arsenide	850
	Gallium antimonide	1,600
	Indium arsenide	3,200

1.12.2.2 Excitation mechanism. The excitation mechanism is the device used to put energy into the active medium. There are three primary types of excitation mechanisms: optical, electrical, and chemical. All three provide the energy necessary to raise the energy state of the atom, ion, or molecule of the active medium to an excited state. The process of imparting energy to the active medium is called *pumping the laser.*

1.12.2.2.1 Optical excitation. An optical excitation mechanism uses light energy of the proper wavelength to excite the active medium. The light may come from any of several sources, including a flash lamp, a continuous arc lamp, another laser, or even the sun. Although most of these use an electric power supply to produce the light, it is not the electrical energy that is used directly to excite the atoms of the active medium but rather the light energy produced by the excitation mechanism.

Optical excitation is generally used with active media that do not conduct electricity—solid lasers like the ruby. Fig. 1.108 is a schematic drawing of a solid laser with an optical pumping source.

The sun is considered a possible optical pumping source for lasers in space. The optical energy from the sun could be focused by curved mirrors onto the laser's active medium. Since the size and weight of an electric power supply is of concern in space travel, solar pumping of lasers is an interesting alternative.

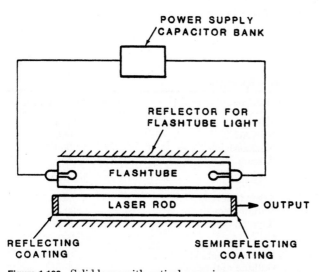

Figure 1.108 Solid laser with optical pumping source.

1.12.2.2.2 Electrical excitation. Electrical excitation is most commonly used when the active medium will support an electric current. This is usually the case with gases and semiconductor materials.

When a high voltage is applied to a gas, current-carrying electrons or ions move through the active medium. As they collide with the atoms, ions, or molecules of the active medium, their energy is transferred and excitation occurs. The atoms, ions, and electrons within the active medium are called *plasma.*

Figure 1.109 is a schematic drawing of a gas laser system with electrical excitation. The gas mixture is held in a gas plasma tube and the power supply is connected to the ends of the plasma tube. When the power supply is turned on, electron movement within the tube is from the negative to the positive terminal.

1.12.2.2.3 Chemical excitation. Chemical excitation is used in a number of lasers. When certain chemicals are mixed, energy is released as chemical bonds are made or broken. This energy can be used as a pumping source. It is most commonly used in hydrogen-fluoride lasers, which are extremely high-powered devices used primarily in military weapons and research. These lasers are attractive for military applications because of the large power-to-weight ratio.

1.12.2.3 Feedback mechanism. Mirrors at each end of the active medium are used as a *feedback mechanism.* The mirrors reflect the light produced in the active medium back into the medium along its longitudinal axis. When the mirrors are aligned parallel to each other, they form a resonant cavity for the light waves produced within the laser. They reflect the light waves back and forth through the active medium.

In order to keep stimulated emission at a maximum, light must be

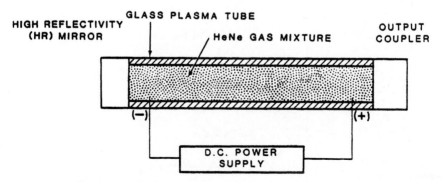

Figure 1.109 Gas laser with electrical excitation.

kept within the amplifying medium for the greatest possible distance. In effect, mirrors increase the distance traveled by the light through the active medium. The path that the light takes through the active medium is determined by the shape of the mirrors. Figure 1.110 shows some of the possible mirror combinations. Curved mirrors are often used to alter the direction in which the reflected light moves.

1.12.2.4 Output coupler. The feedback mechanism keeps the light inside the laser cavity. In order to produce an output beam, a portion of the light in the cavity must be allowed to escape. However, this escape must be controlled. This is most commonly accomplished by using a partially reflective mirror in the feedback mechanism. The amount of reflectance varies with the type of laser. A high-power laser may reflect as little as 35 percent, with the remaining 65 percent being transmitted through the mirror to become the output laser beam. A low-power laser may require an output mirror reflectivity as high as 98 percent, leaving only 2 percent to be transmitted. The output mirror that is designed to transmit a given percentage of the laser light in the cavity between the feedback mirrors is called the *output coupler.*

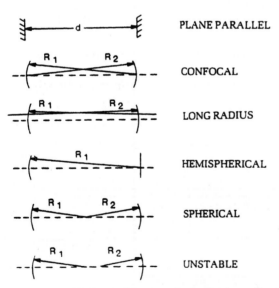

Figure 1.110 Mirror combinations for feedback mechanism.

1.12.3 Semiconductor displacement laser sensors.

Semiconductor displacement laser sensors, consisting of a light-metering element and a position-sensitive detector (PSD), detect targets by using triangulation. A light-emitting diode or semiconductor laser is used as the light source. A semiconductor laser beam is focused on the target by the lens. The target reflects the beam, which is then focused on the PSD, forming a beam spot. The beam spot moves on the PSD as the target moves. The displacement of the workpiece can then be determined by detecting the movement of the beam spot.

1.12.3.1 Industrial applications of semiconductor displacement lasers.
The laser beam emitted from laser diode in the transmitter is converged into a parallel beam by the lens unit. The laser beam is then directed through the slit on the receiver and focused on the light-receiving element. As the target moves through the parallel laser beam, the change in the size of the shadow is translated into the change in received light quantity (voltage). The resulting voltage is used as a comparator to generate an analog output voltage.

1.12.4 Industrial applications of laser sensors*

Electrical and electronics industries:

1. *Warpage and pitch of IC leads.*   The visible beam spot facilitates the positioning of the sensor head for small workpieces. Warpage and pitch can be measured by scanning IC leads with the sensor head (Fig. 1.111).

2. *Measurement of lead pitch of electronic components.*   The sensor performs precise noncontact measurement of pitch using a laser beam (Fig. 1.112).

3. *Measurement of disk head movement.*   The laser sensor is connected to a computer in order to compare the pulse input to the disk head drive unit with actual movement. The measurement is done on-line, thus increasing productivity (Fig. 1.113).

---

*A few nonlaser optical sensors are included, as indicated.

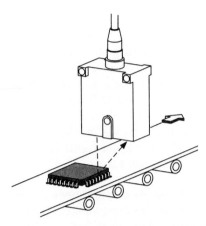

Figure 1.111 Warpage and pitch of IC lead.

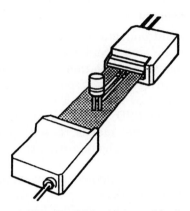

Figure 1.112 Measurement of lead pitch of electronic components.

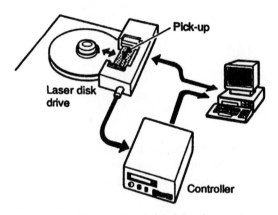

Figure 1.113 Measurement of disk head movement.

4. *Detection of presence/absence of resin coating.* The laser displacement sensor determines whether a resin coating was formed after wire bonding (Fig. 1.114).

5. *Detection of double-fed or mispositioned resistors prior to taping.* Through-beam-type sensor heads are positioned above and below the resistors traveling on a transfer line. A variation on the line changes the quantity of light in the laser beam, thus signaling a defect (Fig. 1.115).

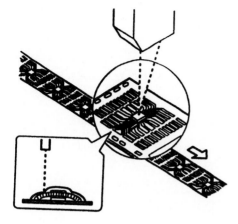

Figure 1.114   Detection of presence/absence of resin coating.

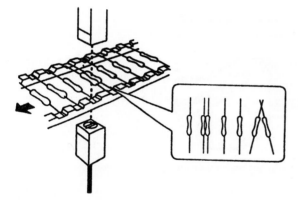

Figure 1.115   Detection of double-fed or mispositioned resistors.

6. *Detection of defective shrink wrapping of videocassette.*   Defective film may wrap or tear during shrink wrapping. The laser sensor detects defective wrapping by detecting a change in the light quantity on the surface of the videocassette (Fig. 1.116).

7. *Measurement of gap between roller and doctor blade.*   Measures the gap between the roller and the doctor blade in submicrometer units. The sensor's automatic measurement operation eliminates reading errors (Fig. 1.117).

8. *Measurement of surface run-out of laser disk.*   The surface run-out of a laser disk is measured at a precision of 0.5 µm. The sensor head enables measurement on a mirror-surface object (Fig. 1.118).

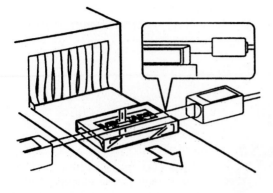

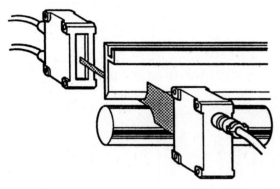

Figure 1.116 Detection of defective shrink wrapping of videocassette.

Figure 1.117 Measurement of gap between roller and doctor blade.

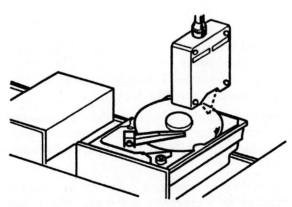

Figure 1.118 Measurement of surface run-out of laser disk.

9. *Displacement of printer impact pins.*   The visible beam spot facilitates positioning of the head of a pin-shaped workpiece, enabling measurement of the vertical displacement of impact pins (Fig. 1.119).

Automotive manufacturing industries:

1. *Measurement of thickness of connecting rod.*   Measures the thickness of the connecting rod by processing the analog inputs in the digital meter relay (Fig. 1.120).

2. *Measurement of depth of valve recesses in piston head.*   Measures the depth of the valve recesses in the piston head so that chamber capacity can be measured. Iron jigs are mounted in front of the sensor head, and the sensor measures the distance the jigs travel when they are pressed onto the piston head (Fig. 1.121).*

3. *Measurement of height of radiator fin.*   Detects improper radiator fin height by comparing the bottom value of the analog output with a stored pair of tolerances (Fig. 1.122).

4. *Measurement of outer diameter of engine valve.*   The laser scan micrometer allows on-line measurement of the outer diameter of engine valves simply by positioning a separate sensor head on either side of the conveyer (Fig. 1.123).

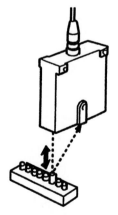

Figure 1.119   Displacement of printer impact pins.

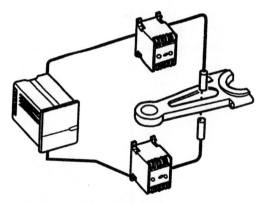

Figure 1.120   Measurement of thickness of connecting rod.

---

*Nonlaser sensor.

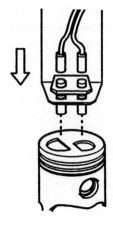

Figure 1.121 Measurement of depth of valve recesses in piston head.

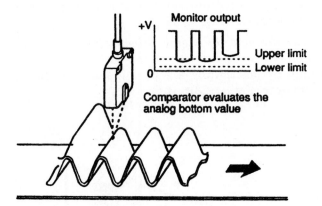

Figure 1.122 Measurement of height of radiator fin.

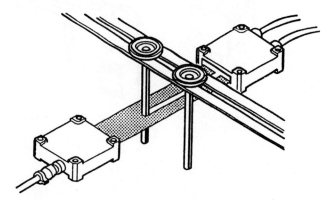

Figure 1.123 Measurement of outer diameter of engine valve.

5. *Positioning of robot arm.*   The laser displacement sensor is used to maintain a specific distance between the robot arm and target. The sensor outputs a plus or minus voltage if the distance becomes greater or less, respectively, than the 100-mm reference distance (Fig. 1.124).

6. *Detection of damage on microdiameter tool.*   Detects a break, chip, or excess swarf from the variation of light quantity received (Fig. 1.125).

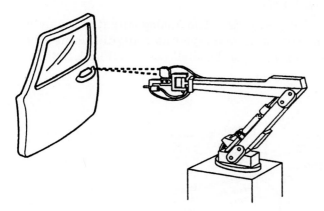

Figure 1.124   Positioning of robot arm.

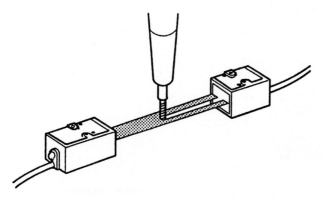

Figure 1.125   Detection of damage on microdiameter tool.

Metal/steel/nonferrous industries:

1. *Detection of misfeeding in high-speed press.*   The noncontact laser sensor, timed by a cam in the press, confirms the material feed by monitoring the pilot holes and outputs the result to an external digital meter relay (Fig. 1.126).

2. *Simultaneous measurement of outer diameter and eccentricity of ferrite core.*   Simultaneously measures the outer diameter and eccentricity of a ferrite core with a single sensor system. The two measured values can then be simultaneously displayed on a single controller unit (Fig. 1.127).

3. *Confirmation of roller centering.*   The analog output of the inductive displacement sensor is displayed as a digital value, thus allowing a numerical reading of the shaft position (Fig. 1.128).

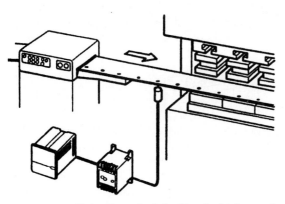

Figure 1.126   Detection of misfeeding in high-speed press.

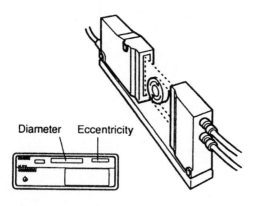

Figure 1.127   Simultaneous measurement of outer diameter and eccentricity of ferrite core.

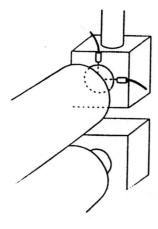

Figure 1.128 Confirmation of roller centering.

4. *Measurement of the height and inner diameter of sintered metal ring.* Determines the height and inner diameter of the metal ring by measuring the interrupted areas of the parallel laser beam (Fig. 1.129).

5. *Measurement of outer diameter of wire in two axes.* Simultaneously measures in *x* axis to determine the average value of the outer diameter, thereby increasing dimensional stability (Fig. 1.130).

6. *Measurement of outer diameter after centerless grinding.* The scanning head allows continuous noncontact measurement of a metal shaft immediately after the grinding process (Fig. 1.131).

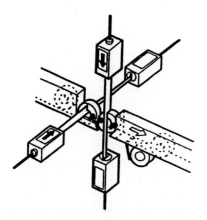

Figure 1.129 Measurement of the height and inner diameter of sintered metal ring.

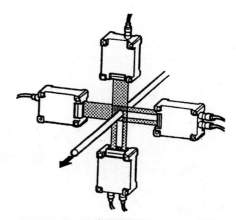

Figure 1.130 Measurement of outer diameter of wire in two axes.

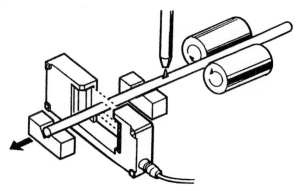

Figure 1.131 Measurement of outer diameter after center-less grinding.

Food processing and packaging:

1. *Detection of material caught during heat sealing.*   Detects materi-al caught in the distance between rollers (Fig. 1.132).

2. *Detection of missing or doubled packing ring in cap.*   Detects a missing or doubled rubber ring in caps by using the comparator to evaluate sensor signals (Fig. 1.133).

3. *Detection of incorrectly positioned small objects.*   The transmitter and the receiver are installed to allow a parallel light beam to scan slightly above the tablet sheets. When a single tablet projects from a line of tablets, the optical axis is interrupted and the light quan-tity changes (Fig. 1.134).

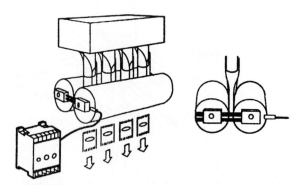

Figure 1.132 Detection of material caught during heat sealing.

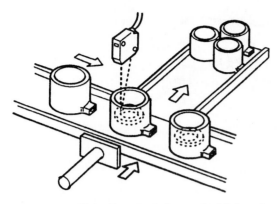

Figure 1.133 Detection of missing or doubled packing ring in cap.

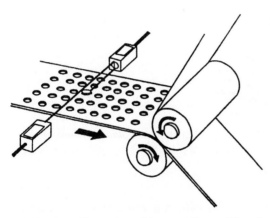

Figure 1.134 Detection of incorrectly positioned small objects.

4. *Measurement of tape width.* Measures the width of running tape to the submicrometer level; 100 percent inspection improves product quality (Fig. 1.135).

5. *Measurement of sheet thickness.* Use of two controllers enables thickness measurement by determining the distance in input values. Thus, thickness measurement is not affected by roller eccentricity (Fig. 1.136).

Automatic machinery:

1. *Detection of surface run-out caused by clamped error.* Improper clamping due to trapped chips will change the rotational speed of the workpiece. The multifractional digital meter relay sensor cal-

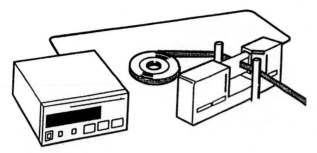

Figure 1.135 Measurement of tape width.

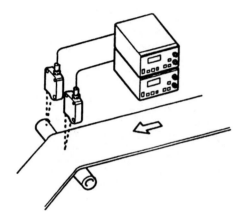

Figure 1.136 Measurement of sheet thickness.

culates the surface run-out of the rotating workpiece, compares it with the stored value, and outputs a detection signal (Fig. 1.137).

2. *Detection of residual resin in injection molding machine.*   When the sensor heads are such that the optical axis covers the surface of the die, any residual resin will interface with this axis (Fig. 1.138).

3. *Measurement of travel of camera lens.*   A separate sensor can be installed without interfering with the camera body, thus assuring a highly reliable reading of lens travel (Fig. 1.139).

4. *Measurement of rubber sheet thickness.*   With the segment function that allows the selection of measuring points, the thickness of a rubber sheet (i.e., the distance between the rollers) can be easily measured (Fig. 1.140).

5. *Measurement of stroke of precision table.*   Detects even minute strokes at a resolution of 0.05 μm. In addition, the AUTO ZERO function allows indication of relative movement (Fig. 1.141).

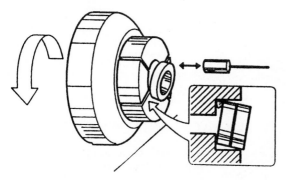

Figure 1.137 Detection of surface run-out caused by clamped error.

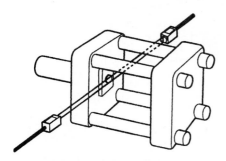

Figure 1.138 Detection of residual resin.

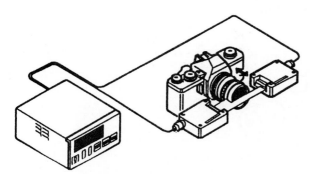

Figure 1.139 Measurement of travel of camera lens.

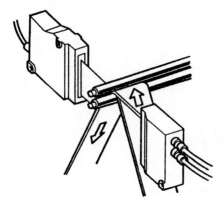

Figure 1.140 Measurement of rubber sheet thickness.

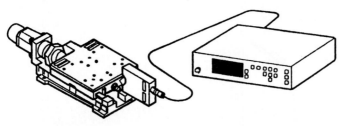

Figure 1.141 Measurement of stroke of precision table.

6. *Measurement of plasterboard thickness.* A sensor head is placed above and below the plasterboard, and its analog outputs are fed into a digital meter relay. The meter relay indicates the absolute thickness value (Fig. 1.142).

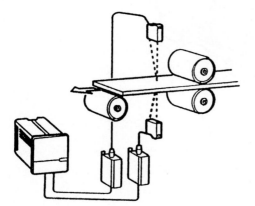

Figure 1.142 Measurement of plasterboard thickness.

Further Reading

Chappel, A. (ed.), *Optoelectronics: Theory and Practice,* McGraw-Hill, New York, 1978.

Doebelin, E. O., *Measurement Systems: Application and Design,* 4th ed., McGraw-Hill, New York, 1990.

Holliday, D., and R. Resnick, *Physics,* Wiley, New York, 1975.

International Organization for Standardization, "Statistical Interpretation of Data: Comparison of Two Means in the Case of Paird Observations," ISO 3301-1975.

Lion, K. L., *Elements of Electrical and Electronic Instrumentation,* McGraw-Hill, New York, 1975.

Neubert, H. K. P., *Instrument Transducers,* 2d ed., Clarendon Press, Oxford, 1975.

Ogata, K., *Modern Control Engineering,* 2d ed., Prentice-Hall, Englewood Cliffs, N.J, 1990.

Rock, I., *Lightness Constancy, Perception,* W. H. Freeman, New York, 1984.

Seippel, R. G., *Optoelectronics,* Reston Publishing Co., Reston, Va., 1981.

Shortley, G., and D. Williams, *Quantum Property of Radiation,* Prentice-Hall, Englewood Cliffs, N.J., 1971.

Todd, C. D. (Bourns Inc.), *The Potentiometer Handbook,* McGraw-Hill, New York, 1975.

White, R. M., "A Sensor Classification Scheme," *IEEE Trans. Ultrasonics, Ferroelectrics, and Frequency Control,* March, 1987.

2

Fiber Optics in Sensors and Control Systems

2.0 Introduction

Accurate position sensing is crucial to automated motion control systems in manufacturing. The most common components used for position sensing are photoelectric sensors, inductive proximity sensors, and limit switches. They offer a variety of options for manufacturing implementation from which highly accurate and reliable systems can be created. Each option has its features, strengths, and weaknesses that manufacturing personnel should understand for proper application. There are three types of sensors used in manufacturing applications:

1. *Photoelectric sensors.* Long-distance detection

2. *Inductive proximity sensors.* Noncontact metal detection

3. *Limit switches.* Detection with traditional reliability

2.1 Photoelectric Sensors—Long-Distance Detection

A photoelectric sensor is a switch that is turned on and off by the presence or absence of receiving light (Fig. 2.1). The basic components of a photoelectric sensor are a power supply, a light source, a photodetector, and an output device. The key is the photodetector, which is made of silicon, a semiconductor material that conducts current in the presence of light. This property is used to control a variety of output devices vital for manufacturing operation and control, such as mechanical relays, triacs, and transistors, which in turn control machinery.

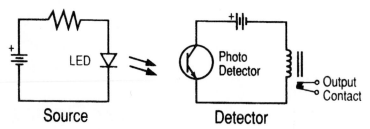

Figure 2.1 Photoelectric sensor.

Early industrial photoelectric controls used focused light from incandescent bulbs to activate a cadmium sulfide photocell (Fig. 2.2). Since they were not modulated, ambient light such as that from arc welders, sunlight, or fluorescent light fixtures could easily false-trigger these devices. Also, the delicate filament in the incandescent bulbs had a relatively short life span, and did not hold up well under high vibration and the kind of shock loads normally found in an industrial environment. Switching speed was also limited by the slow response of the photocell to light/dark changes (Fig. 2.1).

2.1.1 Light-emitting diodes

Photoelectric sensors use an effective light source, light-emitting diodes (LEDs), which were developed in the early 1960s. LEDs are solid-state devices that emit light when current is applied (Fig. 2.3). This is the exact opposite of the photodetector, which emits current when light is received.

LEDs have several advantages over incandescent bulbs and other light sources. LEDs can be turned on and off very rapidly, are

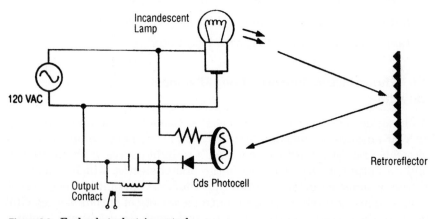

Figure 2.2 Early photoelectric control.

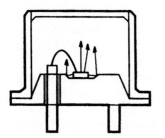

Light Emitting Diodes (LED's)
- Fast turn-on and turn-off
- No warm-up
- Small
- Rugged
- Low power consumption
- High radiant efficiency
- Long life

Figure 2.3   Light-emitting diode.

extremely small, consume little power, and last as long as 100,000 continuous hours. Also, since LEDs are solid-state devices, they are much more immune to vibration than incandescent bulbs.

LEDs emit light energy over a narrow wavelength (Fig. 2.4a). Infrared (ir) gallium arsenide LEDs emit energy only at 940 nm (Fig. 2.4b). As this wavelength is at the peak of a silicon photodiode's response, maximum energy transfer between source and detector is achieved.

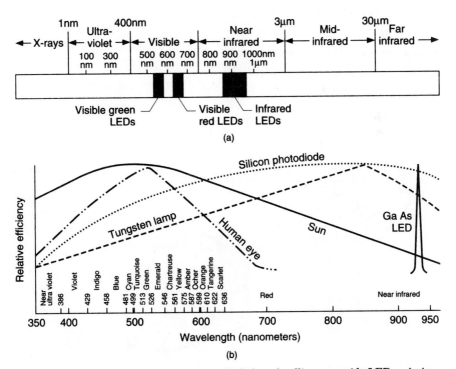

Figure 2.4   (a) LED emission wavelengths. (b) Infrared gallium arsenide LED emission.

A silicon photodetector's sensitivity to light energy also peaks in the infrared light spectrum. This contributes to the high efficiency and long range possible when silicon photodetectors are used in conjunction with gallium arsenide LEDs.

In recent years, visible LEDs have been introduced as light sources in photoelectric controls. Because the beam is visible to the naked eye, the principle advantage of visible LEDs is ease of alignment. Visible beam photoelectric controls usually have lower optical performance than ir LEDs.

The modes of detection for optical sensors are (1) through-beam and (2) reflection (diffuse reflection and reflex detection).

2.1.2 Through-beam sensors

Through-beam sensors have separate source and detector elements aligned opposite each other, with the beam of light crossing the path that an object must cross (Fig. 2.5). The effective beam area is that of the column of light that travels straight between the lenses (Fig. 2.6).

Because the light from the source is transmitted directly to the photodetector, through-beam sensors offer the following benefits:

1. Longest range for sensing

2. Highest possible signal strength

3. Greatest light/dark contrast ratio

4. Best trip point repeatability

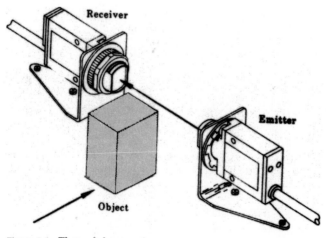

Figure 2.5 Through-beam sensor.

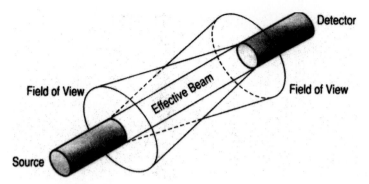

Figure 2.6 Effective beam area.

The limitations of through-beam sensors are as follows:

1. They require wiring of the two components across the detection zone.
2. It may be difficult to align the source and the detector.
3. If the object to be detected is smaller than the effective beam diameter, an aperture over the lens may be required (Fig. 2.7).

2.1.3 Reflex photoelectric controls

Reflex photoelectric controls (Fig. 2.8) position the source and detector parallel to each other on the same side of the target. The light is directed to a retroreflector and returns to the detector. The switching and output occur when an object breaks the beam.

Since the light travels in two directions (hence twice the distance),

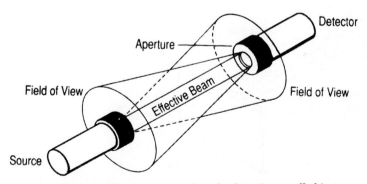

Figure 2.7 Sensor with aperture over lens for detecting small objects.

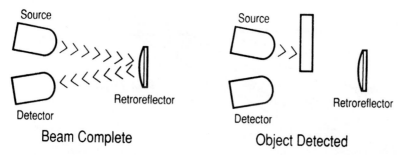

Source

Detector

Beam Complete

Retroreflector

Source

Detector

Object Detected

Retroreflector

Figure 2.8 Reflex photoelectric controls.

reflex controls will not sense as far as through-beam sensors. However, reflex controls offer a powerful sensing system that is easy to mount and does not require that electrical wire be run on both sides of the sensing area. The main limitation of these sensors is that a shiny surface on the target object can trigger false detection.

2.1.4 Polarized reflex detection

Polarized reflection controls use a polarizing filter over the source and detector that conditions the light such that the photoelectric control sees only light returned from the reflector (Fig. 2.9). A polarized reflex sensor is used in applications where shiny surfaces such as metal or shrink-wrapped boxes may false-trigger the control.

Polarized reflex sensing is achieved by combining some unique properties of polarizers and retroreflectors. These properties are (1) polarizers pass light that is aligned along only one plane and (2) cor-

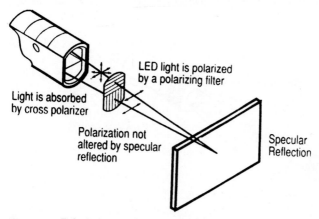

LED light is polarized by a polarizing filter

Light is absorbed by cross polarizer

Polarization not altered by specular reflection

Specular Reflection

Figure 2.9 Polarization reflection controls.

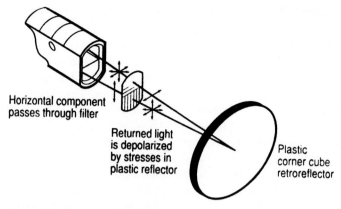

Horizontal component
passes through filter

Returned light
is depolarized
by stresses in
plastic reflector

Plastic
corner cube
retroreflector

Figure 2.10   Corner-cube reflector.

ner-cube reflectors depolarize light as it travels through the face of
the retroreflector (Fig. 2.10).

Light from the source is aligned by a polarizer. When this light
reflects off the retroreflector, it is depolarized. The returning light
passes through another polarizing filter in front of the detector. The
detector's polarizer is oriented at 90° to the source's polarizer. Only
the light which has been rotated by the corner cube retroreflector can
pass through the detector's polarizer. Light that bounces off other
shiny objects, and has not been rotated 90°, cannot pass through the
detector's polarizer, and will not trigger the control.

Polarized reflex sensors will not work with reflective tape containing glass beads. Also, shiny objects wrapped with clear plastic shrink-
wrap will potentially false-trigger a polarized reflex control, since
under certain conditions these act as a corner-cube reflector.

The polarized reflex detection sensor has the following advantages:

1. It is not confused by the first surface reflections from target
   objects.

2. It has a high dark/light contrast ratio.

3. It is easily installed and aligned. One side of the sensing zone only
   need be wired.

   It also has certain limitations:

1. Operating range is half that of a nonpolarized sensor since much
   of the signal is lost in the polarizing filters.

2. The sensor can be fooled by shiny objects wrapped with shrink-
   wrap material.

2.1.5 Proximity (diffuse-reflection) detection

Proximity detection is similar to reflex detection, because the light source and detector elements are mounted on the same side (Fig. 2.11). In this application, the sensors detect light that is bounced off the target object, rather than the breaking of the beam. The detection zone is controlled by the type, texture, and composition of the target object's surface.

Focused proximity sensors are a special type of proximity sensor where the source and the detector are focused to a point in front of the sensor (Fig. 2.12). Focused proximity sensors can detect extremely small objects, or look into holes or cavities in special applications. Background objects will not false-trigger a focused proximity sensor since they are "cross-eyed" and cannot see past a certain point.

Advantages of the focused proximity sensor are

1. Installation and alignment are simple. The control circuit can be wired through only one side of the sensing zone.

2. It can detect differences in surface reflectivity.

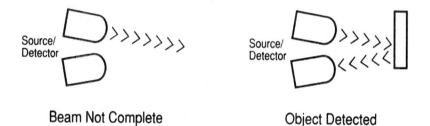

Beam Not Complete	**Object Detected**

Figure 2.11 Proximity detection.

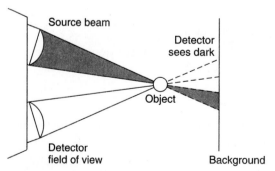

Figure 2.12 Focused proximity sensor.

It also has certain limitations:

1. It has limited sensing range.
2. The light/dark contrast and sensing range depend on the target object's surface reflectivity.

2.1.6 Automated guided vehicle system

The wide sensing field of diffuse reflective photoelectric sensors makes them ideally suited for use as obstacle detection sensors. Sensing distance and field width are adjustable to form the obstacle detection zone vital for manufacturing operations.

Two outputs, *near* and *far,* provide switching at two separate sensing distances which are set by corresponding potentiometers. The sensing distance of the far output is adjustable up to 3 m maximum. The sensing distance of the near output is adjustable from 30 to 80 percent of the far output. Indicators include a red LED which glows with the near output ON and a yellow LED which glows with the far output ON.

An ideal application for this family of sensors is the automated guided vehicle (AGV), which requires both slow-down and stop controls to avoid collisions when obstacles enter its path (Fig. 2.13).

A modulated infrared light source provides immunity to random operation caused by ambient light. Additionally, unwanted sensor operation caused by adjacent sensor interference (crosstalk) is also eliminated through the use of multiple-position modulated frequency adjustments.

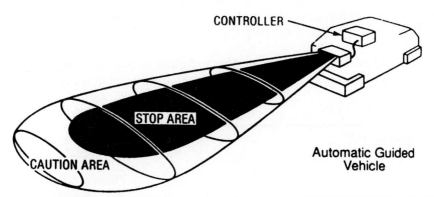

Figure 2.13 Application of diffuse reflective photoelectric sensor in automated guided vehicle system.

2.2 Fiber Optics

Fiber optics has greatly expanded the applications of photoelectric sensors. Fiber optics uses bundles of thin plastic or glass fibers that operate on a principle discovered in 1854 by John Tyndahl. When Tyndahl shined a beam of light through a stream of water, instead of emerging straight from the stream of water as might be expected, the light tended to bend with the water as it arced towards the floor. Tyndahl discovered that the light was transmitted along the stream of water. The light rays inside the water bounced off the internal walls of the water and were thereby contained (Fig. 2.14). This principle has come to be known as *total internal reflection.*

Industry has since discovered that the principle of total internal reflection also applies to small-diameter glass and plastic fibers, and this has lead to rapid growth of applications throughout the industry. Because optical fibers are small in diameter and flexible, they can bend and twist into confined places. Also, because they contain no electronics, they can operate in much higher temperatures—as high as 400°F—and in areas of high vibration. They are limited by sensing distances, which typically are 80 mm in the proximity mode and 400 mm in the through-beam mode. Also, because of their small sensing area, optical fibers can be fooled by a small drop of water or dirt over the sensing area.

Fiber optics is used to transmit data in the communication field and to transmit images or light in medicine and industry. Photoelectric controls use fiber optics to bend the light from the LED source and return it to the detector so sensors can be placed in locations where common photoelectric sensors could not be applied.

Fiber optics used with photoelectric controls consists of a large number of individual glass or plastic fibers which are sheathed in suitable material for protection. The optical fibers used with photoelectric controls are usually covered by either PVC or stainless-steel jackets. Both protect the fibers from excessive flexing and the environment (Fig. 2.15).

Optical fibers are transparent fibers of glass or plastic used for conducting and guiding light energy. They are used in photoelectric sen-

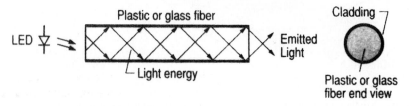

Figure 2.14 Total internal reflection.

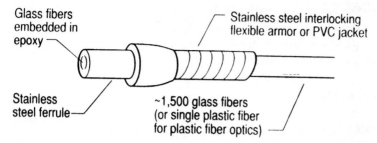

Figure 2.15 Jacketed glass fibers.

sors as "light pipes" to conduct sensing light into and out of a sensing area.

Glass optical-fiber assemblies consist of a bundle of 0.05-mm-diameter discrete glass optical fibers housed within a flexible sheath. Glass optical fibers are also able to withstand hostile sensing environments. Plastic optical-fiber assemblies consist of one or two acrylic monofilaments in a flexible sheath.

There are two basic styles of fiber-optic assemblies: (1) individual fiber optics (Fig. 2.16) and (2) bifurcated fiber optics (Fig. 2.17).

Individual fiber-optic assemblies guide light from an emitter to a sensing location, or to a receiver from a sensing location. Bifurcated fibers use half their fiber area to transmit light and the other half to receive light.

2.2.1 Individual fiber optics

A fiber-optic assembly having one control end and one sensing end is used for piping photoelectric light from an emitter to the sensing loca-

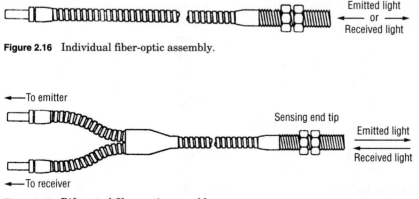

Figure 2.16 Individual fiber-optic assembly.

Figure 2.17 Bifurcated fiber-optic assembly.

tion or from the sensing location back to a receiver. It is usually used in pairs in the opposed sensing mode, but can also be used side by side in the diffuse proximity mode or angled for the specular reflection or mechanical convergent mode.

2.2.2 Bifurcated fiber optics

A bifurcated fiber-optic assembly is branched to combine emitted light with received light in the same assembly. Bifurcated fibers are used for diffused (divergent) proximity sensing, or they may be equipped with a lens for use in the retroreflective mode.

There are three types of sensing modes used in positioning of a sensor so that the maximum amount of emitted energy reaches the receiver sensing element:

1. Opposed sensing mode (Fig. 2.18)

2. Retroreflective sensing mode (Fig. 2.19)

3. Proximity (diffused) sensing mode (Fig. 2.20)

Opposed sensing is the most efficient photoelectric sensing mode and offers the highest level of optical energy to overcome lens contamination, sensor misalignment, and long scanning ranges. It is also often referred to as *direct scanning* and sometimes called the *beam break* mode.

The addition of fiber optics to photoelectric sensing has greatly expanded the application of photoelectric devices. Because they are small in diameter and flexible, optical fibers can bend and twist into tiny places formerly inaccessible to bulky electronic devices.

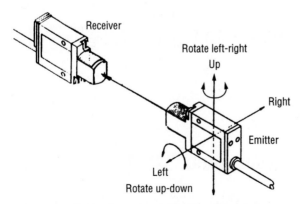

Figure 2.18 Opposed sensing mode. For alignment, move emitter or receiver up-down, left-right, and rotate.

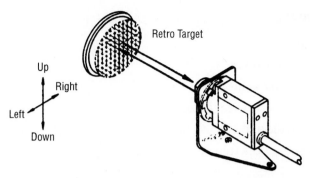

Figure 2.19 Retroreflective sensing mode. For alignment, move target up-down, left-right.

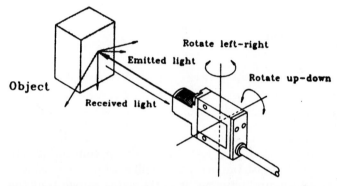

Figure 2.20 Proximity sensing mode. For alignment, rotate up-down, left-right.

Optical fibers operate in the same sensing modes as standard photoelectric controls—through-beam, proximity, and reflex. The sizes and shapes of sensing tips have been developed to accommodate many applications.

Optical fibers have a few drawbacks:

1. Limited sensing distance. Typical sensing distance in the proximity mode is 80 mm; 380 mm for the through-beam mode.

2. Typically more expensive than other photoelectric sensing controls.

3. Easily fooled by a small drop of water or dirt over the sensing surface.

Optical fibers' advantages are

1. Sensing in confined places
2. Ability to bend around corners
3. No electronics at sensing point
4. Operation at high temperatures (glass)
5. Total immunity from electrical noise and interference
6. Easily cut to desired lengths (plastic)

2.3   Optical Fiber Parameters

The most important parameters affecting optical-fiber performance are

1. Excess gain
2. Background suppression
3. Contrast
4. Polarization

2.3.1   Excess gain

Excess gain is the measure of energy available between the source and the detector to overcome signal loss due to dirt or contamination. Excess gain is the single most important consideration in choosing a photoelectric control in manufacturing. It is the extra punch that the sensor has available within its detecting region.

By definition, excess gain is the ratio of the amount of light the detector sees to the minimum amount of light required to trip the sensor. This ratio is depicted graphically for all photoelectric sensors; in Fig. 2.21, excess gain is plotted along the vertical logarithmic axis, starting at 1, the minimum amount of light required to trigger the

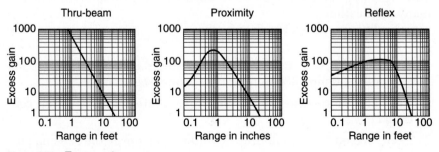

Figure 2.21   Excess gain curves.

detector. Every point above 1 represents the amount of light above that required to trigger the photoelectric control—the excess gain.

Often, the standard of comparison for choosing between different photoelectric sensors is range. Actually, more important to most applications is the excess gain. For a typical application, the higher the excess gain within the sensing region, the more likely the application will work. It is the extra margin that will determine whether the photoelectric control will continue to operate despite the buildup of dirt on the lens or the presence of contamination in the air.

Example An application requires detecting boxes on a conveyer in a filthy industrial environment (Fig. 2.22). The boxes will pass about 2 to 5 mm from the sensor as they move along the conveyer at the sensing location. Given a choice
between the two proximity sensors whose excess gain curves appear in Figs. 2.23
and 2.24, which photoelectric control should be selected for this application?

If the decision were based solely on specified range, the unit described in Fig. 2.23 would be selected. However, if units were

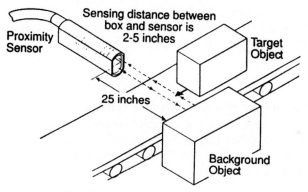

Figure 2.22 Box detection.

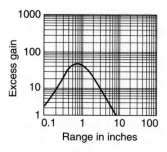

Figure 2.23 Excess gain curve for sensor 1.

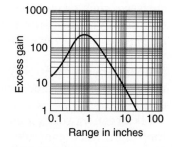

Figure 2.24 Excess gain curve for sensor 2.

installed in this application, it might fail after a short time in operation. Over time, contaminants from the environment would settle on the lens, decreasing the amount of light the sensor sees. Eventually, enough lens contamination would accumulate that the photoelectric control would not have enough excess gain to overcome the signal loss created by the coating, and the application would fail.

A better choice for this application would be the unit represented in Fig. 2.24. It delivers much more excess gain in the operating region required for this application and will therefore work much more successfully than the other unit.

2.3.2 Background suppression

Background suppression enables a diffuse photoelectric sensor to have high excess gain to a predetermined limit and insufficient excess gain beyond that range, where it might pick up objects in motion and yield a false detection. By using triangular ranging, sensor developers have created a sensor that emits light that reflects on the detector from two different target positions. The signal received from the more distant target is subtracted from that of the closer target, providing high excess gain for the closer target.

2.3.3 Contrast

Contrast measures the ability of a photoelectric control to detect an object; it is the ratio of the excess gain under illumination to the excess gain in the dark. All other things being equal, the sensor that provides the greatest contrast ratio should be selected. For reliable operation, a ratio 10:1 is recommended.

2.3.4 Polarization

Polarization is used in reflection sensors in applications where shiny surfaces, such as metal or shrink-wrapped boxes, may trigger the control falsely. The polarizer passes light along only one plane (Fig. 2.25), and the corner-cube reflectors depolarize the light as it passes through the plastic face of the retroreflector (Fig. 2.10). Only light that has been rotated by the corner-cube retroreflector can pass through the polarizer, whereas light that bounces off other shiny objects cannot.

Like regular reflex photoelectric sensors, polarized sensors have a high light/dark contrast ratio and are simple to install and align. However, the polarizers do limit the sensor's operating range because light is lost passing through them.

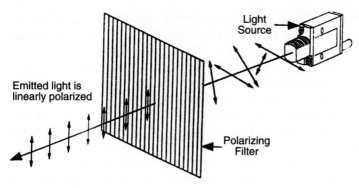

Figure 2.25 Polarization.

2.4 Inductive Proximity Sensors— Noncontact Metal Detection

Inductive proximity sensors are another common choice for position sensing. An inductive proximity sensor consists of four basic elements:

1. The sensor, which comprises a coil and ferrous core

2. An oscillator circuit

3. A detector circuit

4. A solid-state output

In the circuitry in Fig. 2.26, the oscillator generates an electromagnetic field that radiates from the sensor's face. This field is centered

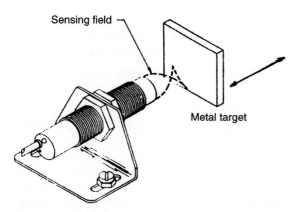

Figure 2.26 Inductive proximity sensor.

around the axis and detected by the ferrite core to the front of the sensor assembly. When a metal object enters the electromagnetic field, eddy currents are induced in the surface of the target. This loads in the oscillator circuit, which reduces its amplitude.

The detector circuit detects the change in the oscillator amplitude and, depending on its programming, switches ON and OFF at a specific oscillator amplitude. The sensing circuit returns to its normal state when the target leaves the sensing area and the oscillator circuit regenerates.

The nominal sensing range of inductive proximity sensors is a function of the diameter of the sensor and the power that is available to generate the electromagnetic field. This is subject to a manufacturing tolerance of ±10 percent, as well as a temperature drift tolerance of ±10 percent. The target size, shape, and material will have an effect on the sensing range. Smaller targets will reduce the sensing range, as will targets that are not flat or are made of nonferrous material.

There are basically two types of inductive proximity sensors: (1) shielded and (2) nonshielded. The shielded version has a metal cover around the ferrite core and coil assembly. This focuses the electromagnetic field to the front of the sensor and allows it to be imbedded in metal without influencing the sensing range. The nonshielded sensor can sense on the side as well as in front of a sensor. It requires a nonmetallic area around the sensor to operate correctly.

Inductive proximity sensors have several benefits:

1. High repeatability. Visibility of the environment is not an issue, since inductive proximity sensors can sense only electromagnetic fields. Therefore, environments from dirt to sunlight pose no problem for inductive proximity sensors. Also, because they are non-contact sensors, nothing wears.

2. Shock and vibration resistance.

3. Versatile connections. They can connect two or three wires with an ac, ac/dc, or dc power supply and up to four-wire dc connections.

4. Wide operating temperature range. They operate between −20 to +70°C, ±10 percent.

5. Very fast response, particularly in the dc models. These sensors can detect presence, send a signal, and reset in 50 μs (2000 times per second) in dc models.

Inductive proximity sensors are generally limited by their sensing distances and the material they sense. The effective range is limited to about 25 mm for most models and can be extended to only about 100 mm with the large models.

2.5   Limit Switches—Traditional Reliability

Limit switches are mechanical position-sensing devices that offer simplicity, robustness, and repeatability to processes. Mechanical limit switches are the oldest and simplest of all presence- or position-sensing devices: contact is made and a switch is engaged. This simplicity contributes generally to the cost advantage of limit switches. Yet, they can provide the control capabilities and versatility demanded in today's error-free manufacturing environment. The key to their versatility is the various forms they can take in the switch, actuating head, and lever operator. There are two-step, dual-pole limit switches which can detect and count two products of different sizes and can provide direct power control to segregate or process the items differently. The lever operator will rotate 10° to activate one set of contacts and 20° to activate another set. Because of the high amperage they can handle, limit switches can control up to 10 contacts from the movement of a single lever.

They are easy to maintain because the operator can hear the operation of the switch and can align it easily to fit the application. They are also robust. Limit switches are capable of handling an inrush current 10 times that of their steady-state current rating. They have rugged enclosures and they have prewiring that uses suitable strain-relief bushings to enable the limit switch to retain cables with 500 to 600 pounds of force on them. Limit switches can also handle direct medium-power switching for varying power factors and inrush stresses. For example, they can control a multihorsepower motor without any interposing starter, relay, or contactor.

Reliability is another benefit. Published claims for repeat accuracy for standard limit switches vary from within 0.03 mm to within 0.001 mm over the temperature range of −4 to +200°F. Limit switches dissipate energy spikes and rarely break down under normal mode surges. They will not be affected by electromagnetic interferences (EMI); there are no premature responses in the face of EMI. However, because they are mechanical devices, limit switches face physical limitations that can shorten their service life even though they are capable of several million operations. Also, heavy sludge, chips, or coolant can interfere with their operation.

2.6   Factors Affecting the Selection of Position Sensors

In selecting a position sensor, there are several key factors that should be considered:

1. *Cost.*   Both initial purchase price and life-cycle cost must be considered.

2. *Sensing distance.*   Photoelectric sensors are often the best selection when sensing distances are longer than 25 mm. Photoelectric sensors can have sensing ranges as long as 300,000 mm for outdoor or extremely dirty applications, down to 25 mm for extremely small parts or for ignoring background. Inductive proximity sensors and limit switches, on the other hand, have short sensing distances. The inductive proximity sensors are limited by the distance of the electromagnetic field—less than 25 mm for most models—and limit switches can sense only as far as the lever operator reaches.

3. *Type of material.*   Inductive proximity sensors can sense only ferrous and nonferrous materials, whereas photoelectric and limit switches can detect the presence of any solid material. Photoelectric sensors, however, may require polarizer if the target's surface is shiny.

4. *Speed.*   Electronic devices using dc power are the fastest—as fast as 2000 cycles per second for inductive proximity models. The fastest-acting limit switches can sense and reset in 4 ms or about 300 times per second.

5. *Environment.*   Proximity sensors can best handle dirty, gritty environments, but they can be fooled by metal chips and other metallic debris. Photoelectric sensors will also be fooled or left inoperable if they are fogged or blinded by debris.

6. *Types of voltages, connections, and requirements of the device's housing.*   All three types can accommodate varying requirements, but the proper selection must be made in light of the power supplies, wiring schemes, and environments.

7. *Third-party certification.*   Underwriters Laboratories (UL), National Electrical Manufacturers Association (NEMA), International Electrotechnical Commission (IEC), Factory Mutual, Canadian Standards Association (CSA), and other organizations impose requirements for safety, often based on the type of application. The certification will ensure the device has been tested and approved for certain uses.

8. *Intangibles.*   These can include the availability of application support and service, the supplier's reputation, local availability, and quality testing statements from the manufacturer.

2.7 Wavelengths of Commonly Used Light-Emitting Diodes

An LED is a semiconductor that emits a small amount of light when current flows through it in the forward direction. In most photoelectric sensors, LEDs are used both as emitters for sensing beams and as visual indicators of alignment or output status for a manufacturing process. Most sensor manufacturers use visible red, visible green, or infrared (invisible) LEDs (Fig. 2.4b). This simple device plays a significant part in industrial automation. It provides instantaneous information regarding an object during the manufacturing operation. LEDs, together with fiber optics, allow a controller to direct a multitude of tasks, simultaneously or sequentially.

2.8 Sensor Alignment Techniques

A sensor should be positioned so that the maximum amount of emitted energy reaches the receiver element in one of three different modes:

1. Opposed sensing mode
2. Retroreflective sensing mode
3. Proximity (diffuse) sensing mode

2.8.1 Opposed sensing mode

In this photoelectric sensing mode, the emitter and receiver are positioned opposite each other so that the light from the emitter shines directly at the receiver. An object then breaks the light beam that is established between the two. Opposed sensing is always the most reliable mode.

2.8.2 Retroreflective sensing mode

Retroreflective sensing is also called the *reflex* mode or simply the *retro* mode. A retroreflective photoelectric sensor contains both emitter and receiver. A light beam is established between the sensor and a special retroreflective target. As in opposed sensing, an object is sensed when it interrupts this beam.

Retro is the most popular mode for conveyer applications where the objects are large (boxes, cartons, etc.), where the sensing environment is relatively clean, and where scanning ranges are typically a few meters in length. Retro is also used for code-reading applications. Automatic storage and retrieval systems and automatic conveyer routing systems use retroreflective code plates to identify locations and/or products.

2.8.3 Proximity (diffuse) sensing mode

In the proximity (diffuse) sensing mode, light from the emitter strikes a surface of an object at some arbitrary angle and is diffused from the surface at all angles. The object is detected when the receiver captures some small percentage of the diffused light. Also called the *direct reflection* mode or simply the photoelectric *proximity* mode, this method provides direct sensing of an object by its presence in front of a sensor. A variation is the ultrasonic proximity sensor, in which an object is sensed when its surface reflects a sound wave back to an acoustic sensor.

2.8.4 Divergent sensing mode

This is a variation of the diffuse photoelectric sensing mode in which the emitted beam and the receiver's field of view are both very wide. Divergent mode sensors (Fig. 2.27), have loose alignment requirements, but have a shorter sensing range than diffuse mode sensors of the same basic design. Divergent sensors are particularly useful for sensing transparent or translucent materials or for sensing objects with irregular surfaces (e.g., webs that flutter). They are also used

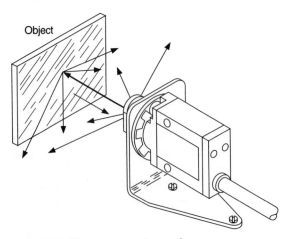

Figure 2.27 Divergent sensing mode.

effectively to sense objects with very small profiles, such as small-diameter thread or wire, at close range.

All unlensed bifurcated optical fibers are divergent. The divergent mode is sometimes called the *wide-beam diffuse* (or *proximity*) mode.

2.8.5   Convergent sensing mode

This is a special variation of diffuse mode photoelectric proximity sensing which uses additional optics to create a small, intense, and well-defined image at a fixed distance from the front surface of the sensor lens (Fig. 2.28). Convergent beam sensing is the first choice for photoelectric sensing of transparent materials that remain within a sensor's depth of field. It is also called the *fixed-focus proximity* mode.

2.8.6   Mechanical convergence

In mechanical convergence (Fig. 2.29), an emitter and a receiver are simply angled toward a common point ahead of the sensor. Although less precise than the optical convergent-beam sensing mode, this approach to reflective sensing uses light more efficiently than diffuse sensing and gives a greater depth of field than true optical convergence.

Mechanical convergence may be customized for an application by mounting the emitter and the receiver to converge at the desired distance. Depth of field is controlled by adjusting the angle between the emitter and the receiver.

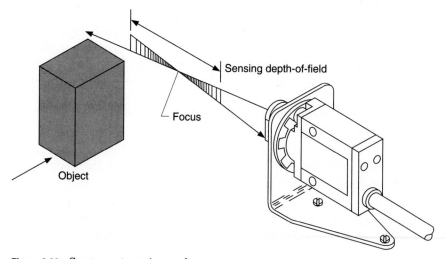

Figure 2.28 Convergent sensing mode.

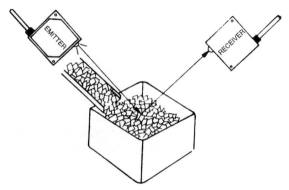

Figure 2.29 Mechanical convergence.

2.9 Fiber Optics in Industrial Communication and Control

The application of fiber optics to industrial information transfer is a natural extension of the current commercial uses of this technology in high-data-rate communications. While the primary advantage of fiber optics in traditional application areas has been extremely reliable communication at high rates, exceeding 1 Gbit/s over distances exceeding 100 km, other intrinsic features of the technology are more important than data rate and distance capability in industrial uses.

The physical mechanism of light propagating through a glass fiber has significant advantages that enable sensors to carry data and plant communications successfully and in a timely manner—a fundamental condition that must be constantly maintained in a computer-integrated manufacturing environment:

1. The light signal is completely undisturbed by electrical noise. This means that the fiber-optic cables can be laid wherever convenient without special shielding. Fiber-optic cables and sensors are unaffected by electrical noise when placed near arc welders, rotating machinery, electrical generators, etc., whereas in similar wired applications, even the best conventional shielding methods are often inadequate.

2. Fiber-optic communication is devoid of any electrical arcing or sparking, and thus can be used successfully in hazardous areas without danger of causing an explosion.

3. The use of a fiber link provides total electrical isolation between terminal points on the link. Over long plant distances, this can avoid troublesome voltage or ground differentials and ground loops.

4. A fiber-optic system can be flexibly configured to provide additional utility in existing hardware.

2.10   Principles of Fiber Optics in Communications

An optical fiber (Fig. 2.30) is a thin strand composed of two layers, an inner core and an outer cladding. The core is usually constructed of glass, and the cladding structure, of glass or plastic. Each layer has a different index of refraction, the core being higher. The difference between the index of refraction of the core material n_1 and that of the surrounding cladding material n_2 causes rays of light injected into the core to continuously reflect back into the core as they propagate down the fiber.

The light-gathering capability of the fiber is expressed in terms of its numerical aperture NA, the sine of the half angle of the acceptance cone for that fiber. Simply stated, the larger the NA, the easier it is for the fiber to accept light (Fig. 2.31).

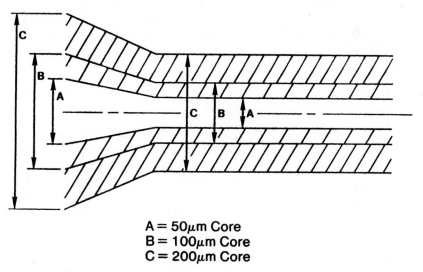

$$A = 50\mu m \text{ Core}$$
$$B = 100\mu m \text{ Core}$$
$$C = 200\mu m \text{ Core}$$

Figure 2.30   Structure of optical fiber.

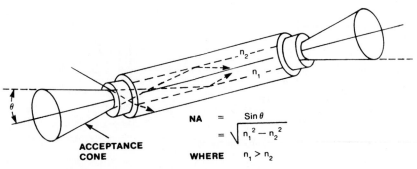

$$NA = \sin\theta$$
$$= \sqrt{n_1^2 - n_2^2}$$
$$\text{WHERE} \quad n_1 > n_2$$

ACCEPTANCE CONE

Figure 2.31   Numerical aperture (NA) of an optical fiber.

2.11   Fiber-Optic Information Link

A fiber-optic communication system (Fig. 2.32) consists of:

1. A light source (LED or laser diode) pulsed by interface circuitry and capable of handling data rates and voltage levels of a given magnitude.
2. A detector (photodiode) which converts light signals to electrical signals and feeds interface circuitry to recreate the original electrical signal.
3. Fiber-optic cables between the light source and the detector (called the *transmitter* and the *receiver,* respectively).

It is usual practice for two-way communication to build a transmitter and receiver into one module and use a duplex cable to communicate to an identical module at the end of the link. The 820- to 850-nm range is the most frequent for low-data-rate communication, but other wavelengths (1300 and 1550 nm) are used in long-distance systems. Fiber selection must always take transmission wavelength into consideration.

2.12   Configurations of Fiber Optics

The selection of optical fiber plays a significant part in sensor performance and information flow. Increased fiber diameter results in higher delivered power, but supports a lower bandwidth. A fiber with a 200-μm core diameter (glass core, silicone plastic cladding), by virtue of its larger diameter and acceptance angle, can transmit 5 to 7 times more light than a 100-μm core fiber, or up to 30 times more light than a 40-μm fiber, the historical telecommunication industry standard.

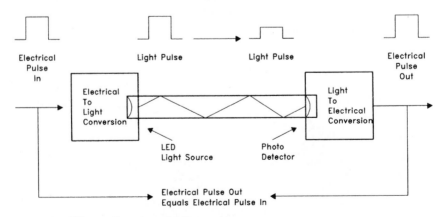

Figure 2.32 Fiber-optic communication system.

Computer data communications systems commonly employ 62.5-μm fiber because of its ability to support very high data rates (up to 100 Mbaud) while offering increased power and ease of handling. Factory bandwidth requirements for most links are typically one or more orders of magnitude less. Therefore, fiber core size may be increased to 200 μm to gain the benefits of enhanced power and handling ease, the decrease in bandwidth being of no consequence (Fig. 2.31).

2.12.1 Optical power budget

An optical power budget examines the available optical power and how it is used and dissipated in a fiber-optic system. It is important to employ the highest possible optical power budget for maximum power margin over the detector requirement. A budget involves four factors:

1. Types of light source

2. Optical fiber acceptance cone

3. Receiver sensitivity

4. Splice, coupling, and connector losses

Laser-diode sources are generally not economically feasible or necessary in industrial systems. Light-emitting diodes are recommended for industrial applications. Such systems are frequently specified with transmitted power at 50 μW or greater if 200-μm-core fiber is used.

Successful communication in industry and commercial applications is determined by the amount of light energy required at the receiver, specified as *receiver sensitivity*. The higher the sensitivity, the less light required from the fiber. High-quality systems require power only in the hundreds of nanowatts to low microwatts range.

Splice losses must be low so that as little light as possible is removed from the optical-fiber system. Splice technology to repair broken cable is readily available, permitting repairs in several locations within a system in a short time (minutes) and causing negligible losses. Couplers and taps are generally formed through the process of glass fusion and operate on the principle of splitting from one fiber to several fibers. New active electronic couplers replenish light as well as distribute the optical signal.

An example of an optical power budget follows:

1. The optical power injected into a 200-μm-core fiber is 200 μW (the same light source would inject approximately 40 μW into a 100-μm-core fiber)

2. The receiver sensitivity is 2 μW

3. The receiver budget (dB) is calculated as:

$$dB = 10 \log [(\text{available light input})/(\text{required light output})]$$

$$= 10 \log [(200 \ \mu W)/(2 \ \mu W)]$$

$$= 10 \log 100$$

$$= 20 \ dB$$

4. Three major sources of loss are estimated as:
   a. 2 to 3 dB loss for each end connector
   b. 1 to 2 dB loss for each splice
   c. 1 dB/150 m loss for fiber of 200 μm diameter

   Most manufacturers specify the optical power budget and translate this into a recommended distance.

2.12.2   Digital links—pulsed

The one-for-one creation of a light pulse for an electrical pulse is shown in Fig. 2.32. This represents the simplest form of data link. It does not matter what format and signal level the electrical data takes (e.g., whether IEEE RS-232 or RS-422 standard format or CMOS or TTL logic level), as long as the interface circuitry is designed to accept them at its input or reproduce them at the output. Voltage conversion may be achieved from one end of the link to the other, if desired, through appropriate interface selection.

The light pulses racing down the fiber are independent of electrical protocol. Several design factors are relevant to these and other types of data links as follows:

1. *Minimum output power.*   The amount of light, typically measured in microwatts, provided to a specific fiber size from the data link's light source

2. *Fiber size.*   Determined from the data link's light source

3. *Receiver sensitivity.*   The amount of light, typically measured in microwatts or nanowatts, required to activate the data link's light detector

4. *Data rate.*   The maximum rate at which data can be accurately transmitted

5. *Bit error rate (BER).*   The frequency with which a light pulse is erroneously interpreted (for example, 10^{-9} BER means no more than one of 10^9 pulses will be incorrect)

6. *Pulse-width distortion.*   The time-based disparity between input and output pulse widths

The simple pulse link is also the basic building block for more complex links. Figure 2.33 provides an example of a 5-Mbit three-channel link used to transmit encoder signals from a servomotor to a remote destination.

2.12.3   Digital links—carrier-based

A carrier-based digital link is a system in which the frequency of the optical carrier is varied by a technique known as *frequency-shift keying* (FSK). Figure 2.34 illustrates the modulation concept; two frequencies are employed to create the logic 0 and 1 states. This scheme is especially useful in systems where electrical "handshaking" (confirmation of reception and acceptance) is employed. Presence of the opti-

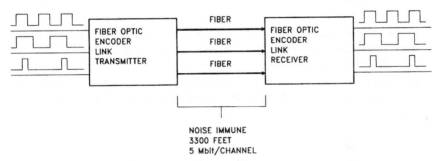

Figure 2.33   Three-channel optical link.

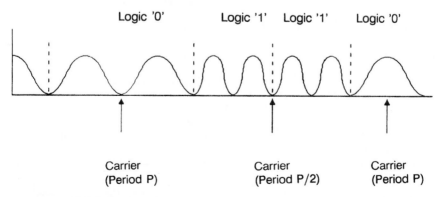

Figure 2.34   Modulation concept.

cal carrier is the equivalent of the handshake signal, with the data signal presented by frequency.

Figure 2.35 illustrates a system where the logic of the fiber-optic line driver recognizes the optical carrier to create a handshake between terminal and processor.

Additionally, since the processor is capable of recognizing only one terminal, the carrier is controlled to deny the handshake to all other terminals once one terminal is actively on line to the processor.

2.12.4   Analog links

It is well-recognized that, in motion control and process measurement and control, transmitting analog information without distortion is important. There are several ways of treating analog information with fiber optics.

Analog data cannot easily be transmitted through light intensity variation. A number of external factors, such as light source variation, bending losses in cable, and connector expansion with temperature, can affect the amount of raw light energy reaching the detector. It is not practical to compensate for all such factors and deliver accurate analog data. A viable method of transmitting data is to use an unmodulated carrier whose frequency depends on the analog signal level. A more advanced means is to convert the analog data to digital data, where accuracy also is determined by the number of bits used,

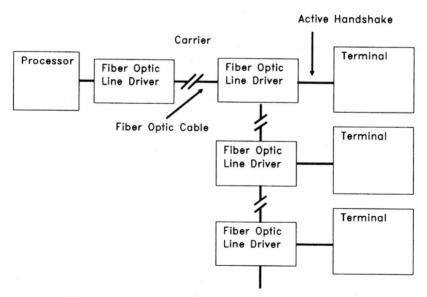

Figure 2.35   System employing optical carrier.

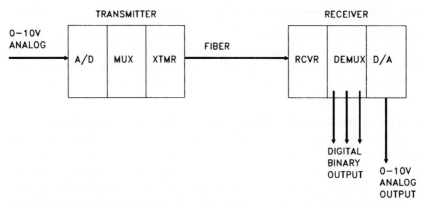

Figure 2.36   Analog and digital data transmission.

multiplex the digital bits into one stream, and use the pulsed digital link approach.

Figure 2.36 illustrates a link in which this last approach is used to produce both digital and analog forms of the data at the output.

2.12.5   Video links

Long-distance video transmission in industrial situations is easily disrupted by radiated noise and lighting. Repeaters and large-diameter coaxial cables are often used for particularly long runs. The use of fiber optics as a substitute for coaxial cable allows propagation of noise-free video over long distances. Either an intensity- or frequency-modulated optical carrier signal is utilized as the transmission means over fiber. With intensity-modulated signals, it is mandatory that some sort of automatic gain control be employed to compensate for light degradation due to varying cable losses, splices, etc. Figure 2.37 illustrates a typical fiber-optic video link in a machine-vision application.

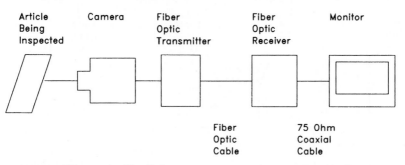

Figure 2.37   Fiber-optic video link.

2.12.6 Data bus networks

Wiring a system often causes serious problems for designers and communication system integrators regarding the choice of topology. The basic difference between fiber and wiring is that one normally does not splice or tap into fiber as one would with coaxial or twin axial cable to create a drop point.

2.12.6.1 Daisy chain data bus. The simplest extension of a point-to-point data link is described in Fig. 2.38. It extends continuously from one drop point (node) to the next by using each node as a repeater. The fiber-optic line driver illustrated in Fig. 2.35 is such a system, providing multiple access points from remote terminals to a programmable controller processor. A system with several repeater points is vulnerable to the loss of any repeater, and with it all downstream points, unless some optical bypass scheme is utilized. Figures 2.38 and 2.39 exhibit such a scheme.

2.12.6.2 Ring coupler. A preferred choice among several current fiber-optic system designs is the token-passing ring structure.

Signals are passed around the ring, with each node serving to amplify and retransmit. Care must be taken to provide for a node becoming nonoperational. This is usually handled by using some type of bypass switching technique, given that the system provides sufficient optical power to tolerate a bypassed repeater. Another contingency method is to provide for the transmitting node to read its own data coming around the ring, and to retransmit in the other direction if necessary, as illustrated in Fig. 2.40. Yet another is to provide for a second pair of fibers paralleling the first, but routed on a physically different path.

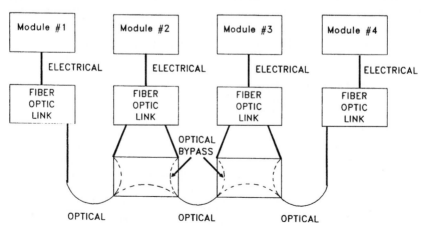

Figure 2.38 Point-to-point data link.

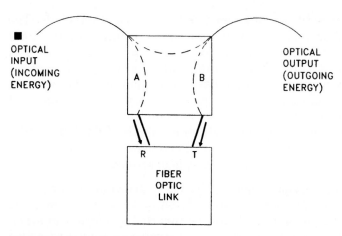

Figure 2.39 Daisy chain data bus.

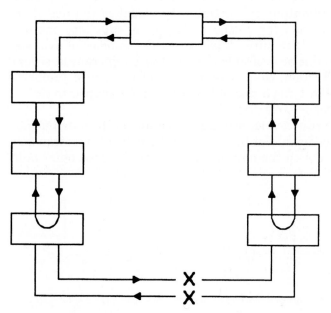

Figure 2.40 Ring coupler.

2.12.6.3 Passive star coupler. Certain systems have attempted to utilize a fiber-optic coupling technology offered from the telecommunications and data communications applications areas. When successful, this technique allows tapping into fiber-optic trunk lines, a direct parallel with coaxial or twin axial systems. Light entering into the tap or coupler is split into a given number of output channels. The amount of light in any output channel is determined by the total amount of light input, less system losses, divided by the number of output channels. Additional losses are incurred at the junction of the main data-carrying fibers with the fiber leads from the tap or star. As such, passive couplers are limited to systems with few drops and moderate distances. Also, it is important to minimize termination losses at the coupler caused by the already diminished light output from the coupler. A partial solution is an active in-line repeater, but a superior solution, the *active star coupler,* is described below.

2.12.6.4 The active star coupler. The basic principle of the active star coupler is that any light signal received as an input is converted to an electrical signal, amplified, and reconverted to optical signals on all other output channels. Figure 2.41 illustrates an eight-port active star coupler, containing eight sets of fiber-optic input/output (I/O) ports. A signal received on the channel 1 input will be transmitted on the channel 2 to 8 output ports. One may visualize the use of the active star coupler as aggregating a number of taps into one box. Should the number of required taps exceed the number of available I/O ports, or should it be desirable to place these tap boxes at several locations in the system, the active star couplers may be jumpered together optically by tying a pair of I/O ports on one coupler to that on another in a hub-and-spoke system.

With the active star coupler serving as the hub of the data bus network, any message broadcast by a unit on the network is retransmitted to all other units on the network. A response of these other units

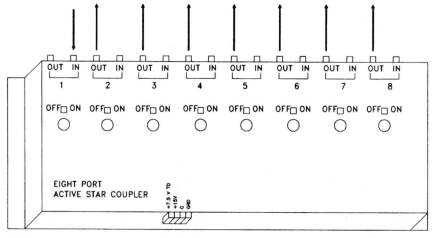

Figure 2.41 Eight-port active star coupler.

is broadcast back to the rest of the network through the star, as in an electrical wired data bus network.

2.13 Configurations of Fiber Optics for Sensors

Fiber-optic sensors for general industrial use have largely been restricted to applications in which their small size has made them convenient replacements for conventional photoelectric sensors. Until recently, fiber-optic sensors have almost exclusively employed standard bundle technology, whereby thin glass fibers are bundled together to form flexible conduits for light.

Recently, however, the advances in fiber optics for data communications have introduced an entirely new dimension into optical sensing technology. Combined with novel but effective transducing technology, they set the stage for a powerful class of fiber-optic sensors.

2.13.1 Fiber-optic bundle

A typical fiber-optic sensor probe, often referred to as a *bundle* (Fig. 2.42), is 1.25 to 3.15 mm in diameter and made of individual fiber elements approximately 0.05 mm in diameter. An average bundle will contain up to several thousand fiber elements, each working on the conventional fiber-optic principle of total internal reflection.

Composite bundles of fibers have an acceptance cone of the light based on the numerical aperture of the individual fiber elements.

$$NA = \sin \Theta$$
$$= \sqrt{n_1^{\,2} - n_2^{\,2}}$$

where $n_1 > n_2$ and Θ = half the cone angle.

Bundles normally have NA values in excess of 0.5 (acceptance cone full angle greater than 60°), contrasted with individual fibers for long-

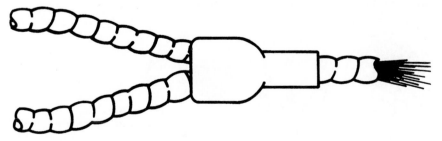

Figure 2.42 Fiber-optic sensor probe.

distance, high-data-rate applications, which have NA values approaching 0.2 (acceptance cone full angle approximately 20°).

The ability of fiber-optic bundles to readily accept light, as well as their large total cross-sectional surface area, have made them an acceptable choice for guiding light to a remote target and from the target area back to a detector element. This has been successfully accomplished by using the pipe as an appendage to conventional photoelectric sensors, proven devices conveniently prepackaged with adequate light source and detector elements.

Bundles are most often used in either opposed beam or reflective mode. In the opposed beam mode, one fiber bundle pipes light from the light source and illuminates a second bundle—placed on the same axis at some distance away—which carries light back to the detector. An object passing between the bundles prevents light from reaching the detector.

In the reflective mode, all fibers are usually contained in one probe but divided into two legs at some junction point in an arrangement known as *bifurcate*. One bifurcate leg is then tied to the light source and the other to the detector (Fig. 2.43). Reflection from a target provides a return path to the detector for the light. The target may be fixed so that it breaks the beam, or it may be moving so that, when present in the probe's field of view, it reflects the beam.

Typical bundle construction in a bifurcate employs one of two arrangements of individual fibers. The sending and receiving fibers are arranged either randomly or hemispherically (Fig. 2.44). As a

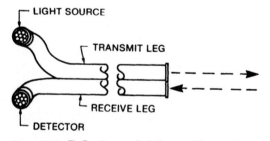

Figure 2.43 Reflective mode bifurcate fiber optics.

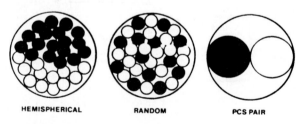

HEMISPHERICAL   RANDOM   PCS PAIR

Figure 2.44 Bundle construction.

practical matter, there is little, if any, noticeable impact on the performance of a photoelectric system in any of the key parameters such as sensitivity and scan range.

Application areas for bundle probes include counting, break detection, shaft rotation, and displacement/proximity sensing.

2.13.2 Bundle design considerations

Microscopic fiber flaws (Fig. 2.45) such as impurities, bubbles, voids, material absorption centers, and material density variations all diminish the ability of rays of light to propagate down the fiber, causing a net loss of light from one end of the fiber to the other. All these effects combine to produce a characteristic absorption curve, which graphically expresses a wavelength-dependent loss relationship for a given fiber (Fig. 2.46). The fiber loss parameter is expressed as attenuation in dB/km as follows:

$$\text{Loss} = -10 \log (p_2/p_1)$$

where p_2 = light power output and p_1 = light power input.

Therefore, a 10-dB/km fiber would produce only 10 percent of the input light at a distance of 1 km.

Because of their inexpensive lead silicate glass composition and relatively simple processing techniques, bundles exhibit losses in the 500-dB/km range. This is several orders of magnitude greater than a communications-grade fiber, which has a loss of 10 dB/km. The maximum practical length for a bundle is thus only about 3 m. Further, the absence of coating on individual fibers and their small diameter make them susceptible to breakage, especially in vibratory environments.

Also, because of fiber microflaws, it is especially important to shield fibers from moisture and contaminants. A fiber exposed to water will

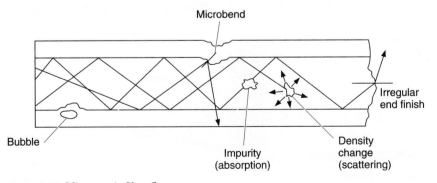

Figure 2.45 Microscopic fiber flaws.

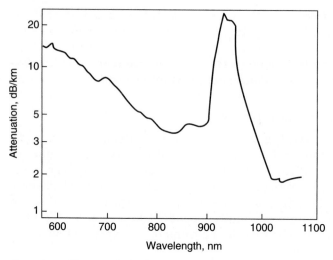

Figure 2.46 Characteristic absorption curve.

gradually erode to the point of failure. This is true of any optical fiber, but is especially true of uncoated fibers in bundles.

2.13.3 Fiber pairs for remote sensing

A viable solution to those design problems that exist with fiber bundles is to use large-core glass fibers that have losses on a par with telecommunication fibers. Although its ability to accept light is less than that of a bundle, a 200- or 400-µm core diameter plastic clad silica (PCS) fiber provides the ability to place sensing points hundreds of meters away from corresponding electronics. The fiber and the cable construction in Fig. 2.47 lend themselves particularly well to conduit pulling, vibratory environments, and general physical abuse. These fibers are typically proof-tested for tensile strength to levels in excess of 50,000 lb/in². A pair of fibers (Fig. 2.44) is used much like a bundle,

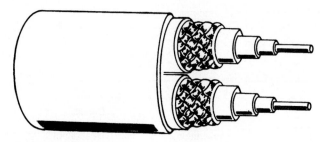

Figure 2.47 Pair of fibers.

where one fiber is used to send light to the sensing point and the other to return light to the detector. The performance limitation of a fiber pair compared to a bundle is reduced scan range; however, lenses may be used to extend the range. A fiber pair may be used in one of two configurations: (1) a single continuous probe, i.e., an unbroken length of cable from electronics to sensing point, or (2) a fiber-optic extension cord to which a standard probe in either a bundle or a fiber pair is coupled mechanically. This allows the economical replacement, if necessary, of the standard probe, leaving the extension cord intact.

The typical application for a fiber pair is object detection in explosive or highly corrosive environments, e.g., ammunition plants. In such cases, electronics must be remote by necessity. Fiber pairs also allow the construction of very small probes for use in such areas as robotics, small object detection, thread break detection, and small target rotation.

2.13.4   Fiber-optic liquid level sensing

Another technique for interfacing with fiber-optic probes involves the use of a prism tip for liquid sensing (Fig. 2.48). Light traveling down

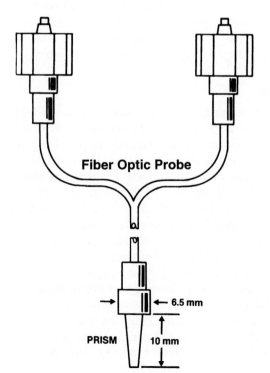

Figure 2.48   Prism tip for liquid sensing.

one leg of the probe is totally internally reflected at the prism-air interface. The index of refraction of air is 1. Air acts as a cladding material around the prism. When the prism contacts the surface of a liquid, light is stripped from the prism, resulting in a loss of energy at the detector. A properly configured system can discriminate between liquid types, such as gasoline and water, by the amount of light lost from the system, a function of the index of refraction of the liquid.

This type of sensor is ideal for set-point use in explosive liquids, in areas where electronics must be remote from the liquid by tens or hundreds of meters, and where foam or liquid turbulence make other level-sensing techniques unusable.

2.14   Flexibility of Fiber Optics

The power of fiber optics is further shown in the flexibility of its system configurations. A master industrial terminal (Fig. 2.49) can access any of a number of remote processors. The flexibility of switching, distance capability, and noise immunity of such a system are its primary advantages.

Figure 2.50 illustrates a passive optical coupler with a two-way fiber-optic link communicating over a single fiber through an on-axis rotary joint. Such a system allows a simple, uninterrupted communication link through rotary tables or other rotating machinery. This is a true challenge for high-data-rate communication in a wire system.

2.14.1   Fiber-optic terminations

Optical fibers are becoming increasingly easier to terminate as rapid advances in termination technology continue to be made. Several manufacturers have connector systems that require no polishing of the fiber end, long a major objection in fiber optics. Products that eliminate epoxy adhesives are also being developed. Field installation times now typically average less than 10 min for large-core fibers (100 and 200 μm) with losses in the 1- to 3-dB range. Further, power budgets for well-designed industrial links normally provide a much greater latitude in making a connection. A 5- to 6-dB-loss connection, while potentially catastrophic in other types of systems, may be quite acceptable in short-haul systems with ample power budgets.

The most popular connector style for industrial communications is the SMA style connector, distinguished by its nose dimensions and configuration, as well as the thread size on the coupling nut. The coupling nut is employed to mechanically join the connector to a mating device on the data link or to a thread splice bushing. Figure 2.51 illustrates an SMA connection to an information link.

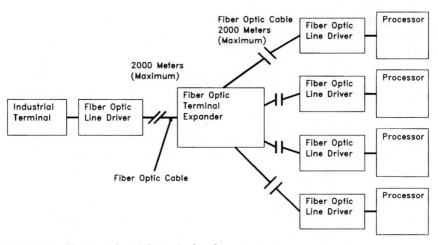

Figure 2.49 Master industrial terminal and remote processors.

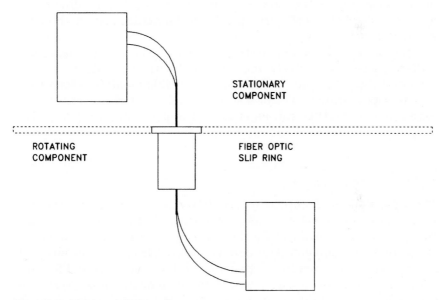

Figure 2.50 Passive optical coupler.

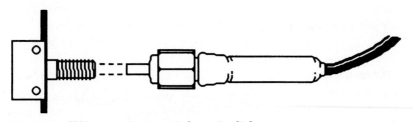

Figure 2.51 SMA connection to an information link.

2.15   Testing of Fiber Optics

Optical measurements, perhaps among the most difficult of all physical measurements, are fundamental to the progress and development of fiber-optic technology. Recently various manufacturers have offered lines of fiber-optic test equipment for use in field and laboratory. Typical field measurement equipment determines the average optical power emitted from the system source, the component and overall system loss, the bit error rate, and the location of breaks in the fiber. Laboratory equipment measures loss through connectors and splicing, characterizes transmitters and receivers, and establishes bit error rate.

Testing of fiber-optic cables or systems is normally done with a calibrated light source and companion power meter. The light source is adjusted to provide a 0-dB reading on the power meter with a short length of jumper cable. The cable assembly under test is then coupled between the jumper and the power meter to provide a reading on the meter, in decibels, that corresponds to the actual loss in the cable assembly.

Alternatively, the power through the cable from the system's transmitter can be read directly and compared with the system's receiver sensitivity specification. In the event of a cable break in a long span, a more sophisticated piece of test equipment, an optical time-domain reflectometer (OTDR), can be employed to determine the exact position of the break.

2.15.1   Test light sources

The Photodyne 9XT optical source driver (Fig. 2.52) is a hand-held unit for driving LED and laser fiber-optic light sources. The test equipment is designed to take the shock and hard wear of the typical work-crew environment. The unit is powered from two rechargeable nicad batteries or from line voltage. The LED series is suited for measurement applications where moderate dynamic range is required or coherent light should be avoided. The laser modules are used for attenuation measurements requiring extreme dynamic range or where narrow spectral width and coherent light are required. The laser modules are the most powerful source modules available.

2.15.2   Power meters

The Photodyne 2285XQ fiber-optic power meter (Fig. 2.53) is a low-cost optical power meter for general-purpose average-power measure-

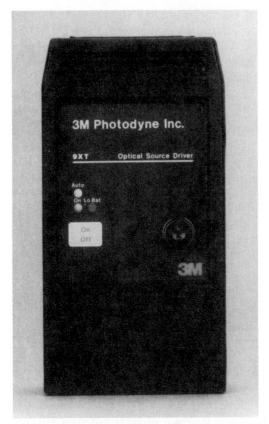

Figure 2.52 Photodyne 9XT optical source driver.
(*Courtesy 3M Corporation*)

Figure 2.53 Photodyne 2285XQ fiber-optic power meter. (*Courtesy 3M Corporation*)

ment, but particularly for fiber-optic applications, in both manual and computer-controlled test setups. A unique and powerful feature of the 2285XSQ is its ratio function. This allows the user to make ratio (A/B) measurements by stacking several instruments together via the interface without the need for a controller.

Another very powerful feature is the built-in data logger. With this function, data may be taken at intervals from 1 to 9999 s. This feature is useful in testing optical devices for short- and long-term operation.

At the heart of the instrument is a built-in large-area high-sensitivity indium gallium arsenide (InGaAs) sensor. All industry connectors may be interfaced to the sensor using any of the Photodyne series 2000 connector adapters. The sensor is calibrated for the all the fiber-optic windows: 820, 850, 1300, and 1550 nm.

2.15.3   Dual laser test sets

The Photodyne 2260XF and 2260XFA are switchable dual laser test sets, with a transmit and receive section in one unit. They measure and display loss in fiber links at both 1300 and 1550 nm simultaneously in one pass. They are designed for use in installing, maintaining, and troubleshooting single-mode wavelength-division multiplexed (WDM) fiber-optic links operating at 1300- and 1550-nm wavelengths. They may also be used for conventional links operating at either 1300 or 1550 nm.

The essential differences between the two models are that the XF version (Fig. 2.54) has a full complement of measurement units, log and linear (dBm, dB, mW, μW, nW), whereas the XFA version has log units only (dBm, dB). In the XF version, laser power output is adjustable over the range 10 to 20 dBm (100 μW to 21 mW). In the XFA version, it is automatically set to a fixed value of 10 dBm. The XF version allows the user to access wavelengths over the ranges 1250 to 1350 nm and 1500 to 1600 nm in 10-nm steps. The XFA version is fixed at 1300 nm to 1550 nm. To control all functions, the XF version has six keys; the XFA has only two. Although both instruments perform identical tasks equally well, the XF may be seen as the more flexible version and the XFA as the simpler-to-use version.

2.15.4   Test sets/talk sets

The hand-held Photodyne 21XTL fiber-optic test set/talk set (Fig. 2.55) is for use in installation and maintenance of fiber cables. This

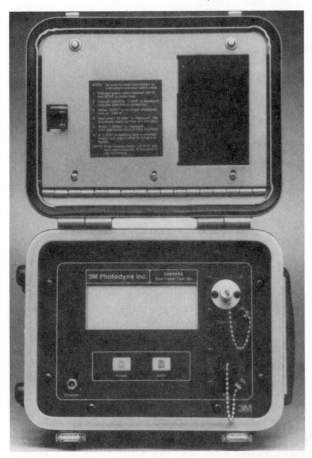

Figure 2.54 Photodyne 2260XF switchable dual laser test set. (*Courtesy 3M Corporation*)

Figure 2.55 Photodyne 21XTL fiber-optic test set/talk set. (*Courtesy 3M Corporation*)

instrument functions as a power meter and test set, as well as a talk set. For maintenance purposes, the user may establish voice communication over the same fiber pair that is being measured.

The 21XTL as a power meter covers an exceptionally wide dynamic range (–80 to +3 dBm). As an option, the receiver may include a silicon or an enhanced InGaAs photodiode. With the InGaAs version, the user can perform measurements and voice communication at short and long wavelengths. The silicon version achieves a superior dynamic range at short wavelengths only.

The highly stabilized LED ensures repeatable and accurate measurements. Precision optics couples the surface-emitter LED to the fiber core. With this technique the fiber end will not wear out or scratch. The transmitter is interchangeable, providing complete flexibility of wavelengths and connector types.

2.15.5 Attenuators

The 19XT optical attenuator (Fig. 2.56) is a hand-held automatic optical attenuator. It provides continuous attenuation of both short- and

Figure 2.56 Photodyne 19XT optical attenuator.(*Courtesy 3M Corporation*)

long-wave optical signals in single-mode and multimode applications. The calibrated wavelengths are 850 nm/1300 nm multimode and/or 1300 nm/1550 nm single mode. An attenuation range of 70 dB is offered with 0.1-dB resolution and an accuracy of ±0.2 dB typical (±0.5 dB maximum). Unique features allow scanning between two preset attenuation values and including the insertion loss in the reading.

The 19XT allows simple front panel operation or external control of attenuation through analog input and output connections.

2.15.6   Fault finders

The Photodyne 5200 series optical fault finders (Fig. 2.57) offer flexible alternatives for localizing faults or trouble areas on any fiber-optic network. The 5200 series fault finders can easily be integrated into the troubleshooting routines of fiber-optic crews. They offer a variety of features, such as:

1. Fast, accurate analysis of faults

2. Autoranging for greatest accuracy

3. Reflective and nonreflective fault detection

4. Multiple-fault detection capabilities

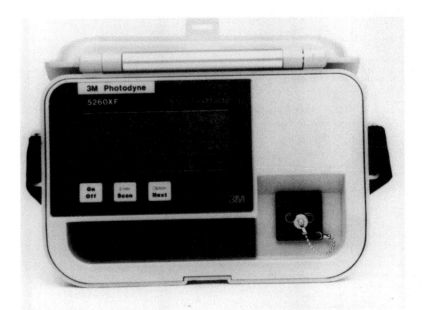

Figure 2.57   Photodyne 5200 series optical fault finder. (*Courtesy 3M Corporation*)

5. Go/no-go splice qualification

6. Variable fault threshold setting (0.5 to 6.0 dB)

7. Fault location up to 82 km

8. Automatic self-test and performance check

9. AC or rechargeable battery operation

10. Large, easy-to-read liquid-crystal display (LCD)

2.15.7 Fiber identifiers

The Photodyne 8000XG fiber identifier (Fig. 2.58) is designed for fast, accurate identification and traffic testing of fiber-optic lines without cutting the fiber line or interrupting normal service. Ideal for use during routine maintenance and line modification, this small, hand-held unit can be used to locate any particular fiber line, identify live fibers, and determine whether or not traffic is present. Features of the 8000XG are

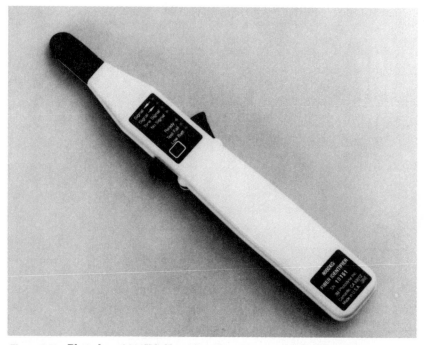

Figure 2.58 Photodyne 8000XG fiber identifier. (*Courtesy 3M Corporation*)

1. Light weight, portability, and battery operation
2. Automatic self-test after each fiber insertion
3. Mechanically damped fiber action
4. Operation over 850- to 1550-nm range
5. Transmission direction indicators
6. 1- and 2-kHz tone detection
7. Low insertion loss at 1300 and 1550 nm
8. Completely self-contained operation

2.16   Networking with Electrooptic links

The following examples describe a number of products utilizing communication through fiber electrooptic modules. The function of the fiber is to replace wire. This can be achieved by interconnecting the electrooptic modules in a variety of ways. Figure 2.59 shows a programmable controller communication network through-coaxial-cable bus branched to four remote input/output stations. The programmable controller polls each of the input/output stations in sequence. All input/output stations hear the programmable controller communication, but only the one currently being addressed responds.

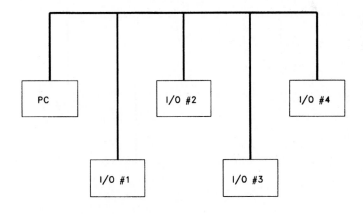

——— WIRE

Figure 2.59 Programmable controller communication network.

2.16.1   Hybrid wire/fiber network

Figure 2.60 shows an electrooptic module used to replace a trouble-some section of coaxial cable subject to noise, grounding problems, lightning, or hazardous environment. When fiber is used in this mode, the electrooptic module should be placed as close to the input/output drop as possible in order to minimize the effect of potential electrical noise problems over the section of coax cable connecting them.

2.16.2   Daisy chain network

The daisy chain configuration (Fig. 2.61) is an economical choice for long, straight-line installations (e.g., conveyer or mine shaft applications). The signal generated at the programmable controller is converted to light and transmitted outward. At each transmitted section, the electrical signal is reconstructed and a light signal is regenerated and sent down the chain.

2.16.3   Active star network

The electrooptical programmable controller links may be joined to an electrooptic module with four-port and eight-port active star couplers

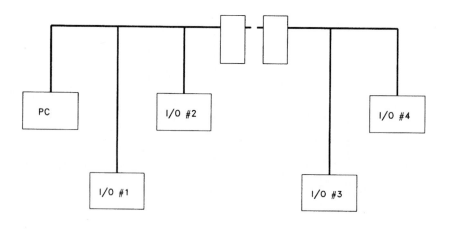

Figure 2.60   Hybrid wire/fiber network.

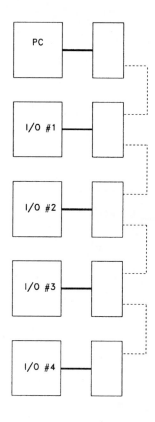

Figure 2.61 Daisy chain network.

—————— FIBER

———— WIRE

for a hub-and-spoke-type system. Figure 2.62 shows two active star couplers joined via fiber, a configuration which might be appropriate where clusters of input/output racks reside in several locations separated by some distance. The star coupler then becomes the distributor of the light signals to the racks in each cluster as well as a potential repeater to a star in another cluster.

2.16.4 Hybrid fiber network

Star and daisy chain network structures can be combined to minimize overall cabling costs (Fig. 2.63). The fiber network configuration

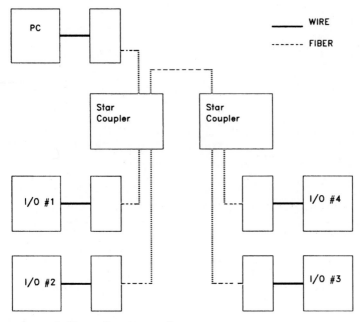

Figure 2.62 Two active star couplers.

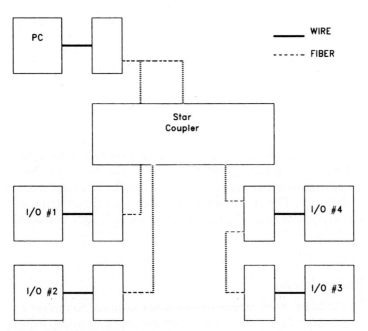

Figure 2.63 Hybrid fiber network.

exactly duplicates the coax network. The final decision on exactly which configuration to choose depends on the following criteria:

1. Economical cabling layout
2. Length of cable runs versus cost of electronics
3. Location of end devices and power availability
4. Power-out considerations

Other considerations specific to the programmable controller being used can be summarized as follows:

- *Pulse width.* Total acceptable pulse-width distortion, which may limit the number of allowable repeater sites, is

Allowable distortion (ns) = allowable percent distortion
$$\times \text{ period of signal (ns)}$$

- *Signal propagation delay.* Allowable system signal propagation delay may limit the overall distance of the fiber network.

Example:

Electrooptic module distortion = 50% allowable distortion
$$\times 1 \text{ Mbaud transmission}$$
$$= 50 \text{ ns}$$

Calculate:

Allowable distortion = 50% allowable distortion $\times 10^{-6}$
$$= 500 \text{ ns}$$

Maximum number of repeater sites = 500 ns/ 50 ns
$$= 10$$

(Note: A star coupler counts as a repeater site.)

Distance Calculation (Signal Propagation Delay)

Delay of light in fiber-optic regeneration	1.5 ns/ft
Delay in module	50 ns

Manufacturers of programmable controllers will provide the value of the system delay. This value must be compared with the calculated allowable delay. If the overall fiber system is longer than published maximum length of the wired system, the system must be reconfigured.

2.16.5 Fiber-optic sensory link for minicell controller

A minicell controller is typically used to coordinate and manage the operation of a manufacturing cell, consisting of a group of automated programmable machine controls (programmable controllers, robots, machine tools, etc.) designed to work together and perform a complete manufacturing or process-related task. A key benefit of a minicell controller is its ability to adjust for changing products and conditions. The minicell controller is instructed to change data or complete programs within the automation work area. A minicell controller is designed to perform a wide variety of functions such as executing programs and data, uploading/downloading from programmable controllers, monitoring, data analysis, tracking trends, generating color graphics, and communicating in the demanding plant floor environment. Successful minicell controllers use fiber-optic links that can interface with a variety of peripheral devices. A minicell controller can be used in a variety of configurations, depending on the optical-fiber lengths, to provide substantial system design and functional flexibility (Fig. 2.64).

2.17 Versatility of Fiber Optics in Industrial Applications

A constant concern in communication is the ever-increasing amount of information that must be sent with greater efficiency over a medium requiring less space and less susceptibility to outside interferences. As speed and transmission distance increase, the problems caused by electromagnetic interference, radio-frequency interference, crosstalk, and signal distortion become more troublesome. In terms of signal integrity, just as in computer-integrated manufacturing data acquisition and information-carrying capacity, fiber optics offers many advantages over copper cables. Furthermore, optical fibers emit no radiation and are safe from sparking and shock. These features make fiber optics the ideal choice for many processing applications where safe operation in hazardous or flammable environments is a requirement.

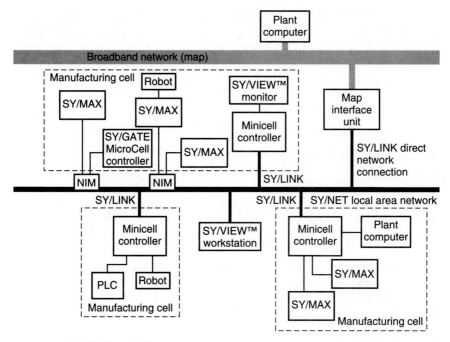

Figure 2.64 Minicell controller.

Accordingly, fiber-optic cables have these advantages in industrial applications:

1. Wide bandwidth
2. Low attenuation
3. Electromagnetic immunity
4. No radio-frequency interference
5. Small size
6. Light weight
7. Security
8. Safety in hazardous environment

2.17.1 High-clad fiber-optic cables

Large-core, multimode, step-index high-clad silica fiber-optic cables make fiber-optic technology user-friendly and help designers of sensors, controls, and communications realize substantial cost saving. The coupling efficiency of high-clad silica fibers allows the use of less expensive transmitters and receivers. High-clad silica polymer technology permits direct crimping onto the fiber cladding. Field terminations can be performed in a few minutes or less, with minimal training. The following lists describe the structure and characteristics of several fiber-optic cables used in industry:

Simplex fiber-optic cables (Fig. 2.65) are used in:

- Light-duty indoor applications
- Cable trays
- Short conduits
- Loose tie wrapping
- Subchannels for breakout cables

Zipcord fiber-optic cables (Fig. 2.66) are used in:

- Light-duty (two-way) transmission
- Indoor runs in cable trays
- Short conduits
- Tie wrapping

Figure 2.65   Simplex fiber-optic cable.

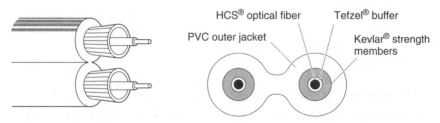

Figure 2.66   Zipcord fiber-optic cable.

Multichannel fiber-optic cables (Fig. 2.67) are used in:

- Outdoor environments
- Multifiber runs where each channel is connectorized and routed separately
- Aerial runs
- Long conduit pulls
- Two, four, and six channels (standard)
- Eight to 18 channels

Heavy-duty duplex fiber-optic cables (Fig. 2.68) are used in:

- Rugged applications
- Wide-temperature-range environments
- Direct burial
- Loose tube subchannel design
- High-tensile-stress applications

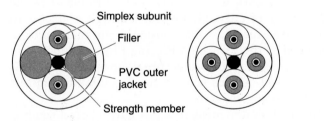

Figure 2.67 Multichannel fiber-optic cable.

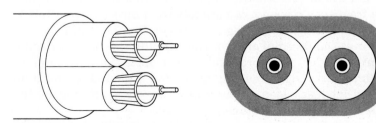

Figure 2.68 Heavy-duty fiber-optic cable.

TABLE 2.1   Fiber-Optic Cable Characteristics

Characteristics	Simplex	Zipcord	Duplex	Channels		
				2	4	6
Cable diameter, mm	2.5	2.5 × 5.4	3.5 × 6	8	8	10
Weight, kg/km	6.5	11	20	41	41	61
Jacket type	PVC	PVC	PVC	PE	PE	PE
Jacket color	Orange	Orange	Orange	Black	Black	Black
Pull tension	330 N	490 N	670 N	890 N	1150 N	2490 N
Max. long-term tension	200 N	310 N	400 N	525 N	870 N	1425 N
Max. break strength	890 N	1340 N	1780 N	2370 N	4000 N	11,000 N
Impact strength at 1.6 N • m	100	100	150	200	200	200
Cyclic flexing, cycles	>5000	>5000	>5000	>2000	>2000	>2000
Minimum bend radius, mm	25	25	25	50	50	75

Cable attenuation	0.6 dB/km at 820 nm
Operating temperature	−40 to +85°C
Storage temperature	−40 to +85°C

Table 2.1 summarizes fiber-optic cable characteristics.

2.17.2   Fiber-optic ammeter

In many applications, including the fusion reactors, radio-frequency systems, and telemetry systems, it is often necessary to measure the magnitude and frequency of current flowing through a circuit in which high dc voltages are present. A fiber-optic current monitor (Fig. 2.69) has been developed at the Princeton Plasma Physics Laboratory (PPPL) in response to a transient voltage breakdown problem that caused failures of Hall-effect devices used in the Tokamak fusion test reactor's natural-beam heating systems.

The fiber-optic current monitor measures low current in a conductor at very high voltage. Typical voltages range between tens of kilovolts and several hundred kilovolts. With a dead band of approximately 3 mA, the circuit derives its power from the conductor being measured and couples information to a (safe) area by means of fiber optics. The frequency response is normally from direct current to 100 kHz, and a typical magnitude range is between 5 and 600 mA.

The system is composed of an inverting amplifier, a current regulator, transorbs, diodes, resistors, and a fiber-optic cable. Around an inverting amplifier, a light-emitting diode and a photodiode form an optical closed feedback loop. A fraction of the light emitted by the LED is coupled to the fiber-optic cable.

As the current flows through the first diode, it splits between the 1.5-mA current regulator and the sampling resistor. The voltage

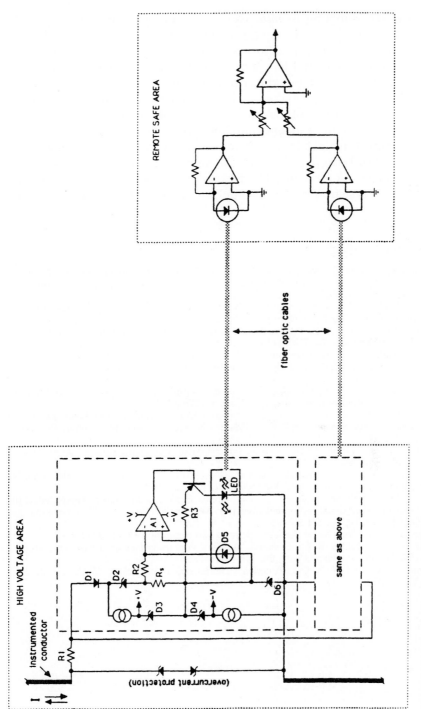

Figure 2.69 Fiber-optic current monitor.

157

across the sampling resistor causes a small current to flow into the inverting amplifier summing junction and is proportional to the current in the sampling resistor. Since photodiodes are quite linear, the light power from the LED is proportional to the current through the sampling resistor. The light is split between the local photodiode and the fiber cable. A photodiode, located in a remote safe area, receives light that is linearly proportional to the conductor current (for current greater than 5 mA and less than 600 mA).

To protect against fault conditions, the design utilizes two back-to-back transorbs in parallel with the monitor circuit. The transorbs are rated for 400 A for 1 ms. The fiber-optic ammeter is an effective tool for fusion research and other applications where high voltage is present.

Further Reading

Berwick, M., J. D. C. Jones, and D. A. Jackson, "Alternating Current Measurement and Non-Invasive Data Ring Utilizing the Faraday Effect in a Closed Loop Fiber Magnetometer," *Optics Lett.,* **12**(294) (1987).

Cole, J. H., B. A. Danver, and J. A. Bucaro, "Synthetic Heterodyne Interferometric Demodulation," *IEEE J. Quant. Electron.,* **QE-18**(684) (1982).

Dandridge, A., and A. B. Tveten, "Phase Compensation in Interferometric Fiber Optic Sensors," *Optics Lett.,* **7**(279) (1982).

Desforges, F. X., L. B. Jeunhomme, Ph. Graindorge, and G. L. Baudec, "Fiber Optic Microswitch for Industrial Use," presented at SPIE O-E Fiber Conf., San Diego, no. 838–41 (1987).

Favre, F., and D. LeGuen, "High Frequency Stability in Laser Diode for Heterodyne Communication Systems," *Electron. Lett.* **16**(179) (1980).

Giallorenzi, T. G., "Optical Fiber Interferometer Technology and Hydrophones in Optical Fiber Sensors," NATO ASI Series E, No. 132, Martinus Nijhoff Dordrecht, 35–50 (1987).

Hocker, G. B., "Fiber Optic Sensing for Pressure and Temperature," *Appl. Optics,* **18**(1445) (1979).

Jackson, D. A., A. D. Kersey, and A. C. Lewin, "Fiber Gyroscope with Passive Quadrature Demodulation," *Electron. Lett.,* **20**(399) (1984).

Optical Society of America, *Optical Fiber Optics.* Summaries of papers presented at the Optical Fiber Sensors Topical Meeting, January 27–29, 1988, New Orleans, La. (IEEE, Catalog No. 88CH2524–7).

Popovic, R. S., "Hall Effect Devices," *Sensors and Actuators,* **17,** 39–53 (1989).

Saxena, S. C., and S. B. Lal Seksena, "A Self-Compensated Smart LVDT Transducer," *IEEE Trans. Instrum. Meas.,* **38,** 748–783 (1989).

Wong, Y. L., and W. E. Ott, *Function Circuits and Applications,* McGraw-Hill, New York, 1976.

3

Networking of Sensors and Control Systems in Manufacturing

3.0 Introduction

Central to the development of any computer-integrated manufacturing facility is the selection of the appropriate automated manufacturing system and the sensors and control systems to implement it. The degree to which a CIM configuration can be realized depends on the capabilities and cost of available equipment and the simplicity of information flow.

When designing an error-free manufacturing system, the manufacturing design group must have an appreciation for the functional limits of the automated manufacturing equipment of interest and the ability of the sensors to provide effective information flow, since these parameters will constrain the range of possible design configurations. Obviously, it is not useful to design a manufacturing facility that cannot be implemented because it exceeds the equipment's capabilities. It is desirable to match automated manufacturing equipment to the application. Although sensors and control systems are—by far—less costly than the automated manufacturing equipment, it is neither useful nor cost-effective to apply the most sophisticated sensors and controls, with the highest performance, to every possible application. Rather, it is important that the design process determines the preferred parameter values.

The preferred values must be compatible with available equipment and sensors and control systems, and should be those appropriate for the particular factory. The parameters associated with the available equipment and sensors and control systems drive a functional process of modeling the manufacturing operation and facility. The parameters determine how the real-world equipment constraints will be incorporated into the functional design process. In turn, as many different functional configurations are considered, the cost-benefit relations of these alternatives can be evaluated and preferred parameter values determined. So long as these preferred values are within the limits of available automated manufacturing equipment and sensory and control systems, the design group is assured that the automated manufacturing equipment can meet its requirements. To the degree that optimum design configurations exceed present equipment capabilities, original equipment manufacturers (OEMs) are motivated to develop new equipment designs and advanced sensors and control systems.

Sensors and control systems, actuators/effectors, controllers, and control loops must be considered in order to appreciate the fundamental limitations associated with manufacturing equipment for error-free manufacturing. Many levels of factory automation are associated with manufacturing equipment; the objective at all times should be to choose the levels of automation and information flow that are appropriate for the facility being designed, as revealed through cost-benefit studies. Manufacturing facilities can be designed by describing each manufacturing system—and the sensors and controls to be used in it—by a set of functional parameters. These parameters are

1. The number of product categories for which the automated manufacturing equipment, sensors, and control systems can be used (with software downloaded for each product type)
2. The mean time between operator interventions (MTOI)
3. The mean time of intervention (MTI)
4. The percentage yield of product of acceptable quality
5. The mean processing time per product

An *ideal equipment unit* would be infinitely flexible so that it could handle any number of categories desired, would require no operator intervention between setup times, would produce only product of acceptable quality, and would have unbounded production capabilities.

The degree to which real equipment containing sensors and control systems can approach this ideal depends on the physical constraints associated with the design and operation of the equipment and the ability to obtain instantaneous information about equipment performance through sensors and control systems. The performance of the equipment in each of the five parameters above is related to details of the equipment's operation in an error-free environment. Relationships must be developed between the physical description of the equipment's operation and the functional parameters that will be associated with this operation. The objective is to link together the physical design of the equipment and its functional performance through sensory and control systems in the factory setting.

This concept provides insight into an area in which future manufacturing system improvements would be advantageous, and also suggests the magnitude of the cost-benefit payoffs that might be associated with various equipment designs. It also reveals the operational efficiency of such systems.

An understanding of the relationships between the equipment characteristics and the performance parameters based on sensors and control systems can be used to select the best equipment for the parameter requirements associated with a given factory configuration. In this way, the manufacturing design team can survey alternative types of available equipment and select the units that are most appropriate for each potential configuration.

3.1   Number of Products in a Flexible System

The first parameter listed above, the number of product categories for which the manufacturing system can be used, represents the key concern in flexible manufacturing. A unit of automated manufacturing equipment is described in terms of the number of product categories for which it can be used with only a software download to distinguish among product types. A completely fixed automated manufacturing system that cannot respond to computer control might be able to accommodate only one product category without a manual setup. On the other hand, a very flexible manufacturing system would be able to accommodate a wide range of product categories with the aid of effective sensors and control systems. This parameter will thus be defined by the breadth of the processes that can be performed by an automated manufacturing equipment unit and the ability of the unit to respond to external control data to shift among these operations.

The most effective solution will depend on the factory configuration that is of interest. Thus OEMs are always concerned with anticipating future types of factories in order to ensure that their equipment will be an optimum match to the intended configuration. This also will ensure that the *concept of error-free manufacturing can be implemented with a high degree of spontaneity*. There is a continual trade-off between flexibility and cost. In general, more flexible and "smarter" manufacturing equipment will cost more. Therefore, the objective in a particular setting will be to achieve just the required amount of flexibility, without any extra capability built into the equipment unit.

3.2   Sensors Tracking the Mean Time between Operator Interventions

The MTOI value should be matched to the factory configuration in use. In a highly manual operation, it may be acceptable to have an operator intervene frequently. On the other hand, if the objective is to achieve operator-independent manufacturing between manual setups, then the equipment must be designed so that the MTOI is longer than the planned duration between manual setups. The manufacturer of automated equipment with adequate sensors and control systems must try to assess the ways in which factories will be configured and produce equipment that can satisfy manufacturing needs without incurring any extra cost due to needed features.

3.3   Sensors Tracking the Mean Time of Intervention

Each time an intervention is required, it is desirable to compare the intervention interval with that for which the system was designed. If the intervention time becomes large with respect to the planned mean time between operator interventions, then the efficiency of the automated manufacturing equipment drops rapidly in terms of the fraction of time it is available to manufacture the desired product.

3.4   Sensors Tracking Yield

In a competitive environment, it is essential that all automated manufacturing equipment emphasize the production of quality product. If the automated manufacturing equipment produces a large quantity of product that must be either discarded or reworked, then the operation of the factory is strongly affected, and costs will increase rapidly. The objective, then, is to determine the product yields that are required for given configurations and to design automated manufacturing equipment containing sensors and control systems that can achieve these levels of yield. Achieving higher yield levels will, in general, require additional sensing and adaptability features for the equipment. These features will enable the equipment to adjust and monitor itself and, if it gets out of alignment, to discontinue operation.

3.5 Sensors Tracking the Mean Processing Time

If more product units can be completed in a given time, the cost of automated manufacturing equipment with sensors and control systems can be more widely amortized. As the mean processing time is reduced, the equipment can produce more product units in a given time, reducing the manufacturing cost per unit. Again, automated manufacturing equipment containing sensory and control systems generally becomes more expensive as the processing time is reduced. Tradeoffs are generally necessary among improvements in the five parameters and the cost of equipment. If high-performance equipment is to be employed, the factory configuration must make effective use of the equipment's capabilities to justify its higher cost. On the other hand, if the factory configuration does not require the highest parameters, then it is far more cost-effective to choose equipment units that are less sophisticated but adequate for the purposes of the facility. This interplay between parameter values and equipment design and cost is an essential aspect of system design.

Table 3.1 illustrates the difference between available parameter values and optimum parameter values, where the subscripts for equipment E represent increasing levels of complexity. The table shows the type of data that can be collected to evaluate cost and benefits. These data have significant impact on system design and perfor-

TABLE 3.1 Values of Available Parameters

Equipment	MTOI, min	MTI, min	Yield, %	Process, min	R&D expense, thousands of dollars	Equipment cost, thousands of dollars
Production function A						
E_1	0.1	0.1	90	12		50
E_2	1.0	0.1	85	8		75
E_3	10	1.0	80	10		85
E_4	18	1.0	90	8	280	155
Production function B						
E_1	1.0	0.1	95	10		150
E_2	10	0.5	90	2		300
Production function C						
E_1	0.1	0.1	98	3		125
E_2	5.0	1.0	98	2		250
E_3	8.0	2.0	96	1		300
E_4	20	2.0	96	1	540	400

mance which, in turn, has direct impact on product cost. Given the type of information in Table 3.1, system designers can evaluate the effects of utilizing various levels of sensors and control systems on new equipment and whether they improve performance enough to be worth the research and development and production investment.

One of the difficulties associated with manufacturing strategy in the United States is that many companies procure manufacturing equipment only from commercial vendors and do not consider modifying it to suit their own needs. Custom modification can produce pivotal manufacturing advantage, but also requires the company to expand both its planning scope and product development skills. The type of analysis indicated in Table 3.1 may enable an organization to determine the value and return on investment of customizing manufacturing equipment to incorporate advanced sensors and control systems. Alternatively, enterprises with limited research and development resources may decide to contract for development of the optimum equipment in such a way that the sponsor retains proprietary rights for a period of time.

3.6   Network of Sensors Detecting Machinery Faults

A comprehensive detection system for automated manufacturing equipment must be seriously considered as part of the manufacturing strategy. A major component of any effort to develop an intelligent and flexible automatic manufacturing system is the concurrent development of automated diagnostic systems, with a network of sensors, to handle machinery maintenance and process control functions. This will undoubtedly lead to significant gains in productivity and product quality. Sensors and control systems are one of the enabling technologies for the "lights-out" factory of the future.

A flexible manufacturing system often contains a variety of manufacturing work cells. Each work cell in turn consists of various workstations. The flexible manufacturing cell may consist of a CNC lathe or mill whose capabilities are extended by a robotic handling device, thus creating a highly flexible machining cell whose functions are coordinated by its own computer. In most cases, the cell robot exchanges workpieces, tools (including chucks), and even its own gripping jaws in the cell (Fig. 3.1).

3.6.1   Diagnostic systems

A diagnostic system generally relies on copious amounts of *a priori* and *a posteriori* information. *A priori* information is any previously

Figure 3.1 Flexible machining cell.

established fact or relationship that the system can exploit in making a diagnosis. *A posteriori* information is the information concerning the problem at hand for which the diagnosis will be made. The first step in collecting data is to use sensors and transducers to convert physical states into electrical signals. After processing, a signal will be in an appropriate form for analysis (perhaps as a table of values, a time-domain waveform, or a frequency spectrum). Then the analysis, including correlations with other data and trending, can proceed.

After the data have been distilled into information, the deductive process begins, leading finally to the fault diagnosis. Expert systems have been used effectively for diagnostic efforts, with the diagnostic system presenting either a single diagnosis or a set of possibilities with their respective probabilities, based on the *a priori* and *a posteriori* information.

3.6.2 Resonance and vibration analysis

Resonance and vibration analysis is a proven method for diagnosing deteriorating machine elements in steady-state process equipment such as turbomachinery and fans. The effectiveness of resonance and vibration analysis in diagnosing faults in machinery operating at variable speed is not proved, but additional study has indicated good potential for its application in robots. One difficulty with resonance and vibration analysis is the attenuation of the signal as it travels through a structure on the way to the sensors and transducers. Moreover, all motion of the machine contributes to the motion measured by the sensors and transducers, so sensors and transducers must be located as close as possible to the component of concern to maximize the signal-to-noise ratio.

3.6.3 Sensing motor current for signature analysis

Electric motors generate back electromotive force (emf) when subjected to mechanical load. This property makes a motor a transducer for measuring load vibrations via current fluctuations. Motor current signature analysis uses many of the same techniques as vibration analysis for interpreting the signals. But motor current signature analysis is nonintrusive because motor current can be measured anywhere along the motor power cables, whereas a vibration sensor or transducer must be mounted close to the machine element of interest. The limited bandwidth of the signals associated with motor drive signature analysis, however, may restrict its applicability.

3.6.4 Acoustics

A good operator can tell from the noise that a machine makes whether a fault is developing or not. It is natural to extend this concept to automatic diagnosis. The operator, obviously, has access to subtle, innate pattern recognition techniques, and thus is able to discern sounds within a myriad of background noises. Any diagnostic system based on sound would have to be able to identify damage-related sounds and separate the information from the ambient noise. Acoustic sensing (looking for sounds that indicate faults) is a nonlocal, noncontact inspection method. Any acoustic technique is subject to outside disturbances, but is potentially a very powerful tool, provided that operating conditions are acoustically repeatable and that the diagnostic system can effectively recognize acoustic patterns.

3.6.5 Temperature

Using temperature as a measurement parameter is common, particularly for equipment running at high speed, where faults cause enough waste heat to raise temperature significantly. This method is generally best for indicating that a fault has occurred, rather than the precise nature of the fault.

3.6.6 Sensors for diagnostic systems

Assuming an automated diagnostic system is required, the necessary sensors are normally mounted permanently at their monitoring sites. This works well if data are required continuously, or if there are only a few monitoring locations. However, for those cases where many sites must be monitored and the data need not be continuously received during operation of the flexible manufacturing cell, it may be possible to use the same sensor or transducer, sequentially, in the many locations.

The robot is well-suited to gathering data at multiple points with a limited number of sensors and transducers. This would extend the mandate of the robot from simply moving workpieces and tools within the flexible manufacturing cell (for production) to include moving sensors (for diagnostic inspection).

Within the flexible manufacturing cell, a robot can make measurements at sites inside its work space by taking a sensor or transducer from a tool magazine, delivering the sensor or transducer to a sensing location, detaching it during data collection, and then retrieving it before moving to the next sensing position.

Sensor mobility does, however, add some problems. First, the robot will not be able to reach all points within the flexible manufacturing cell because its work space is only a subspace of the volume taken up by the flexible manufacturing cell. The manipulator may be unable to assume an orientation desired for a measurement even inside its work space. Also, the inspection procedure must limit the robot's influence on the measurement as much as possible. Finally, sensors require connectors on the robot end effectors for signals and power. The end effector would have to be able to accommodate all the types of sensors to be mounted on it.

3.6.7 Quantifying the quality of a workpiece

If workpiece quality can be quantified, then quality can become a process variable. Any system using product quality as a measure of its performance needs tight error checks so as not to discard product unnecessarily while the flexible manufacturing cell adjusts its operating parameters. Such a system would depend heavily, at first, on the continued supervision of an operator who remains in the loop to assess product quality. Since it is forbidden for the operator to influence the process while it is under automatic control, it is more realistic for the operator to look for damage to product after each stage of manufacture within the cell. In that way, the flexible manufacturing cell receives diagnostic information about product deficiencies close to the time that improper manufacture occurred.

In the future, these quality assessments will be handled by the flexible manufacturing cell itself, using sensors and diagnostic information for process control. Robots, too, will be used for maintenance and physical inspection as part of the regular operation of the flexible manufacturing cell. In the near term, the flexible manufacturing cell robot may be used as a sensor-transfer device, replacing inspectors who would otherwise apply sensors to collect data.

3.6.8   Evaluation of an existing flexible manufacturing cell using a sensing network

A study was conducted at the Mi-TNO in the Netherlands of flexible manufacturing cells for low-volume orders (often called job production, ranging from 1 to 100 parts per order). The automated manufacturing equipment used in the study consisted of two free-standing flexible manufacturing cells. The first cell was a turning-machine cell; the second, a milling-machine cell. The turning cell contained two armed gantry robots for material handling. The study was mainly conducted to assess the diagnostics for flexible manufacturing systems (FMS). In considering the approach to setting up diagnostics for an FMS, it was decided to divide development of the diagnostics program into three major blocks (Fig. 3.2):

1. Analyzing the existing design
2. Setting up diagnostics that are machine-based

   a. Choosing important points in the flexible manufacturing cells where critical failure can occur and where sensors are mounted
   b. Setting up a diagnostic decision system for the hardware system

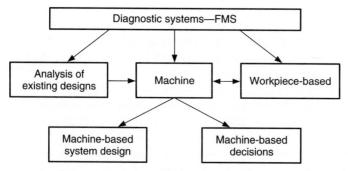

Figure 3.2 Diagnostics for an FMS.

3. Establishing a workpiece-based diagnostic system that is actually part of quality control

The flexible manufacturing cells were divided into control volumes (as shown in Fig. 3.3 for the turning cell). For the turning cell, for example, the hardware control volumes were denoted *a, b, c, d,* and *e,* and software control volumes *p1, p2, p3,* and *p4.* The failures within each of these control volumes were further categorized as:

1. Fault detected by an existing sensor
2. Fault that could have been detected if a sensor had been present
3. Operator learning phase problem
4. Failure due to manufacturer problem
5. Software logic problem
6. Repeat of a problem
7. System down for an extended period

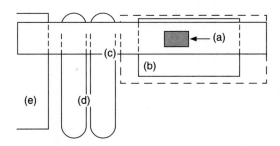

CV Identification
(a) Tool/workpiece interface
(b) Machining volume
(c) Robot volume
(d) Workpiece pallet volume
(e) Tool magazine volume
   1. VAX, 8600
   2. Microvax
   3. PDP 11/231
   4. Turning FMC controller—NC

Figure 3.3 Hardware control volumes in a manufacturing cell.

3.6.8.1 Software problems. It is often assumed that disturbances in the cell software control system can be detected and evaluated relatively easily. Software diagnostics are common in most turnkey operations; however, it has been shown that software diagnostics are far from perfect. Indeed, software problems are of particular concern when a revised program is introduced into a cell which has been running smoothly (existing bugs having been ironed out). The availability of the cell plummets in these situations, with considerable loss of production capabilities and concomitant higher cost. The frequency of software faults increases dramatically when revised packages are introduced to upgrade the system (Fig. 3.4). This is mainly due to human error on the part of either the vendor or the operator. Two possible approaches to software defect prevention and detection are

1. Investigate software methodologies and procedures and recommend alternative languages or more tools as defect prevention measurements. This is tedious and uses ambiguous results because such investigations are not based on data.

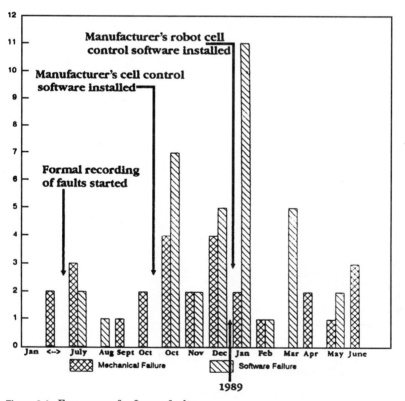

Figure 3.4   Frequency of software faults.

2. Analyze the problems that result from the current design and develop a solution for each class of problem. This produces less ambiguous solutions and is typically used to solve only immediate problems, thereby producing only short-term solutions.

To identify the types of faults that occur in programs, it is necessary to know what caused the problem and what remedial actions were taken. Program faults can be subdivided into the categories shown below and restated in Figure 3.5:

Matching faults:

- Wrong names of global variables or constants
- Wrong type of structure or module arguments
- Wrong number of hardware units
- Wrong procedures for writing data to hardware

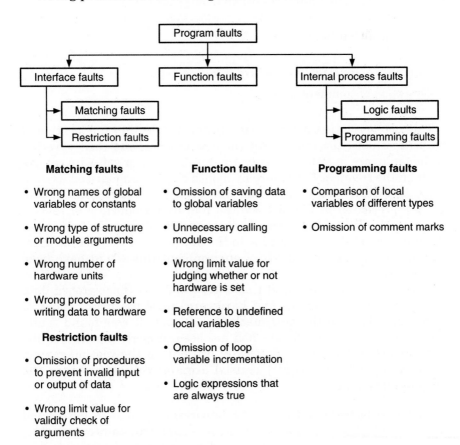

Figure 3.5 Program fault categories.

Restriction faults

- Omission of procedures to prevent invalid input or output of data
- Wrong limit value for validity check of arguments

Function faults

- Omission of saving data to global variables
- Unnecessary calling modules
- Wrong limit value for judging whether or not hardware is set
- Reference to undefined local variables
- Omission of loop variable incrementation
- Logic expressions that are always true

Programming faults

- Comparison of local variables of different types
- Omission of comment marks

This categorization provides significant insight into the location of fault conditions, the reasons for their occurrence, and their severity. If the faults are in either a hardware (mechanical) or software category, then the frequency of failure by month can be summarized as indicated in Fig. 3.4. Two major and unexpected milestones in the program represented in Fig. 3.4 are the routine introduction of revised cell control software and revised robot control software. In both cases there was a substantial increase in the downtime of the flexible manufacturing cell. In an industrial environment this would have been very costly.

In this study, it was found that interface faults (mismatched data transfer between modules and hardware) were the major cause of downtime. Also, in the new machine software, it was found faults occurred because the software had not been properly matched to the number of tool positions physically present on the tool magazine. Once, such a fault actually caused a major collision within the machining volume.

3.6.8.2 Detecting tool failure. An important element in automated process control is real-time detection of cutting tool failure, including both wear and fracture mechanisms. The ability to detect such failures on line allows remedial action to be undertaken in a timely fashion, thus ensuring consistently high product quality and preventing potential damage to the process machinery. The preliminary results from a study to investigate the possibility of using vibration signals generated during face milling to detect both progressive (wear) and catastrophic (breakage) tool failure are discussed below.

3.6.8.2.1 Experimental technique. The experimental studies were carried out using a 3-hp vertical milling machine. The cutting tool was a 381-mm-diameter face mill employing three Carboloy TPC-322E grade 370 tungsten carbide cutting inserts. The standard workpiece was a mild steel plate with a length of 305 mm, a height of 152 mm, and a width of 13 mm. While cutting, the mill traversed the length of the workpiece, performing an interrupted symmetric cut. The sensor sensed the vibration generated during the milling process on the workpiece clamp. The vibration signals were recorded for analysis. Inserts with various magnitudes of wear and fracture (ranging from 0.13 mm to 0.78 mm) were used in the experiments.

3.6.8.2.2 Manufacturing status of parts. Figure 3.6 shows typical acceleration versus time histories. Figure 3.7a is the acceleration for three sharp inserts. Note that the engagement of each insert in the workpiece is clearly evident and that all engagements share similar characteristics, although they are by no means identical.

Figure 3.7b shows the acceleration for the combination of two sharp inserts and one insert with a 0.39-mm fracture. The sharp inserts produce signals consistent with those shown in Fig. 3.6, while the fractured insert produces a significantly different output.

The reduced output level for the fractured insert is a result of the much smaller depth of cut associated with it. It would seem from the time-domain data that use of either an envelope detection or a threshold crossing scheme would provide the ability to automate the detection of tool fracture in a multi-insert milling operation.

Figure 3.7 shows typical frequency spectra for various tool conditions. It is immediately apparent that, in general, fracture phenomena are indicated by an increase in the level of spectra components within the range of 10,000 to 17,000 Hz. A comparison of Fig. 3.7a and b indicates a notable increase in the spectra around 11 kHz when a single insert fracture of 0.39 mm is present. For two fractured inserts (Fig. 3.7c), the peak shifts to around 13.5 kHz. For three fractured inserts (Fig. 3.7d), both the 13.5-kHz peak and an additional peak at about 17 kHz are apparent.

Comparing Figs. 3.7e and c, it is seen that, in general, increasing the size of insert fracture results in an increase in the spectral peak associ-

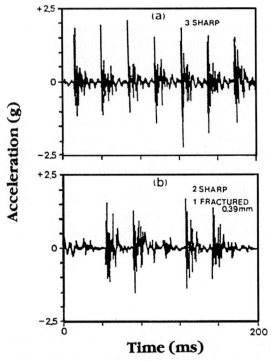

Figure 3.6 Typical acceleration level versus time plot.

ated with the failure condition. Usually, the assemblies used to obtain the spectral data are not synchronously averaged. Therefore, it may be possible to improve the spectral data by utilizing a combination of synchronous averaging and delayed triggering to ensure that data representative of each insert in the cutter is obtained and processed.

In general, the acceleration–time histories for the worn inserts do not produce the noticeably different engagement signals evident in a case of fracture. However, by processing the data in a slightly different manner, it is possible to detect evidence of tool wear.

Figure 3.8 shows the amplitude probability density (APD) as a function of time for several tool conditions. The data are averaged for eight assemblies. It thus seems possible that insert wear could be detected using such features as the location of the peak in the APD, the magnitude of peak, and the area under specific segments of the distribution.

As with fracture, the presence of insert wear resulted in a significant increase in the spectral components within the 10- to 13-kHz band. Although this would seem to indicate that the presence of flank

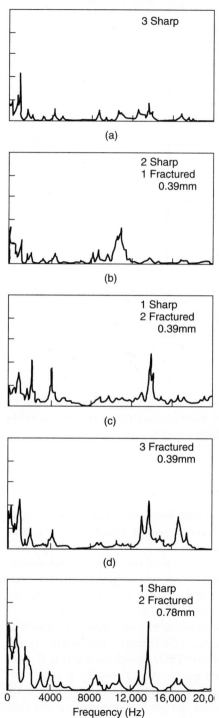

Figure 3.7 Acceleration level versus time for various inserts.

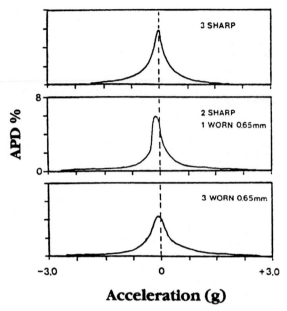

Figure 3.8 Amplitude probability density (APD) for several tool conditions.

wear could be detected by simple spectral analysis, it is not yet clear if this method would be sufficiently discriminating to permit reliable determination of the magnitude of flank wear.

3.7 Understanding Computer Communications and Sensors' Role

The evolution in computer software and hardware has had a major impact on the capability of computer-integrated manufacturing concepts. The development of smart manufacturing equipment and sensors and control systems, as well as methods of networking computers, has made it feasible to consider cost-effective computer applications that enhance manufacturing. In addition, this growth has changed approaches to design of manufacturing facilities.

Messages are exchanged among computers according to various protocols. The open system interconnect (OSI) model developed by the International Standards Organization (ISO) provides such a framework. Figure 3.9 illustrates how two users might employ a computer network. As illustrated, the transfer of information takes place from User 1 to User 2. Each message passes through a series of layers that are associated with message processing:

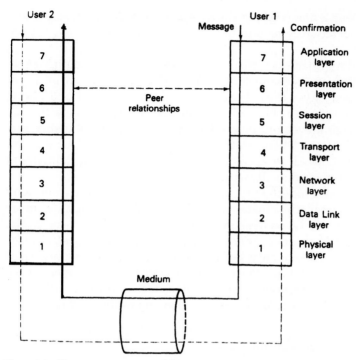

Figure 3.9   Message movement within a computer network.

Sequence of layers—User 1	Sequence of layers—User 2
(PhL)1 Physical layer	(PhL)1 Physical layer
(DL)2 Data link layer	(DL)2 Data link layer
(NL)3 Network layer	(NL)3 Network layer
(TL)4 Transport layer	(TL)4 Transport layer
(SL)5 Session layer	(SL)5 Session layer
(PL)6 Presentation layer	(PL)6 Presentation layer
(AL)7 Application layer	(AL)7 Application layer

A message is sent from User 1 to User 2, with message acknowledgment sent back from User 2 to User 1. The objective of this communication system is to transfer a message from User 1 to User 2 and to confirm message receipt. The message is developed at application layer (AL)7, and passes from there to presentation layer (PL)6, from there to session layer (SL)5, and so forth until the message is actually transmitted over the communication path. The message arrives at physical layer (PhL)1 for User 2 and then proceeds from physical layer (PhL)1 to data link layer (DL)2, to network layer (NL)3, and so forth, until User 2 has received the message. In order for Users 1 and 2 to communicate with one another, every message must pass though all the layers.

The layered approach provides a structure for the messaging procedure. Each time a message moves from User 1 [application layer (AL)7 to physical layer (PhL)1] additional processing and addressing information is added to the beginning or end of the message. As the original message moves on, new information is added to ensure correct communication. Then, as the message moves to User 2 [from physical layer (PhL)1 to application layer (AL)7], this additional information is removed as illustrated in Fig. 3.9.

The layers work together to achieve "peer" communication. Any information or operation instruction that is added at session layer (SL)5 for User 1 will be addressing session layer (SL)5 for User 2. The layers thus work as peers; each layer has a given operational or addressing task to make sure the message is correctly communicated from User 1 to User 2. Each layer associates only with the layers above and below itself. The layer receives messages from one direction, processes the messages, and then passes them on to the next layer.

3.7.1   Application layer communication

Communication begins when User 1 requests that a message be transferred from location 1 to location 2. In developing this message, it may be necessary to have application software that will provide supporting services to the users. This type of service is provided by application layer (AL)7. Common software application tools enable, for example, the transfer of files or the arranging of messages in standard formats.

3.7.2   Presentation layer communication

The initial message is passed down from the application layer (AL)7 to presentation layer (PL)6, where any necessary translation is performed to develop a common message syntax that User 2 will understand. If the two different users apply different computer or equipment languages, it will be necessary to define the differences in such a way that the users can communicate with one another. The basic message that began with User 1 is translated to a common syntax that will result in understanding by User 2. Additional information is added to the message at presentation layer (PL)6 to explain to User 2 the nature of the communication that is taking place. An extended message begins to form.

3.7.3 Session layer communication

The message now passes from presentation layer (PL)6 to session layer (SL)5. The objective of the session layer is to set up the ability for the two users to converse, instead of just passing unrelated messages back and forth. The session layer will remember that there is an ongoing dialog and will provide the necessary linkages between the individual messages so that an extended transfer of information can take place.

3.7.4 Transport layer communication

The message then passes from session layer (SL)5 to transport layer (TL)4, which controls the individual messages as part of the communication sequence. The purpose of the transport layer is to make sure that individual messages are transferred from User 1 to User 2 as part of the overall communication session that is defined by session layer (SL)5. Additional information is added to the message so that the transport of this particular portion of the communication exchange results.

3.7.5 Network layer communication

The message is then passed to network layer (NL)3, divided into packets, and guided to the correct destination. The network layer operates so that the message (and all accompanying information) is tracked and routed correctly so that it will end up at User 2. This includes making sure that addresses are correct and that any intermediate stops between User 1 and User 2 are completely defined.

3.7.6 Data link layer communication by fiber optics or coaxial cable

The system now passes a message to data link layer (DL)2, which directs each frame of the message routing in transit. Each frame is loaded into the communication system in preparation for transmission. The message is prepared to leave the User 1 environment and to move into the communication medium. The next step is from data link layer (DL)2 to physical layer (PhL)1. At this point, the frame is converted into series of digital or analog electronic signals that can be placed on the communication medium itself—wire, fiber optics, coaxial cables, or other means—to achieve the transmission of the message from User 1 to User 2.

3.7.7   Physical layer communication

The electronic signal arrives at the correctly addressed location for User 2 and is received by physical layer (PhL)1. The physical layer then converts the electronic signal back to the original frame that was placed on the medium. This extended message is passed up to the data link layer, which confirms that error-free communication has taken place and that the frame has been received at the correct location. Data link layer (DL)2 directs the frame routing in transit. When the full frame is received at data link layer (DL)2, the routing information is removed and the remaining information is transferred to network layer (NL)3. Network layer (NL)3 confirms the appropriate routing and assembles the packets. Then the routing information is stripped from the message. The message is then passed to transport layer (TL)4, which controls the transport of the individual messages. Transport layer (TL)4 confirms that the correct connection has been achieved between User 1 and User 2 and receives the information necessary to achieve the appropriate connection for this particular exchange.

The remaining information is often passed to session layer (SL)5, which interprets whether the message is part of a continuing communication and, if so, identifies it as part of an ongoing conversation. The information is then passed to presentation layer (PL)6, which performs any necessary translation to make sure that User 2 can understand the message as it has been presented. The message is then passed to application layer (AL)7, which identifies the necessary support programs and software that are necessary for interpretation of the message by User 2. Finally, User 2 receives the message and understands its meaning.

3.7.8   Adding and removing information in computer networks based on open system interconnect (OSI)

This step-by-step passing of the message from User 1 "down" to the actual communication medium and "up" to User 2 involves adding information to the original message prior to the transfer of information and removing this extra information on arrival (Fig. 3.10). At User 1, additional information is added step by step until the communication medium is reached, forming an extended message. When this information arrives at User 2, the additional information is removed step by step until User 2 receives the original message. As noted, a peer relationship exists between the various levels. Each level communicates only with the level above or below it, and the levels perform a matching service. Whatever information is added to a message by a

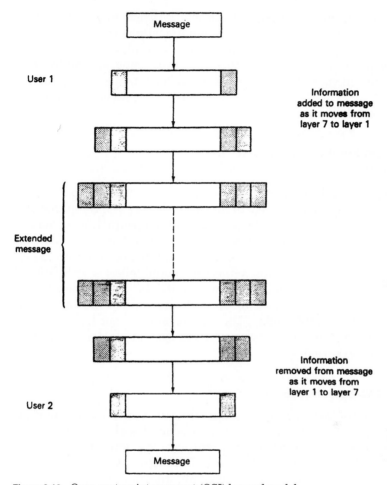

Figure 3.10 Open system interconnect (OSI) layered model.

given level at User 1 is removed from the message by the matching level associated with User 2. Communication is made possible by having User 1's physical layer (PhL)1 communicate with User 2's physical layer (PhL)1, data link layer (DL)2 communicate with data link (DL)2, network layer (NL)3 communicate with network layer (NL)3, and so forth to achieve an exchange between User 1 and User 2.

The two users see only the message that is originated and delivered; they do not see all of the intermediate steps. This is analogous to the steps involved in making a telephone call or sending a letter, in which the two users know only that they have communicated with one another, but do not have any specific knowledge of the details involved in passing a message from one location to another. This orderly and structured approach to communications is useful because it separates the various tasks that must take place. It provides a means for assuring that the methods for processing and addressing messages are always the same at every node. Whether a message is being sent or is being received, a sequential processing activity always takes place.

3.8   Understanding Networks in Manufacturing

A seemingly simple problem in the development of computer networks is to establish the ability to interconnect *any* two computer system elements. This might involve a computer terminal, a modem, a bar-code scanner, a printer, sensors, and other system elements that must exchange information in manufacturing. It might seem reasonable that such one-to-one interconnection would follow a well-defined strategy and make maximum use of standards. Unfortunately, for historical reasons and because of the wide diversity of the equipment units that are available today, this situation does not hold. In fact, achieving interconnection between typical system elements can be a disconcerting experience.

3.8.1   RS-232–based networks

One of the most common approaches used to interconnect computer system elements in manufacturing is associated with a strategy that was never really intended for this purpose. As noted by Campbell (1984), "In 1969, the EIA (Electronic Industries Association), Bell Laboratories, and manufacturers of communication equipment cooperatively formulated and issued 'EIA RS-232,' which almost immediately underwent minor revisions to become 'RS-232-C.' The RS-232 interface was developed to allow data equipment terminals to be connected to modems so that data could be transmitted over the telephone network. The entire purpose of this standard was to assure that the use of telephone lines for computer communications would be handled in a way acceptable to the telephone company.

Thus, in its general application today, RS-232 is not a standard. It is more a guideline for addressing some of the issues involved in interfacing equipment. Many issues must be resolved on an individual basis, which leads to the potential for difficulties. Essentially, a vendor's statement that a computer system element is RS-232–compatible provides a starting point to consider how the equipment unit might be interconnected. However, the detailed aspects of the interconnection require further understanding of the ways in which equipment units are intended to communicate. Campbell (1984) is a helpful introduction to applying RS-232 concepts.

In a sense, the history of the RS-232 interface illustrates the difficulties associated with creating a well-defined means for allowing the various elements of a computer network to interface. Past experience also indicates how difficult it is to develop standards that will apply in the future to all the difficult situations that will be encountered. As it has evolved, the RS-232 approach to the communications interface is an improvement over the "total anarchy" at position A in Fig. 3.11, but it still leads to a wide range of problems.

An entire computer network can be configured by using combinations of point-to-point RS-232 connections. In fact, a number of networks of this type are in common use. Such networks require that multiple RS-232 interfaces be present on each equipment unit, as is often the case. Each particular interconnection must be customized for the two equipment units being considered. Thus, a system integra-

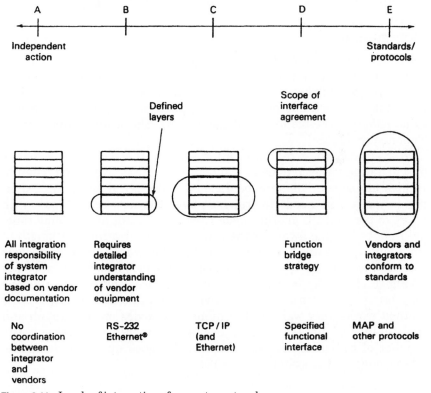

Figure 3.11 Levels of integration of computer networks.

tor not only must decide on the elements of the system and how they should perform in a functional sense, but also must develop a detailed understanding of the ways in which RS-232 concepts have been applied to the particular equipment units that are used for the network. The system integrator incurs a substantial expense in achieving the required interconnections.

It is essential to realize that the RS-232 pseudo-standard only addresses the ability to transfer serially information bit by bit from one system element to another. The higher level communications protocol in Fig. 3.11 is not considered. RS-232 provides the means for running wires or fiber-optic cables from one element to another in order to allow digital signals to be conveyed between system elements. The meaning associated with these bits of information is completely dependent on the hardware and software that is implemented in the system elements. RS-232 is a widely used approach for computer elements to transfer information. Using RS-232 is certainly much better than using no guidelines at all. However, because RS-232 does not completely define all of the relationships that must exist in communication links, it falls far short of being a true standard or protocol.

3.8.2   Ethernet[1]

As illustrated in Fig. 3.11, one approach to local-area networks (LANs) is to define a protocol for the first two layers of a communication strategy and then allow individual users to define the upper layers. This approach has been widely applied using a method referred to as *Ethernet* [Metcalfe and Boggs (1976), Shock and Hupp (1980), Tanenbaum (1988)].

In every computer communication system, there must be a means of scheduling for each node to transmit onto the network and listen to receive messages. This may be done on a statistical basis. For example, when a unit needs to transmit over the network, it makes an effort to transmit. If another node tries to transmit at the same time, both nodes become aware of the conflict, wait for a random length of time, and try again. It might seem that this would be an inefficient means of controlling a network, since the various nodes are randomly trying to claim the network for their own use, and many collisions may occur. As it turns out, for lower communication volumes, this method works very well. As the number of nodes on the system and the number of messages being exchanged increases, however, the number of collisions between active nodes goes up and reduces the effectiveness of the system (Fig. 3.12).

---

[1]Ethernet is a trademark of Xerox Corp.

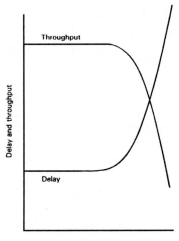

Network load (number of users)

Figure 3.12 Delay and throughput versus network load (number of users).

This type of access control for a computer network is referred to as *carrier-sense multiple-access with collision detection* (CSMA/CD). Ethernet and similar solutions are widely applied to create CSMA/CD networks, particularly in settings in which a maximum waiting time for access to the network does not have to be guaranteed. This type of network is simple to install, and a wide range of hardware and software products are available for support. On the other hand, as indicated in Fig. 3.12, network performance can degrade significantly under high load; therefore, the utility of an Ethernet-oriented network will depend on the particular configuration and loads that are expected for the network.

3.8.3 Transmission control protocol (TCP)/Internet protocol (IP)

TCP/IP applies to the transport and network layers indicated in Fig. 3.9. TCP/IP thus provides a means for addressing intermediate protocol levels, and in fact is often combined with Ethernet in a communication approach that defines both the lower and middle aspects of the system. TCP/IP functions by dividing any message provided to these middle layers into *packets* of 64 kbytes and then sending packets one at a time to the communication network. TCP/IP must also reassemble the packets in the correct order at the receiving user.

TCP/IP provides a common strategy to use for networking. It allows extension of the Ethernet lower layers to a midlayer protocol on which the final application and presentation layers may be constructed.

3.9   Manufacturing Automation Protocol

The manufacturing automation protocol (MAP) is one of the protocols that has been developed for computer communication systems. MAP was developed specifically for use in a factory environment. General Motors Corp. has been the leading advocate of this particular protocol. When faced with a need for networking many types of equipment in its factory environment, General Motors decided that a new type of protocol was required. Beginning in 1980, General Motors began to develop a protocol that could accommodate the high data rate expected in its future factories and provide the necessary noise immunity expected for this environment. In addition, the intent was to work within a mature communications technology and to develop a protocol that could be used for all types of equipment in General Motors factories. MAP was developed to meet these needs. The General Motors effort has drawn on a combination of Institute of Electrical and Electronics Engineers (IEEE) and ISO standards, and is based on the open system interconnect (OSI) layered model as illustrated in Fig. 3.10.

Several versions of MAP have been developed. One difficulty among many has been obtaining agreement among many different countries and vendor groups on a specific standard. Another problem is that the resulting standards are so broad that they have become very complex, making it difficult to develop the hardware and software to implement the system and consequently driving up related costs. The early version of MAP addressed some of the OSI layers to a limited degree, and made provision for users to individualize the application layer for a particular use. The latest version of MAP makes an effort to define more completely all the application layer software support as well as the other layers. This has led to continuing disagreements and struggles to produce a protocol that can be adopted by every vendor group in every country to achieve MAP goals.

Because of its complexity, MAP compatibility among equipment units, or *interoperability,* has been a continuing difficulty. MAP has not been applied as rapidly as was initially hoped for by its proponents because of the complexity, costs, and disagreements on how it should be implemented. Assembling a complete set of documentation for MAP is a difficult activity that requires compiling a file of standards organization reports, a number of industry organization reports, and documentation from all the working committees associated with ISO.

3.9.1   Broadband system for MAP protocol

The MAP protocol was developed with several alternatives for physical layer (PhL)1. MAP can be implemented through what is called a *broadband* system (Fig. 3.13). So that the manufacturing units can talk to one another, transmitted messages are placed on the cable; a *head-end remodulator* retransmits these messages and directs them to the receiving station. The broadband version of MAP has the highest capabilities because it allows several types of communications to take place on the same cabling at the same time. On the other hand, because of its greater flexibility, a broadband system is more complex and more expensive to install. It requires modems and MAP interface

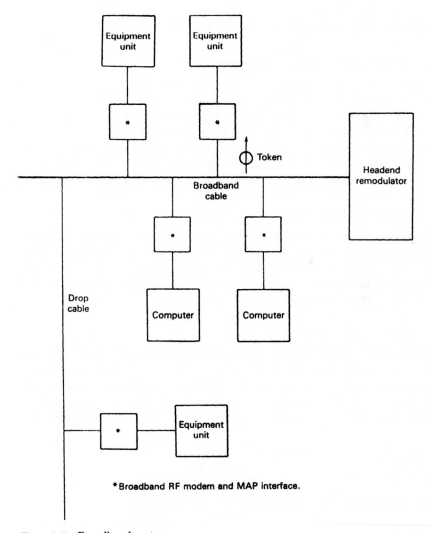

Figure 3.13 Broadband system.

equipment for each item of equipment and a head-end remodulator to serve the entire network. The main cable used for broadband is unwieldy (approximately 25 mm in diameter) and is appropriate only for wiring very large factories. Multiple drop cables can be branched off the main cable for the different MAP nodes. Broadband communication can achieve a high data rate of 10 Mb/s and can split the frequency spectrum to allow several different communications to take place simultaneously. As indicated in Fig. 3.14, the three transmit frequencies and the three receive frequencies are separated from one another in the frequency domain. Three different channels can coexist on the MAP network. The head-end remodulator transfers messages from the low frequencies to the high frequencies.

3.9.2   Carrier-band system for MAP protocol

Another type of MAP network makes use of a carrier-band approach, which uses somewhat less expensive modems and interface units and does not require heavy duty cable (Fig. 3.15). For a small factory, a carrier-band version of MAP can be much more cost-effective. The carrier-band communication can also achieve a high data rate of 5 to 10 Mb/s, but only one channel can operate on the cable at a given time. A single channel is used for both transmission and reception.

3.9.3   Bridges MAP protocol

It is possible to use devices called *bridges* to link a broadband factory-wide communication network to local carrier-band networks (Fig. 3.16). The bridge transforms the message format provided on one side to the message formats that are required on the other. In this sense, a bridge transforms one operating protocol to another.

3.9.4   Token system for MAP protocol

In developing MAP, General Motors was concerned with assuring that every MAP mode would be able to claim control of the network and communicate with other nodes within a certain maximum waiting time. Within this waiting time, every node would be able to initiate its required communications. To do this, MAP implements a *token*

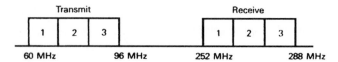

Figure 3.14 Broadband frequencies.

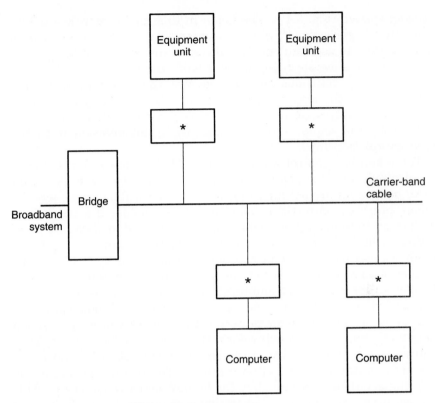

Figure 3.15 Carrier-band system for MAP protocol.

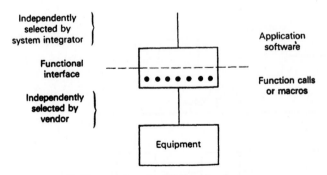

Figure 3.16 Bridges system for MAP protocol.

passing system (Fig. 3.17). The token in this case is merely a digital word that is recognized by the computer. The token is rotated from node address to node address; a node can claim control of the network and transmit a message only when it holds the token. The token prevents message collisions and also ensures that, for a given system configuration, the maximum waiting time is completely defined (so long as no failures occur). The token is passed around a logic ring defined by the sequence of node addresses, not necessarily by the physical relationships.

The token is a control word, and each MAP node can initiate communication only if it passes the token. It is interesting to note that MAP nodes that are not a part of the logic ring will never possess the token, but they may still respond to a token holder if a message is addressed to them. Token management is handled at data link layer (DL)2 of the OSI model. This layer controls the features of the token application with respect to how long a token can be held, the sequence of addresses that is to take place, and the amount of time that is allowed for retrying communications before failure is assumed. If the logic ring is broken at some point—for example, if one equipment unit is no longer able to operate—the other nodes will wait a certain length of time and then will reform the token passing scheme. They will do this by an algorithm through which the token is awarded to the highest station address that contends. The rest of the stations on the ring are then determined by the highest-address successors. This process is repeated until the token ring is re-formed.

Physical layer (PhL)1 involves encoding and modulation of the message so that the digital data are transferred into analog and digital communication signals. Each MAP application requires a modem for this

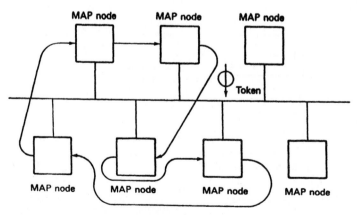

Figure 3.17 Token system for MAP protocol.

purpose. The modem takes the extended message that has been developed at higher layers and modifies it so that it can be used to provide an electronic signal to the communications medium. The medium itself provides the means for transferring the signal from User 1 to User 2.

MAP continues to be an important protocol approach for sensors and control systems in computer-integrated manufacturing. For very large factories the broadband option is available, and for a smaller factories a carrier-band system is also available. A number of vendors now produce the hardware and software necessary to establish a MAP network. However, such networks typically are quite high in cost and, because of the complexity of the protocol, can also be difficult to develop and maintain. Thus MAP is only one of several solutions that are available to planning and implementation teams.

3.10   Multiple-Ring Digital Communication Network—AbNET

An optical-fiber digital communication network has been proposed to support the data acquisition and control functions of electric power distribution networks. The optical-fiber links would follow the power distribution routes. Since the fiber can cross open power switches, the communication network would include multiple interconnected loops with occasional spurs (Fig. 3.18). At each intersection a node is needed. The nodes of the communication network would also include power distribution substations and power controlling units. In addition to serv-

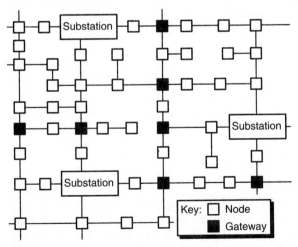

Figure 3.18   Multiple-ring digital communication network—AbNET.

ing data acquisition and control functions, each node would act as a repeater, passing on messages to the next node.

Network topology is arbitrary, governed by the power system. The token ring protocols that are used in single-ring digital communication networks are not adequate. The multiple-ring communication network would operate on the new AbNET protocol, which has already been developed, and would feature fiber optics, for this more complicated network.

Initially, a message inserted anywhere in the network would pass from node to node throughout the network, eventually reaching all connected nodes. On the first reception of a message, each node would record an identifying number and transmit the message to the next node. On second reception of the message, each node would recognize the identifying number and refrain from retransmitting the message. This would prevent the endless repetition and recirculating of messages. This aspect of the protocol resembles the behavior of cells in the immune system, which learn to recognize invading organisms on first exposure and kill them with antibodies when they encounter the organisms again. For this reason, the protocol is called *AbNET* after the microbiologists' abbreviation *Ab* for *antibodies*. The AbNET protocols include features designed to maximize the efficiency and fault-tolerant nature of the approach. Multiple service territories can be accommodated, interconnected by *gateway* nodes (Fig. 3.18).

The AbNET protocol is expected to enable a network to operate as economically as a single ring that includes an *active monitor* node to prevent the recirculation of messages. With AbNET the performance of the proposed network would probably exceed that of a network that relies on a central unit to route messages. Communications would automatically be maintained in the remaining intact parts of the network even if fibers were broken.

For the power system application, the advantages of optical-fiber communications include electrical isolation and immunity to electrical noise. The AbNET protocols augment these advantages by allowing an economical system to be built with topology-independent and fault-tolerant features.

3.11   Universal Memory Network

The universal memory network (UMN) is a modular, digital data communication system that enables computers with differing bus architectures to share 32-bit-wide data between locations up to 3 km apart with less than 1 ms of latency (Fig. 3.19). This network makes it possible to design sophisticated real-time and near-real-time data processing systems without the data transfer bottlenecks that now exist when computers use the usual communication protocols. This

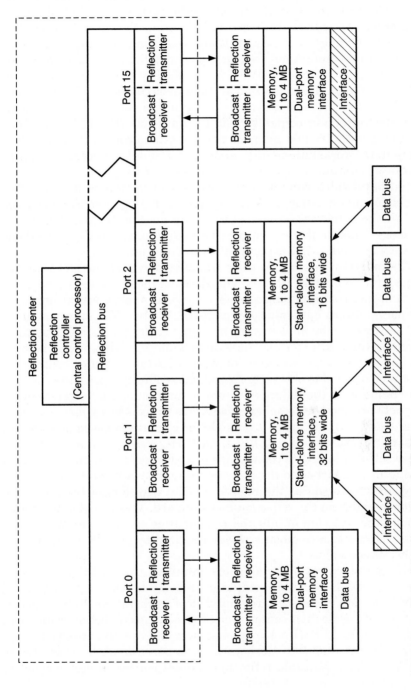

Figure 3.19 Universal memory network (UMN).

enterprise network permits the transmission of the volume of data equivalent to an average encyclopedia each second (40 Mbyte/s). Examples of facilities that can benefit from the universal memory network include telemetry stations, real-time-monitoring through laser sensors, simulation facilities, power plants, and large laboratories (e.g., particle accelerators), or any facility that shares very large volumes of data. The main hub of the universal memory network uses a *reflection center*—a subsystem containing a central control processor (the reflection controller) and a data bus (the reflection bus) equipped with 16 dual memory parts. Various configurations of host computers, workstations, file servers, and small networks or subnetworks of computers can be interconnected by providing memory speed-bandwidth connectivity. The reflection center provides full duplex communication between the ports, thereby effectively combining all the memories in the network into dual-ported, random-access memory. This dual-port characteristic eliminates the CPU overhead on each computer that is incurred with Ethernet.

The reflection bus carries write transfers only and operates at a sustained data rate of 40 Mbyte/s. This does not include address, error correction, and coordination information, which makes actual universal memory network bus traffic approach 100 Mbyte/s. The universal memory network can be implemented in copper cables for distances up to 15 m and in fiber optics for distances up to 3 km. A combination of both for media can be used in the same network. Multiple reflection centers can be interconnected to obtain configurations requiring more ports.

In addition to the reflection center of main hub, the universal memory network includes smaller subsystems called *shared memory interfaces* (SMIs), which make it possible for computers based on different bus architectures (e.g., SELBus, DEC BI, Multi Bus, and VME, or other selected buses) to communicate via the reflection bus. Each host computer is attached to the reflection center by a bus-interface circuit card, which translates the read and write transfers of the host computer to and from the reflection-bus standard. This translation centers around the ordering of bits and conversation used by various vendor architectures to a common strategy required by the 100-ns cycle time of the reflection bus.

The standard memory interface enhances the modular nature of the network. It provides computer memory access to processors of lower cost and enables a large number of workstations to be supported from one reflection center. For example, one reflection center can support up to 12 SMI memory interfaces, each with capacity to support between 8 and 16 workstations, depending on local hardware configurations. Multiple reflection centers can be interconnected to support even more workstations.

Further Reading

Ames, J. G., "Which Network Is the Right One," *Manufacturing Engineering,* **56,** May 1988.

Borggraaf, P., "Semiconductor Factory Automation: Current Theories," *Semiconductor International,* **88,** October 1985.

Bux, W., "Local Area Subnetworks: A Performance Comparison," *IEEE Trans. Communications,* **COM-29**(10):1465 (1981).

Campbell, J., *The RS-232 Solution,* Alameda, California: Sybex, Inc. (1984).

Kaminski, A. M., Jr., "Protocols for Communicating in the Factory," *IEEE Spectrum,* **65,** April 1986.

Kleinrock, L., and S. S. Lam, "Packet Switching in a Multiaccess Broadcast Channel: Performance Evaluation," *IEEE Trans. Communications,* **COM-23**(4):410 (1975).

Metcalfe, R. M., and D. R. Boggs, "Ehternet: Distributed Packet-Switching for Local Computer Networks," *Communications of the ACM* **19:**395 (1976).

Shock, J. F., and J. A. Hupp, "Measured Performance of an Ethernet Local Network," *Communications of the ACM,* **23:**711 (1980).

Tanenbaum, A. S., *Computer Networks,* 2d ed., Englewood Cliffs, N.J.: Prentice-Hall (1988).

Talvage, J., and R. G. Hannam, *Flexible Manufacturing Systems in Practice: Application Design and Simulation,* New York: Marcel Decker (1988).

Voelcker, J., "Helping Computers Communicate," *IEEE Spectrum,* **61,** March 1986.

Warndorf, P. R., and M. E. Merchant, "Development and Future Trends in Computer Integrated Manufacturing in the USA," *International Journal in Technology Management,* **1**(1–2):162 (1986).

Wick, C., "Advances in Machining Centers," *Manufacturing Engineering,* **24,** October 1987.

Wittry, E. J., *Managing Information Systems: An International Approach,* Dearborn, Mich.: Society of Manufacturing Engineers (1987).

4

The Role of Sensors and Control Technology in Computer-Integrated Manufacturing

4.0 Introduction

According to various studies conducted in the United States, nearly 50 percent of the productivity increase during the period 1950–1990 was due to technological innovation. That is, the increase was due to the introduction of high-value-added products and more efficient manufacturing processes, which in turn have caused the United States to enjoy one of the highest living standards in the world. However, unless the United States continues to lead in technological innovation, the relative living standard of the country will decline over the long term.

This clearly means that the United States has to invest more in research and development, promote scientific education, and create incentives for technological innovation. In the R&D arena, the United States has been lagging behind other nations: about 1.9 percent of the U.S. gross national product (GNP) (versus about 2.6 percent of the GNP in Japan and West Germany) goes for R&D. The introduction of computer-integrated manufacturing (CIM) strategy in U.S. industry has begun to provide a successful flow of communication which may well lead to a turnaround. *Sensors and control systems in manufacturing* are a powerful tool for implementing CIM. Current world business leaders view CIM as justifying automation to save our standard of living. Successful implementation of CIM depends largely on creative information gathering through sensors and control systems, with information flow as feedback response. Information gathering through sensors and control systems is *imbedded* in CIM. CIM provides manufacturers with the ability to react more quickly to market demands and to achieve levels of productivity previously unattainable.

Effective implementation of sensors and control subsystems within the CIM manufacturing environment will enable the entire manufacturing enterprise to work together to achieve new business goals.

4.1 CIM Plan

This chapter will address implementation of a CIM plan through the technique of modeling.

A model can be defined as a tentative description of a system or theory; the model accounts for many of the system's known properties. An *enterprise model* can be defined (in terms of its functions) as the function of each area, the performance of each area, and the performance of these areas interactively. The creation of a model requires an accurate description of the needs of an enterprise.

In any manufacturing enterprise there is a unique set of business processes that are performed in order to design, produce, and market the enterprise's products. Regardless of how unique an enterprise or its set of processes is, it shares with others the same set of high-level objectives. To attain the objectives, the following criteria must be met:

1. Management of manufacturing finances and accounting
2. Development of enterprise directives and financial plans
3. Development and design of product and manufacturing processes utilizing adequate and economical sensors and control systems
4. Management of manufacturing operations
5. Management of external demands

4.1.1 CIM plan in manufacturing

In manufacturing, CIM promotes customer satisfaction by allowing order entry from customers, faster response to customer enquiries and changes, via electronic sensors, and more accurate sales projections.

4.1.2 CIM plan in engineering and research

In engineering and research, CIM benefits include quicker design, development, prototyping, and testing; faster access to current and historical product information; and paperless release of products, processes, and engineering changes to manufacturing.

4.1.3   CIM plan in production planning

In production planning, CIM offers more accurate, realistic production scheduling that requires less expediting, canceling, and rescheduling of production and purchase orders.

In plant operations, CIM helps to control processes, optimize inventory, improve yields, manage changes to product and processes, and reduce scrap and rework. CIM also helps in utilizing people and equipment more effectively, eliminating production crises, and reducing lead time and product costs.

4.1.4   CIM plan in physical distribution

In physical distribution, where external demands are satisfied with products shipped to the customer, CIM helps in planning requirements; managing the flow of products; improving efficiency of shipping, vehicle, and service scheduling; allocating supplies to distribution centers; and expediting processing of returned goods.

4.1.5   CIM plan for business management

For business management activities such as managing manufacturing, finance, and accounting, and developing enterprise directives and financial plans, CIM offers better product cost tracking, more accuracy in financial projections, and improved cash flow.

4.1.6   CIM plan for the enterprise

For the enterprise as a whole, these advantages add up to faster release of new products, shorter delivery times, optimized finished goods inventory, shorter production planning and development cycles, reduced production lead time, improved product quality, reliability and serviceability, increased responsiveness, and greater competitiveness. In effect, CIM replaces an enterprise's short-term technical improvements with a long-term strategic solution.

The advantages of CIM with sensors and control systems are not just limited to the four walls of an enterprise. It can also deliver real productivity gains in the outside world. For example, suppliers will be able to plan production, schedule deliveries, and track shipments more efficiently. Customers will benefit from shorter order-to-delivery times, on-time deliveries, and less expensive, higher-quality products.

4.2 Manufacturing Enterprise Model

The integration and productivity gains made possible by CIM with sensors and control systems are the key to maintaining a competitive edge in today's manufacturing environments. The enterprise model defines an enterprise in terms of its functions. In a traditional enterprise that relies on a complex organization structure, operations and functional management are divided into separate departments, each with its own objectives, responsibilities, resources, and productivity tools.

Yet, for the enterprise to operate profitably, these departments must perform in concert. Sensors and control systems that improve one operation at the expense of another, and tie up the enterprise's resources, are counterproductive. New sensors and control systems in CIM can create a systematic network out of these insulated pockets of productivity. But to understand how, one must examine the elements of an enterprise model and see how its various functional areas work—independently and with each other.

Creating a model of the enterprise can help to expose operations that are redundant, unnecessary, or even missing. It can also help determine which information is critical to a successful implementation once effective sensors and control systems are incorporated.

Obviously, this model is a general description. There are many industry-unique variations to the model. Some enterprises may not require all of the functions described, while others may require more than are listed. Still other enterprises may use the same types of functions, but group them differently.

For example, in the aerospace industry, life-cycle maintenance of products is an essential requirement and may require extensions to the model. In the process industry, real-time monitoring and control of the process must be included in the model.

Computer integrated manufacturing harnesses sensors and control system technology to integrate these manufacturing and business objectives. When implemented properly, CIM can deliver increased productivity, cost-efficiency, and responsiveness throughout the enterprise. CIM can accomplish these objectives by addressing each of the major functional areas of the manufacturing enterprise:

1. Marketing
2. Engineering research and development of sensors in CIM strategy
3. Production planning
4. Plant operations incorporating sensors and shop floor control systems
5. Physical distribution
6. Business management

Integrating these functions and their resources requires the ability to share and exchange information about the many events that occur during the various phases of production; manufacturing systems must be able to communicate with the other information systems throughout the enterprise (Fig. 4.1). There must also be the means to capture data close to the source, then distribute the data at the division or corporate level, as well as to external suppliers, subcontractors, and even customers.

To meet these needs, the CIM environment requires a dynamic network of distributed functions. These functions may reside on independent system platforms and require data from various sources. Some may be general-purpose platforms, while others are tailored to specific environments. But the result is an environment that encompasses the total information requirements of the enterprise, from developing its business plans to shipping its products.

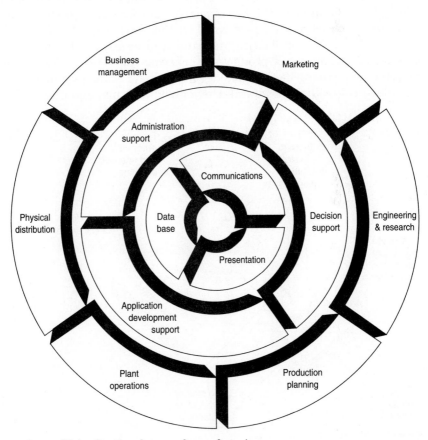

Figure 4.1   Major functional areas of manufacturing.

With this enterprise-wide purpose, CIM can deliver its benefits to all types of manufacturing operations, from plants that operate one shift per day to processes that must flow continuously, from unit fabrication and assembly to lots with by-products and coproducts. These benefits can also be realized in those enterprises where flexible manufacturing systems are being used to produce diversified products over shorter runs, as well as in those that use CIM with sensors and control systems to maintain an error-free environment.

By creating more efficient, more comprehensive management information systems through sensors and control systems, CIM supports management efforts to meet the challenges of competing effectively in today's world markets.

4.2.1 Marketing

Marketing acts as an enterprise's primary contact with its customer (Fig. 4.2). To help meet the key objectives of increasing product sales, a number of functions are performed within marketing. These include market research; forecasting demand and sales; analyzing sales; tracking performance of products, marketing segments, sales personnel, and advertising campaigns; developing and managing marketing channels; controlling profits and revenues; and managing sales per-

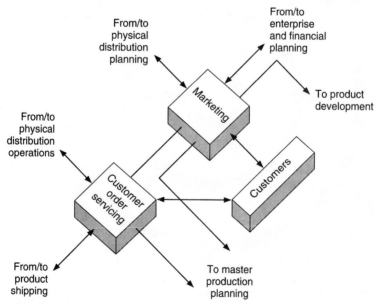

Figure 4.2 Marketing relationships.

sonnel, sales plans, and promotion. Input comes from business management and customers. Output goes to customers, product development, customer order servicing, and master production planning.

The information handling requirements of marketing include monumental texts and graphics as well as queries and analysis of internal and external data. The internal data is gathered through a variety of software routines.

Customer order servicing involves entering and tracking customer orders. This can be for standard products or custom-designed products. Other customer order servicing activities include providing product quotations, checking customer credit, pricing product, allocating order quantities, and selecting shipments from distribution centers.

Input to this area includes order and forecast data from marketing or directly from customers as well as available-to-promise data from production planning. Output can include allocation of all orders, quotations for custom products, communication with production engineering regarding custom products, order consideration, and shipping releases. Customer service will significantly improve through electronic data interchange (EDI).

4.2.2   Engineering and research

The engineering and research areas of an enterprise can be broken down into separate activities (Fig. 4.3). Each of these has its own special needs, tools, and relationships to other areas.

The research activities include investigating and developing new materials, products, and process technology. Information processing needs include complex analyses, extensive texts, imaging, graphics, and videos.

Research input comes from such outside research sources as universities, trade journals, and laboratory reports. Then research must communicate its findings to three other functional areas—product development, process development, and facilities engineering.

4.2.2.1   Product development.   In today's increasingly competitive manufacturing markets, creation of new material and production technologies is essential for the successful development of products. The product development area uses these materials and production technologies to design, model, simulate, and analyze new products.

Product development activities include preparing product specifications and processing requirements, drawings, materials or parts lists, and bills of material for new products or engineering changes.

In this area, laboratory analysis tools, computer-aided design/computer-aided engineering (CAD/CAE) tools, and group technology (GT)

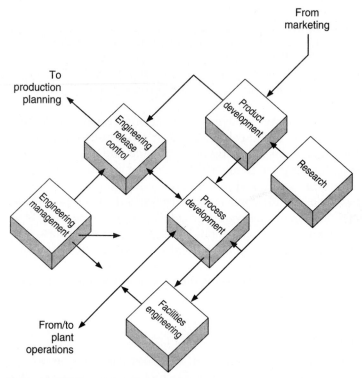

Figure 4.3 Engineering and research.

applications are helping to reduce product development time, increase productivity, and improve product quality.

Product development comes from marketing, research, and plant operations. Its output—product specifications, manufacturing control requirements, drawings, text, and messages—is directed to process development and engineering release control.

4.2.2.2 Process development. This functional area creates process control specifications, manufacturing routings, quality test and statistical quality control specifications, and numerical control (NC) programming. It also validates the manufacturability of product designs. Computer-aided process planning (CAPP) programs and group technology applications for routing similar parts have helped streamline these functions. Expert systems have also been used to supplement traditional product testing and defect-analysis processes. This area is also responsible for the application of new manufacturing technologies such as work cells, conveyer systems, and robotics. Sensors and control systems play a fundamental role within this work-cell structure.

Process development receives input from research and product development as well as statistical process data from plant operations. Output includes providing routings, process control algorithms, and work-cell programming to plant operations by way of engineering release control.

4.2.2.3 Facilities engineering.   The chief responsibility of facilities engineering is the installation of plant automation incorporating new equipment with sensors and control systems, material flow, inventory staging space, and tools. Tools may include driverless material handling equipment, conveyers, and automated storage and retrieval systems. This area is also responsible for plant layout and implementation of such plant services and utilities as electricity, piping, heat, refrigeration, and light.

Input to facilities engineering is from research and process development. Output, such as plant layouts, changes of schedule, and forecasts of new equipment availability, goes to plant operations.

4.2.2.4 Engineering release control.   This function involves the coordination of the release of new products, processes, tools, and engineering changes to manufacturing. A major check point in the product cycle is obtaining assurance that all necessary documentation is available, after which the information is released to manufacturing.

Input is received from product and process development activities. Specific output, including product and tool drawings, process and quality specifications, NC programs, and bills of material, is transferred to production planning and plant operations.

4.2.2.5 Engineering management.   Among the activities of engineering management are introducing engineering releases, controlling product definition data, estimating cost, entering and controlling process data, and defining production resources.

Input and output for engineering management include exchanging designs and descriptions with engineering design; reviewing product data, routings, and schedules with production planning; and accepting engineering change requests from plant operations.

4.2.3 Production planning

Production planning can be viewed as five related functional areas (Fig. 4.4).

4.2.3.1 Master production planning.   In master production planning, information is consolidated from customer order forecasts, distribution centers, and multiple plans in order to anticipate and satisfy demands for the enterprise's products. Output includes time-phased requirements sent to the material planning function as well as the final assembly schedule.

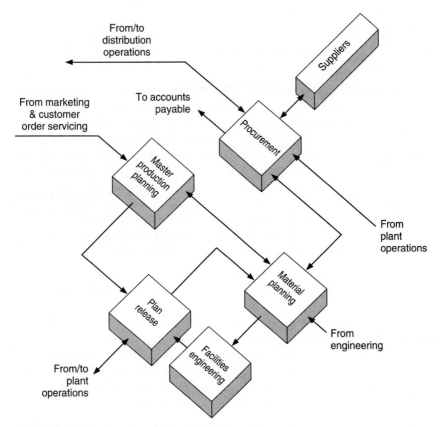

Figure 4.4 Production planning.

4.2.3.2 Material planning and resource planning.
These two areas require timely and accurate information—demand schedules, production commitment, inventory and work-in-progress status, scrap, actual versus planning receipts, shortages, and equipment breakdowns—in order to keep planning up to date with product demands.

Product and process definition data come from the engineering areas. Output is to plant operation and procurement and includes production schedules, order releases, and plans for manufactured and purchased items.

4.2.3.3 Procurement.
Procurement involves selecting suppliers and handling purchase requirements and purchase orders for parts and materials. Among the input is material requirements from material planning and just-in-time delivery requests from plant operations. Other input includes shipping notices, invoices, and freight bills.

Output to suppliers includes contracts, schedules, drawings, purchase orders, acknowledgments, requests for quotations, release of vendor payments, and part and process specifications. In order to streamline this output, as well as support just-in-time concepts, many enterprises rely on sensors for electronic data interchange with vendors.

4.2.3.4 Plant release. The functions of this area can vary, depending on the type of manufacturing environment. In continuous-flow environments, for example, this area produces schedules, recipes to optimize use of capacity, specifications, and process routings.

For job-shop fabrication and assembly environments, this area prefers electronic or paperless-shop documents consisting of engineering change levels; part, assembly, setup, and test drawings and specifications; manufacturing routings; order and project control numbers; and bar codes, tags, or order-identification documents.

However, large-volume industries are evolving from typical job-shop operations to continuous-flow operations, even for fabrication and assembly-type processes.

Input—typically an exploded production plan detailing required manufacturing items—comes from material planning. Output—schedules, recipes or shop packets—is used in plant operations for scheduling.

4.2.4 Plant operations

Plant operations can be described in terms of nine functions (Fig. 4.5).

4.2.4.1 Production management. This area provides dynamic scheduling functions for the plant floor by assigning priorities, personnel, and machines. Other activities include sending material and tool requests for just-in-time delivery.

Input for production management includes new orders and schedules from production planning and real-time feedback from plant operation. Output includes dynamic schedules and priorities that are used to manage operations on the plant floor.

4.2.4.2 Material receiving. The material receiving function includes accepting and tracking goods, materials, supplies, and equipment from outside suppliers or other locations within the enterprise. Input to this area includes receiving reports and purchase order notices.

Output includes identifying receipts with appropriate documentation, then routing materials to the proper destination. Data are also sent to accounting, procurement, and production management. Production management can also direct materials based on more immediate needs.

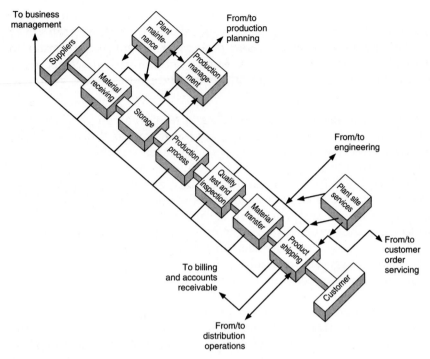

To business
management

From/to
production
planning

From/to
engineering

From/to
customer
order
servicing

To billing
and accounts
receivable

From/to
distribution
operations

Figure 4.5 Plant operations.

4.2.4.3 Storage. This represents inventory management—where
materials are stored and are accessible to the proper production locations. The materials include finished goods, raw materials, parts, supplies, work-in-progress, and tools, as well as nonproduction materials
and equipment.

Storage functions include preparing item identification and storage
tags, managing storage locations, processing pick requests, reporting
picks and kit activities, and planning physical inventory cycles and
reporting counts.

Storage input includes storage and picking requests from production management scheduling functions. Output includes receiving
and disbursement reports for use in production management and
accounting.

4.2.4.4 Production process. Production process functions include
managing the production process, processing materials, fabricating
parts, grading or reworking components, assembling final products,
and packaging for distribution.

One of today's trends in fabrication and assembly processes is toward continuous processing, such as continuous movement with minimal intermediate inventories. This can be described with such terms as continuous-flow manufacturing, flexible manufacturing cells, just-in-time logistics, and group technology. Unfortunately, in many instances, these automation efforts are autonomous, without regard to the other functions of the enterprise.

The information handling needs of the production process can include analog and digital data, text, graphics, geometries, applications programs—even images, video, and voice. Processing this information may require subsecond access and response time.

Input to this area includes shop documents, operator instructions, recipes, and schedules from production management as well as NC programs from process development. Output consists of material and tool requests, machine maintenance requests, material transfer requests, production, and interruption reports for production management, production and labor reports for cost accounting and payroll, and statistical process reports for production management and process development.

4.2.4.5 Quality test and inspection. Testing items and products to assure the conformity of specifications is the main activity in quality test and inspection. This includes analyzing and reporting results quickly by means of metrological sensors and control systems, in order to reduce scrap and rework costs.

Quality test and product specifications are input from engineering. Chief output includes purchased-item inspection results to procurement, manufactured-item inspection and product test results to production process and production management, quality test and inspection activity reports to cost accounting, and rejected part and product dispositions to material handling.

4.2.4.6 Material transfer. Material transfer involves the movement of materials, tools, parts, and products among the functional areas of the plant. These activities may be manual. Or they may be semiautomated, using control panels, forklifts, trucks, and conveyers. Or they may be fully automated, relying on programmable logical controllers, distribution control systems, stacker cranes, programmed conveyers, automated guided vehicles, and pipelines.

Input in this area may be manual requests or those generated by the system. Output includes reporting completed moves to production management.

4.2.4.7 Product shipping. This area supports the movement of products to customers, distributors, warehouses, and other plants. Among the activities are selecting shipment and routing needs, consolidating products for a customer or carrier order, preparing shipping lists and bills of lading, reporting shipments, and returning goods to vendors.

The primary input is from customer order servicing, and it includes the type of product and the method and date of shipment. Output includes reporting shipment dates to customer order servicing, billing, and accounts receivable.

4.2.4.8 Plant maintenance. Plant maintenance includes those functions that ensure the availability of production equipment and facilities. Maintenance categories include routine, emergency, preventive, and inspection services. In addition, many of today's advanced users of systems are moving toward diagnostic tools based on expert systems, which reduce equipment downtime. Input (maintenance requests) can be initiated by plant personnel, a preventive maintenance and inspection system, or a process and equipment monitoring system. Output includes requests for purchase of maintenance items, schedules for maintenance for use in production management, requests for equipment from facilities engineering, and maintenance work order costs to cost accounting.

4.2.4.9 Plant site services. The final area of plant operation is plant site services. Input received and output provided cover such functions as energy supply and utilities management, security, environmental control, grounds maintenance, and computer and communications installations.

4.2.5 Physical distribution

Physical distribution can be viewed as having two functional areas (Fig. 4.6).

4.2.5.1 Physical distribution planning. This involves planning and control of the external flow of parts and products through warehouses, distribution centers, other manufacturing locations, and points of sale. These functions may also include allocating demand, planning finished goods inventory, and scheduling vehicles. Some industries require another major set of functions that relate to logistics reports. This includes spare parts, maintenance, training, and technical documentation.

Input and output are exchanged with marketing and physical distribution operations. The information handling requirements of physical distribution planning are usually medium to heavy, especially if multiple distribution centers exist.

4.2.5.2 Physical distribution operations. This area includes receiving, storing, and shipping finished goods at the distribution center or warehouse. Receipts arrive from plants or other suppliers, and ship-

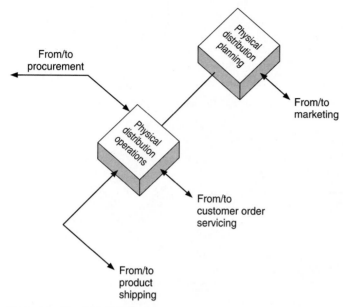

Figure 4.6 Physical distribution.

ments are made to customers and dealers. Other functions can include scheduling and dispatching vehicles, processing returned goods, and servicing warranties and repairs.

Input is received from plant site product shipping, procurement, physical distribution planning, and customer order servicing. Output includes acknowledgments to plant site product shipping and procurement, as well as data for updates to physical distribution planning and customer order servicing.

4.2.6   Business management

Within the enterprise model, a business management function may be composed of seven areas (Fig. 4.7).

4.2.6.1   Financial planning and management.   In financial planning and management, financial resource plans are developed and enterprise goals are established. Among the functions are planning costs, budgets, revenues, expenses, cash flow, and investments.

Input includes financial goals and objectives established by management as well as summarized financial data received from all areas of the enterprise. Output includes financial reports to stockholders, departmental budgets, and general ledger accounting.

4.2.6.2   Accounts payable.   Accounts payable primarily involves paying vendors. Input includes vendor invoices and goods-received reports. The output includes discount calculations to vendors and

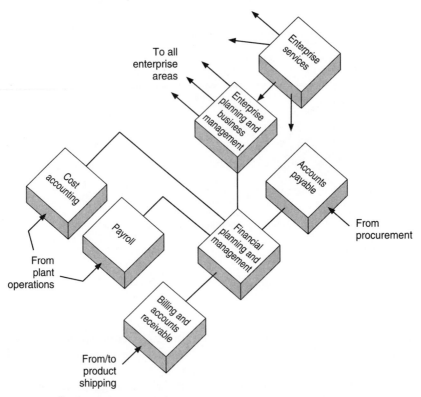

Figure 4.7 Business management.

payment checks. This last function lends itself to the use of electronic data interchange to electronically transfer funds to vendors.

4.2.6.3 Billing and accounts receivable. This area prepares invoices to customers and manages customer account collections. Input to this area consists of product shipping data and cash received. Output includes customer invoices, delinquent account reports, and credit ratings. Transferring funds electronically through EDI can simplify this area significantly.

4.2.6.4 Cost accounting. Cost accounting activity supports product pricing and financial planning by establishing product costs. These costs can include those of materials, direct and indirect labor, fixed production elements (machinery and equipment), variable production elements (electricity, fuels, or chemicals), and overhead. Other functions can include establishing standard costs, reporting variances to standards, costing job orders, and determining accrued costs.

Cost accounting input is acquired primarily from plan operations. Output is sent to financial management and planning.

4.2.6.5   Payroll. This area computes payments, taxes, and other deductions for employees. It also includes reporting and paying employee tax withholdings to government agencies. Input includes time and attendance information and production data from plant operations. Output includes payroll checks, labor distribution, and government reports and payments.

4.2.6.6   Enterprise planning and business management. These functions include establishing goals and strategies for marketing, finance, engineering and research, plant automation, sensors and control systems, and information systems. Input and output are exchanged through sensors and control systems with virtually every other area of the enterprise.

4.2.6.7   Enterprise services. Enterprise services include such support services as office personnel, management information services, personnel resources, and public relations. These services require extensive administrative support tools, such as text processing, decision support, and graphic tools. But since input and output will be exchanged throughout the entire enterprise, it is imperative that these tools be integrated with the enterprise's other systems.

4.3   Design of CIM with Sensors and Control Systems

With the advent of low-priced computers and sensors and control systems, there have been a number of technological developments related to manufacturing that can be used to make production more efficient and competitive. The primary purpose is to develop several computer-concepts as related to the overall manufacturing plan of CIM systems.

In order for the manufacturing enterprise to succeed in the future, it is imperative that it adopt a manufacturing strategy that integrates its various functions. CIM systems have the potential to accomplish this task. The implementation of CIM with sensory and control systems on the shop floor represents a formidable, albeit obtainable, objective. To accomplish this goal, enterprises must have access to information on what is available in CIM. A secondary purpose of obtaining access to information is to provide a framework that can aid in the search for information. Once the information is obtained, it becomes necessary to look at the current system objectively and decide how to approach the problem of implementing CIM with sensors and control systems.

While many of the ideas associated with CIM are new and untried, progressive enterprises, with the realization that old methods are ineffective, are doing their part in implementing this new technology. Some of the concepts currently being implemented are flexible manufacturing systems, decision support systems (DSS), artificial intelligence (AI), just-in-time (JIT) inventory management, and group technology. While all of these concepts are intended to improve efficiency, each one alone can only accomplish so much. For example, an FMS may reduce work-in-process (WIP) inventory while little is accomplished in the area of decision support systems and artificial intelligence to relate all aspects of manufacturing management and technology to each other for FMS. The advent of inexpensive sensors and control systems enables the concept of CIM to be implemented with greater confidence.

Recent advances in computer technology and sensors in terms of speed, memory, and physical space have enabled small, powerful, personal computers to revolutionize the manufacturing sector and become an essential part of design, engineering, and manufacturing, through, for example, database management systems (DBMSs) and local-area networks (LANs). The coordination of the various aspects of a manufacturing environment means that complex systems inherently interact with one another. Due to a lack of standards and poor communication between departments, many components and databases are currently incompatible.

Table 4.1 describes some benefits of CIM.

The potential for CIM, according to Table 4.1, is overwhelming, but the main issue is how to analyze and design CIM that incorporates sensors, control systems, and decision support so that it is utilized effectively.

Manufacturing problems are inherently multiobjective. For example, improving quality usually increases cost and/or reduces productivity. Furthermore, one cannot maximize quality and productivity simultaneously; there is a tradeoff among these objectives. These conflicting objectives are treated differently by different levels and/or units of production and management. Obviously, without a clear understanding of objectives and their interdependence at different levels, one cannot successfully achieve CIM with sensors and control systems.

TABLE 4.1 CIM Benefits

Application	Improvement with CIM, %
Manufacturing productivity	120
Product quality	140
Lead time (design to sale)	60
Lead time (order to shipment)	45
Increase in capital equipment utilization	340
Decrease in WIP inventory	75

4.3.1 Components of CIM with sensors and control systems

The decision making in design of CIM with effective sensors and control systems can be classified into three stages.

1. *Strategic level.* The strategic level concerns those decisions typically made by the chief executive officer (CEO) and the board of directors. Upper management decisions of this type are characterized by a relatively long planning horizon, lasting anywhere from 1 to 10 years. Implementing CIM with sensors and control systems has to begin at this level. Even though small enterprises may not have as many resources at their disposal, they have the added advantage of fewer levels of management to work through while constructing CIM.

2. *Tactical level.* At the tactical level, decisions are made that specify how and when to perform particular manufacturing activities. The planning horizon for these decisions typically spans a period from 1 to 24 months. Activities at this level include such intermediate functions as purchasing and inventory control. They affect the amount of material on the shop floor but do not control the use of the material within the manufacturing process.

3. *Operational level.* Day-to-day tasks, such as scheduling, are performed at the operational level. The primary responsibility at this level is the effective utilization of the resources made available through the decisions made on the strategic and tactical levels. Because of the variability in demand or machine down time, the planning horizon at this level must be relatively short, normally 1 to 15 days.

While each of these levels has certain responsibilities in a manufacturing plant, the objectives are often conflicting. This can be attributed to inherent differences between departments, e.g., sales and marketing may require a large variety of products to serve every customer's needs while the production department finds its job easier if there is little product variation. One of the main causes of conflicting decisions is a lack of communication due to ineffective sensors and control systems between levels and departments. CIM with adequate sensors and control systems provides the ability to link together technological advances, eliminate much of the communication gap between levels, and bring all elements into a coherent production system.

4.3.2 CIM with sensors and control systems at the plant level

Some of the important emerging concepts related to CIM with effective sensors and control systems are flexible manufacturing systems, material handling systems, automated storage and retrieval systems (AS/RS), computer-aided design (CAD), computer-aided engineering (CAE), computer-aided manufacturing (CAM), and microcomputers. These components of CIM can be classified into three major groups (Fig. 4.8).

4.3.2.1 Flexible manufacturing systems incorporating sensors and control systems. An FMS can link several elements on the shop floor through sensors in order to coordinate those elements. While CIM can be applied to any manufacturing industry, FMSs find their niche in the realm of discrete production systems such as job shops.

The most important FMS elements are numerical control machines and an automated material handling network to transport the product from raw material inventory, through the NC operations, and finally to the finished goods inventory.

Numerical control technology has made major advances with the advent of computer numerical control and direct numerical control (CNC/DNC). Microprocessors and sensors located on the machine itself can now provide the codes necessary for the parts to be operated on.

4.3.2.2 Material handling. Material handling is the means of loading, unloading, and transporting workpieces among different machines and departments. There are several ways in which material handling is accomplished:

1. *Transfer lines* consist of fixed automation machinery such as conveyor belts. Their advantages are high speed and low cost. Their major disadvantage is their lack of flexibility. Dedicated transfer lines can handle only a limited number of parts and cannot be easily changed once in place, thus defeating the goals of an FMS.

2. *Robots* provide another alternative for moving workpieces. Generally robots can be made very flexible because of their programmability, but they are limited to their region of access.

3. *Automated guided vehicles* can move workpieces a great distance, but they lack the speed found in both robot and transfer lines. Yet, because of their ability to be programmed to different routes, they are more flexible than transfer lines.

4.3.2.3 Automated storage and retrieval systems. By means of AGVs, raw materials can be taken from the loading dock and placed in a designated place in inventory. By means of an AS/RS, inventory can be tracked throughout the manufacturing process and optimized for strategically locating items in storage. Because the process is computerized, data on what exactly is in inventory assist planners in determining order and production schedules.

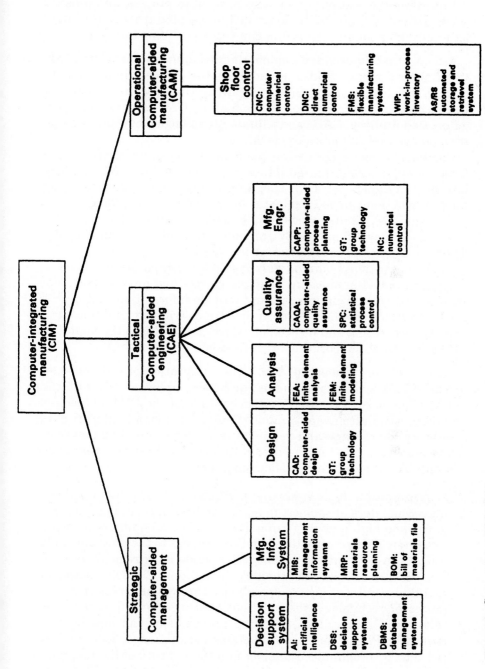

Figure 4.8 Components within CIM.

Inventories consist of raw materials, work in process, and finished goods. Inventories should be controlled by keeping track of inventory locations and amounts. An AS/RS can accomplish this task.

4.3.2.4 Computer-aided engineering/design/manufacturing (CAE/CAD/CAM).

Computer-aided design helps the engineer in many ways during the design stage. Simply drawing the part on a computer increases the productivity of designers, but CAD is more than just automated drafting. CAD can facilitate group technology and the construction of a bill of materials (BOM) file.

Computer-aided engineering consists of the many computerized facets of engineering that go into a particular product. When a part has been designed, the CAE subgroup is responsible for generating the NC code that can be used by the NC machines on the floor.

By using GT, similar parts can be classified by similar attributes and placed in part families. By grouping parts this way, much redundancy in design and manufacturing is eliminated.

In *computer-aided manufacturing* (CAM), computers are used to plan and conduct production. They can correct and process data, control the manufacturing process, and provide information that can be used in decision making. CAM can involve distributed quality control and product testing and inspection, which are built into the manufacturing process to support the larger functional relationships.

4.3.2.5 Microcomputers for CIM.

The integration of different elements of CIM and sensor systems can be accomplished only with the aid of a computer. Because of the magnitude of the CIM and sensor database, a mainframe (or a minicomputer for a small enterprise) is necessary for storage and retrieval of information. The power of microcomputers has increased so dramatically, however, that they are suitable for most other applications, such as:

1. *Programmable logic controllers (PLCs).* On the factory floor, microcomputers are subject to harsh conditions. Dust, high temperature, and high humidity can quickly destroy an ordinary personal computer. Traditionally the shop floor belongs to the programmable logic controller. The components can operate in extreme conditions, e.g., 55°C and 90 percent humidity. In addition, PLCs are geared to real-time control of factory operations. Through advances in microprocessor technology, the functions that were once controlled by relays and mechanical switching mechanisms can now be performed by PLCs. The ladder diagrams used to notate logic circuits can now be programmed directly into the memory of the programmable controller. Because of microelectronic circuitry, PLCs can process control information quickly and shut down automatically in

case of emergencies. Whereas PLCs have become commonplace on the floor of discrete product operations, process control computers have become indispensable in the control of process plants where conditions must be monitored constantly. They are also used in the control of office and factory environments. For example, a PLC can turn furnaces and air-conditioning units ON and OFF to provide suitable working conditions while optimizing energy use.

2. *Industrial personal computers (PCs).*   Until recently, PCs were commonly found in the protected environments of offices. Now, manufacturers have introduced rugged versions of popular personal computers. For example, IBM has developed the 5531 and 7531/32, industrialized versions of PC/XT and PC/AT, respectively, to withstand the environment of the factory floor. They have the advantage of being able to run any PC-DOS–based software, but are unable to perform real-time control. This problem has been remedied with the advent of the industrial computer disk operating system (IC-DOS), a real-time operating system that is compatible with other IBM software. This allows a shop floor computer to provide real-time control while using software packages previously found only on office models.

3. *Microsupercomputers.*   The hardware of microsupercomputers has increased computing power significantly. Offering high performance, transportability, and low price, the new microsupercomputers compare favorably to mainframes for many applications.

4.4   Decision Support System for CIM with Sensors and Control Systems

With an increase in production volume and efficiency comes a need to have a more effective method of scheduling and controlling resources. Herein lies a connection between CAE and computer-aided management. The long-range plans of a company must include forecasts of what the demand will be for various products in the future. Through these forecasts, the enterprise determines what strategy it will take to ensure survival and growth.

For the enterprise to make intelligent decisions, reliable information must be available. In regard to the three levels of decision making, it is also important that the information be consistent throughout each level. The best way to assure this availability and consistency is to make the same database available to all individuals involved in the production process. Because of lack of good communication between levels and sometimes the reluctance of upper-level managers to commit themselves to CIM, constructing a centralized database represents one of the most difficult problems in the implementation of CIM.

4.4.1 Computer-integrated manufacturing database (CIM DB)

The creation of a CIM DB is at the heart of the effective functioning of CIM. Most manufacturers have separate databases set up for nearly every application. Since data from one segment of an enterprise may not be structured for access by other segments' software and hardware, a serious problem for meeting the CIM goal of having readily available data for all levels occurs. Another problem with multiple databases is in the redundancy of data. Both the strategic and tactical decision makers, for example, may need information concerning a bill of material file. Even with the assumption that the databases contain consistent data (i.e. the same information in each), maintaining them both represents inefficient use of computer time and storage and labor. To install CIM, these databases must be consolidated. Unfortunately, bringing multiple databases into one CIM DB that remains available to everyone and consistent in all levels presents a significant obstacle because of the large investment needed in time and computer hardware and software.

4.4.2 Structure of multiobjective support decision systems

The success of CIM also depends largely on the ability to incorporate sensor technology with a database. The database is utilized in making decisions on all levels—decisions that are used to update to the database. Decision support systems can provide a framework for efficient database utilization by allowing storage and retrieval of information and problem solving through easy communications.

Decision making problems in manufacturing can be grouped into two general classes:

- *Structured decisions* are those that are constrained by physical or practical limitations and can be made almost automatically with the correct input. An example is generating a group technology part code given the geometry of the part.

- *Unstructured decisions* typically are those that contain a large degree of uncertainty. Decisions considered by strategic planners are almost always unstructured. Deciding whether or not to expand a certain product line, for example, may be based on demand forecasting and on the expected growth of competitors.

Due to the predictive nature of these decisions, they inherently contain more uncertainty than structured decisions. Long-range planning consists primarily of unstructured information.

Decision support systems mainly consist of three separate parts:

1. *Language systems.*   The function of a language system (LS) is to provide a means for the user to communicate with the DSS. Some considerations for the choice of a language are that the formulation should be easily understood, implementable, and modifiable. Moreover, processing the language should be possible on a separate level or on the problem processing system (PPS) level. An obvious choice for a language would be the spoken language of the user. This would require little or no training for the user to interact with a computer, but the complexity of sentences and the use of words that have multiple meanings present difficult problems that, when solved, would introduce unwanted inefficiency into the language system. An alternative would be to use a more formal language based on logic, e.g., PROLOG. The advantage here is that a language can be used at all levels of the DSS. In the design and use of an LS for the user interface, one can consider objectives such as ease of communication, the level of complexity that can be presented by the LS, and the time needed for the user to learn it.

2. *Knowledge systems.*   The basic function of a knowledge system (KS) is the representation and organization of the "knowledge" in the system. Two possible approaches are storing in the information in database form or representing the data as a base for artificial intelligence using methods from, for example, predicate calculus. The objective of KS is to ease accessibility of data for the DSS. The KS should be able to organize and classify databases and problem domains according to objectives that are sensible and convenient for the user. Some of the objectives in the design of the KS are to reduce the amount of computer memory required, increase the speed with which the data can be retrieved or stored, and increase the number of classifications of data and problem domains possible.

3. *Problem processing systems.*   The problem processing system of a DSS provides an interface between the LS and the KS. The primary function is to receive the problem from the user via the LS and use the knowledge and data from the KS to determine a solution. Once a solution is found, the PPS sends it through the KS to be translated into a form the user can recognize. More importantly, in the model formulation, analysis, and solution procedure of PPS, the conflicting objectives of stated problems must be considered.

The PPS should provide methodology that can optimize all conflicting objectives and generate a compromise solution acceptable to the user. Some of the objectives in the development of such multiobjective approaches are to reduce the amount of time that the user must spend to solve the problem, increase the degree of interactiveness (e.g., how many questions the user should answer), reduce the difficulty of questions posed to the user, and increase the robustness of the underlying assumptions and procedures.

4.5   Analysis and Design of CIM with Sensors and Control Systems

Many manufacturing systems are complex, and finding a place to begin a system description is often difficult. Breaking down each function of the system into its lowest possible level and specifying objectives for each level and their interactions will be an effective step. The objectives and decision variables, as related to elements or units for all possible levels, are outlined in the following sections.

4.5.1   Structured analysis and design technique (SADT)

The structured analysis and design technique is a structured methodology. It combines the graphical diagramming language of structured analysis (SA) with the formal thought discipline of a design technique (DT). The advantage of SADT is that it contains a formalized notation and procedure for defining system functions.

Psychological studies have shown that the human mind has difficulty grasping more than five to seven concepts at one time. Based on this observation, SADT follows the structured analysis maxim: "Everything worth saying must be expressed in six or fewer pieces." Limiting each part to six elements ensures that individual parts are not too difficult to understand. Even complex manufacturing systems can be subjected to top-down decomposition without becoming overwhelming.

The basic unit for top-down decomposition is the structural analysis box (Fig. 4.9). Each of the four sides represents a specific action for the SA box. The four actions implement:

1. *Input,* measured in terms of different decision variables.

2. *Control,* to represent constraints and limitations.

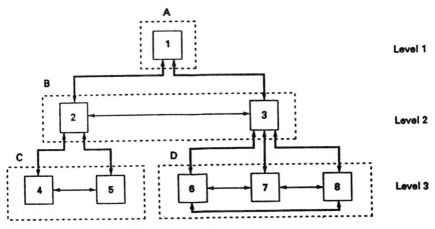

Figure 4.9 Structural analysis box.

3. *Output,* measured in the form of a set of objectives.

4. *Mechanism for translation,* which performs the translations (for mapping) of input into output as constrained by the control action.

Since each box represents an idea in the system, each box contains a detailed diagram. A parent-child relationship exists between the box being detailed and those boxes under it. The same relationship holds for the parent diagram as the diagram directly preceding the child diagrams.

The interfaces linking boxes through their neighbors at the same level are the input, output, and control actions. In graphical terms, these interfaces are designated by arrows. Methodically proceeding through a given network, the entire system can be modeled in terms of boxes and arrows.

While SADT provides a realistic approach to modeling any system, it cannot provide the solution to any problem. The integrated computer-aided manufacturing definition method comes one step closer to the realization of a functional CIM system. It utilizes teamwork and demands that all correspondence and analysis be in written form so that others can obtain a grasp of the situation and errors can be more readily located. Because the written word is required during the implementation phases, the documentation that usually is done at the end of most major projects can be nearly eliminated. Keeping accurate records also plays an important role in debugging the system in the future.

4.5.2   A multiobjective approach for selection of sensors in manufacturing

There are six criteria for the evaluation of sensors in manufacturing:

1. *Cost* is simply the price for the sensor and its integrated circuitry if it should be purchased.

2. *Integrability* is the degree to which the sensor can be used in conjunction with the manufacturing system it serves. This can be usually be measured in terms of compatibility with existing hardware control circuits and software.

3. *Reliability* is the quality of the sensors as indicated by the *mean time between failures* (MTBF), and can be measured by performing a simple stress test on the sensor under severe limits of operation. If the sensor operates under a certain high temperature for a certain period of time, it will assure the user that the system will perform satisfactorily under normal operating conditions. It will also indicate that the electronic control circuits are reliable, according to the *burn-in philosophy*.

4. *Maintenance* involves the total cost to update and maintain the sensor and how often the sensor needs to be serviced.

5. *Expandability* is how readily the sensor can be modified or expanded as new needs arise because of a changing environment.

6. *User friendliness* indicates the ease of using and understanding the unit. It may include the quality of documentation in terms of simplicity, completeness, and step-by-step descriptions of procedures.

4.6   Data Acquisition for Sensors and Control Systems in CIM Environment

The input signals generated by sensors can be fed into an interface board, called an *I/O board*. This board can be placed inside a PC-based system. As personal computers for CIM have become more affordable, and I/O boards have become increasingly reliable and readily available, PC-based CIM data acquisition has been widely implemented in laboratory automation, industrial monitoring and control, and automatic test and measurement.

To create a data acquisition system for sensors and control systems that really meets the engineering requirements, some knowledge of electrical and computer engineering is required. The following key areas are fundamental in understanding the concept of data acquisition for sensors and control systems:

1. Real-world phenomena

2. Sensors and actuators

3. Signal conditioning

4. Data acquisition for sensors and control hardware

5. Computer systems

6. Communication interfaces

7. Software

4.6.1 Real-world phenomena

Data acquisition and process control systems measure real-world phenomena, such as temperature, pressure, and flow rate. These phenomena are sensed by sensors, and are then converted into analog signals which are eventually sent to the computer as digital signals.

Some real-world events, such as contact monitoring and event counting, can be detected and transmitted as digital signals directly. The computer then records and analyzes this digital data to interpret real-world phenomena as useful information.

The real world can also be controlled by devices or equipment operated by analog or digital signals which are generated by the computer (Fig. 4.10).

4.6.2 Sensors and actuators

A sensor converts a physical phenomenon such as temperature, pressure, level, length, position, or presence or absence, into a voltage, current, frequency, pulses, etc.

For temperature measurements, some of the most common sensors include thermocouples, thermistors, and resistance temperature detectors (RTDs). Other types of sensors include flow sensors, pressure sensors, strain gauges, load cells, and optical sensors.

An actuator is a device that activates process control equipment by using pneumatic, hydraulic, electromechanical, or electronic signals. For example, a valve actuator is used to control fluid rate for opening and closing a valve.

4.6.3 Signal conditioning

A signal conditioner is a circuit module specifically intended to provide signal scaling, amplification, linearization, cold junction compensation, filtering, attenuation, excitation, common mode rejection, etc. Signal conditioning improves the quality of the sensor signals that

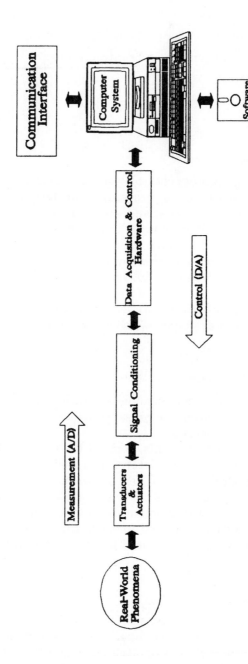

Figure 4.10 Integration of computer-controlled devices.

will be converted into digital signals by the PC's data acquisition hardware.

One of the most common functions of signal conditioning is amplification. Amplifying a sensor signal provides an analog-to-digital (A/D) converter with a much stronger signal and thereby increases resolution. To acquire the highest resolution during A/D conversion, the amplified signal should be equal to approximately the maximum input range of the A/D converter.

4.6.4 Data acquisition for sensors and control hardware

In general, data acquisition for sensors and control hardware performs one or more of the following functions:

1. Analog input
2. Analog output
3. Digital input
4. Digital output
5. Counter/timer

4.6.4.1 Analog input (A/D). An analog-to-digital converter produces digital output directly proportional to an analog signal input, so that it can be digitally read by the computer. This conversion is imperative for CIM (Fig. 4.11).

The most significant aspects of selecting A/D hardware are

1. Number of input channels
2. Single-ended or differential input
3. Sampling rate (in samples per second)
4. Resolution (in bits)
5. Input range (specified as full-scale volts)
6. Noise and nonlinearity

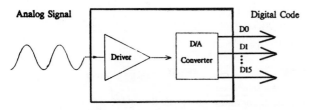

Figure 4.11 Analog-to-digital converter.

4.6.4.2   Analog output (D/A).   A digital-to-analog (D/A) converter changes digital information into a corresponding analog voltage or current. This conversion allows the computer to control real-world events.

Analog output may directly control equipment in a process that is then measured as an analog input. It is possible to perform a closed loop or proportional integral-differential (PID) control with this function. Analog output can also generate waveforms in a function generator (Fig. 4.12).

4.6.4.3   Digital input and output.   Digital input and output are useful in many applications, such as contact closure and switch status monitoring, industrial ON/OFF control, and digital communication (Fig. 4.13).

4.6.4.4   Counter/timer.   A counter/timer can be used to perform event counting, flow meter monitoring, frequency counting, pulse width and time period measurement, etc.

Most data acquisition and control hardware is designed with the multiplicity of functions described above on a single card for maximum performance and flexibility. Multifunction data acquisition for high-performance hardware can be obtained through PC boards specially designed by various manufacturers for data acquisition systems.

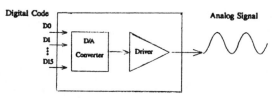

Figure 4.12   Analog output.

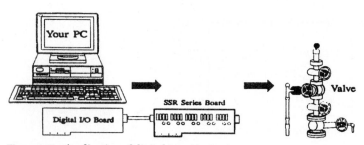

Figure 4.13   Application of digital input/output.

4.6.5 Computer system

Today's rapidly growing PC market offers a great selection of PC hardware and software in a wide price range. Thus, a CIM strategy can be economically implemented.

4.6.5.1 Hardware considerations. Different applications require different system performance levels. Currently, there are 286, 386, and 486 CPUs which will allow a PC to run at benchmark speeds from 20 up to 150 MHz. Measurements and process control applications usually require 80286 systems. But for applications that require high speed, real-time data analysis, an 80386 or 80486 system will be much more suitable.

4.6.5.2 Industrial PCs. An *industrial PC* (IPC) is designed specifically to protect the system hardware in harsh operating environments. IPCs are rugged chasses that protect system hardware against excessive heat, dust, moisture, shock, and vibration. Some IPCs are even equipped with power supplies that can withstand temperatures from -20 to $+85°C$ for added reliability in harsh environments.

4.6.5.3 Passive backplane and CPU card. More and more industrial data acquisition for sensors and control systems are using passive backplane and CPU card configurations. The advantages of these configurations are reduced mean time to repair (MTTR), ease of upgrading the system, and increased PC-bus expansion slot capacity.

A passive backplane allows the user to plug in and unplug a CPU card without the effort of removing an entire motherboard in case of damage or repair.

4.6.6 Communication interfaces

The most common types of communication interfaces used in PC-based data acquisition for sensor and control system applications are RS-232, RS-422/485, and the IEEE-488 general-purpose interface bus (GPIB).

The RS-232 interface is the most widely used interface in data acquisition for sensors and control systems. However, it is not always suitable for distances longer than 50 m or for multidrop network interfaces. The RS-422 protocol has been designed for long distances (up to 1200 m) and high-speed (usually up to 56,000 bit/s) serial data communication. The RS-485 interface can support multidrop data communication networks.

4.6.7 Software

The driving force behind any data acquisition for sensors and control systems is its software control. Programming the data acquisition for sensors and control systems can be accomplished by one of three methods:

1. *Hardware-level programming* is used to directly program the data acquisition hardware's data registers. In order to achieve this, the control code values must determine what will be written to the hardware's registers. This requires that the programmer use a language that can write or read data from the data acquisition hardware connected to the PC. Hardware-level programming is complex, and requires significant time—time that might be prohibitive to spend. This is the reason that most manufacturers of data acquisition hardware supply their customers with either driver-level or package-level programs.

2. *Driver-level programming* uses function calls with popular programming languages such as C, PASCAL, and BASIC, thereby simplifying data register programming.

3. *Package-level programming* is the most convenient technique of programming the entire data acquisition system. It integrates data analysis, presentation, and instrument control capabilities into a single software package. These programs offer a multitude of features, such as pull-down menus and icons, data logging and analysis, and real-time graphic displays.

4.7   Developing CIM Strategy with Emphasis on Sensors' Role in Manufacturing

To develop a comprehensive CIM strategy incorporating sensors and control systems, an enterprise must begin with solid foundation, such as a CIM architecture. A CIM architecture is an information system structure that enables the industrial enterprise to integrate information and business processes. It accomplishes this by (1) establishing the direction integration will take and (2) defining the interfaces between the users and the providers of this integration function.

Figure 4.14 shows how CIM architecture answers the enterprise's integration means. A CIM architecture provides a core of common services. These services support every other area in the enterprise, from its common support function to its highly specialized business processes.

4.7.1   CIM and building blocks

The information environment of an industrial enterprise is subject to frequent changes in system configuration and technologies. A CIM architecture incorporating sensors and control systems can offer a flexible structure that enables it to react to the changes. This struc-

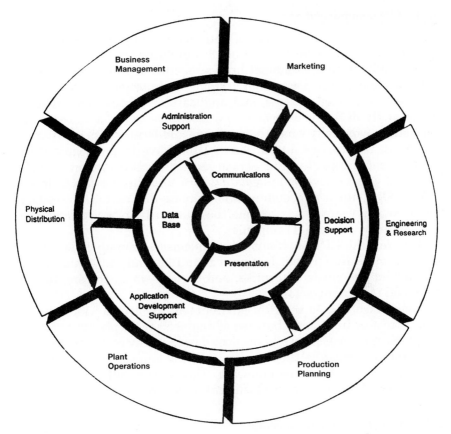

Figure 4.14 CIM architecture.

ture relies on a number of modular elements that allow systems to change easily to grow with the enterprise's needs.

Figure 4.17 (p. 237) shows a modular structure that gives CIM flexibility. It is based on three key building blocks:

1. *Communications.*   The communication and distribution of data

2. *Data management.*   The definition, storage, and use of data

3. *Presentation.*   The presentation of data to people and devices throughout the enterprise

Utilizing the building blocks, CIM can provide a base for integrating the enterprise's products, processes, and business data. It can define the structure of the hardware, software, and services required to support the enterprise's complex requirements. It can also translate information into a form that can be used by the enterprise's people, devices, and applications.

4.7.2 CIM communications

Communications—the delivery of enterprise data to people, systems, and devices—is a critical aspect of CIM architecture. This is because today's industrial environment brings together a wide range of computer systems, data acquisition systems, technologies, system architectures, operating systems, and applications. This range makes it increasingly difficult for people and machines to communicate with each other, especially when they describe and format data differently.

Various enterprises, in particular IBM, have long recognized the need to communicate data across multiple environments. IBM's response was to develop *systems network architecture* (SNA) in the 1970s. SNA supports communication among different IBM systems, and over the years it has become the standard for host communications in many industrial companies.

However, the CIM environment with sensor communications must be even more integrated. It must expand beyond individual areas, throughout the entire enterprise, and beyond—to customers, to vendors, and to subcontractors.

Communications in the CIM environment involves a wide range of data transfer, from large batches of engineering or planning data to single-bit messages from a plant floor device. Many connectivity types and protocols must be supported so that the enterprise's people, systems, and devices can communicate. This is especially true in cases where response time is critical, such as during process alerts.

4.7.3 Plant floor communications

Plant floor communications can be the most challenging aspect of factory control. This is due to the wide range of manufacturing and computer equipment that have been used in production tasks over the decades.

IBM's solution for communicating across the systems is the IBM plant floor series, a set of software products. One of these products, *Distributed Automation Edition* (DAE), is a systems enabler designed to provide communication functions that can be utilized by plant floor applications. These functions include:

1. Defining and managing networks

2. Making logical device assignments

3. Managing a program library for queueing and routing messages

4. Establishing alert procedures

5. Monitoring work-cell status

With these functions, Distributed Automation Edition can (1) help manufacturing engineers select or develop application programs to control work-cell operations and (2) provide communication capabilities between area- and plant-level systems (Fig. 4.15).

DAE supports several communication protocols to meet the needs of a variety of enterprises. For example, it supports SNA for connections to plant host systems and the IBM PC network, as well as the IBM token-ring protocol and manufacturing automated protocol (MAP) for plant floor communications. MAP is the evolving plant floor communication industry standard, adopted by the International Standards Organization for communications among systems provided by different vendors.

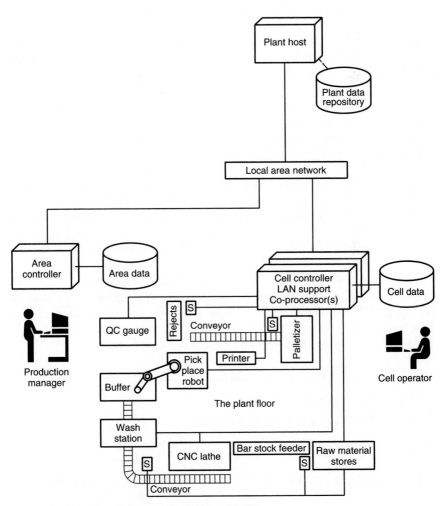

Figure 4.15 Distributed Automation Edition (DAE).

4.7.4 Managing data in the CIM environment

The second building block of a CIM architecture incorporating sensors and control technology is data management. This includes how data are defined, how different data elements are related, where data are stored, and who has access to that data. Data management is particularly critical in today's industrial environment, since there are very many different databases, formats, and storage and access techniques.

Standards are evolving. For example, *Structured Query Language* (*SQL*) provides a medium for *relational database* applications and for users to access a database. Unfortunately, there is a significant amount of data that exists today in other database technologies that are not accessible by current standards.

Data management defines and records the location of the data created and used by the enterprise's business functions. Data management also means enabling users to obtain the data needed without having to know where the data are located.

Relationships among several data elements must be known if data are to be shared by users and applications. In addition, other data attributes are important in sharing data. These include the type of data (text, graphic, image), their status (working, review, completed), and their source (person, application, or machine).

In CIM with sensory architecture, data management can be accomplished through three individual storage functions: (1) the data repository, (2) the enterprise data storage, and (3) the local data files.

Some of the key data management functions—the repository, for example—are already being implemented by the *consolidated design file* (CDF) established through the IBM *Data Communication Service* (DCS).

The consolidated design file operates on a relational database and is built on SQL. One example of its use is as an engineering database to integrate CAD/CAM applications with the business needs of the engineering management function. This environment, IBM's DCS/CDF, provides the following repository functions:

1. Transforming data to a user-selected format
2. Storing CAD/CAM data
3. Adding attributes to CAD/CAM data
4. Enabling users to query data and attributes

DCS/CDF also provides communications functions to transfer data between the repository and CAD/CAM applications (Fig. 4.16).

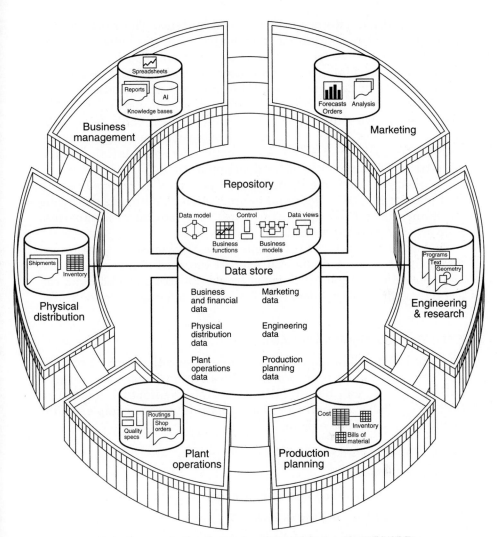

Figure 4.16 Data Communication Service/consolidated design file (DCS/CDF).

4.7.5 CIM environment presentation

Presentation in the CIM environment means providing data to and accepting data from people and devices. Obviously this data must assume appropriate data definitions and screen formats to be usable.

Because today's industrial enterprise contains such a wide array of devices and information needs, it must have a consistent way to distribute and present information to people, terminals, workstations, machine tools, robots, sensors, bar-code readers, automated guided vehicles, and part storage and retrieval systems. The range of this information covers everything from simple messages between people to large data arrays for engineering design and applications (Fig. 4.17). It may originate from a CIM user in one functional area of the enterprise and be delivered to a CIM user or device in another area.

In today's environments, presentation occurs on displays that utilize various technologies. Some are nonprogrammable terminals, some are programmable workstations, and some are uniquely implemented for each application. As a result, the same information is often treated differently by individual applications.

For example, the same manufactured part may be referred to as a part number in a bill of material in production planning, as a drawing in engineering's CAD application, and as a routing in a paperless shop order from plant operations.

As data are shared across the enterprise, they must be transformed into definitions and formats that support the need of individual users and applications. Applications must be able to access shared data, collect the required information, then format that information for delivery.

4.7.6   The requirement for integration

Communication, data management, and presentation each have their own set of technical requirements. In addition, before these three building blocks can be integrated, a CIM architecture must also address a number of enterprise-wide constraints. For example, a CIM architecture should be able to:

1. Utilize standard platforms
2. Integrate data
3. Protect the installed base investment
4. Work with heterogeneous systems
5. Utilize industry-standard operator interfaces
6. Reduce application support cost
7. Provide a customizable solution
8. Offer phased implementation
9. Deliver selectable functions
10. Improve the business process

4.7.6.1   Standard platforms.   Utilizing standard computing platforms is one step industrial enterprises can take toward integration.

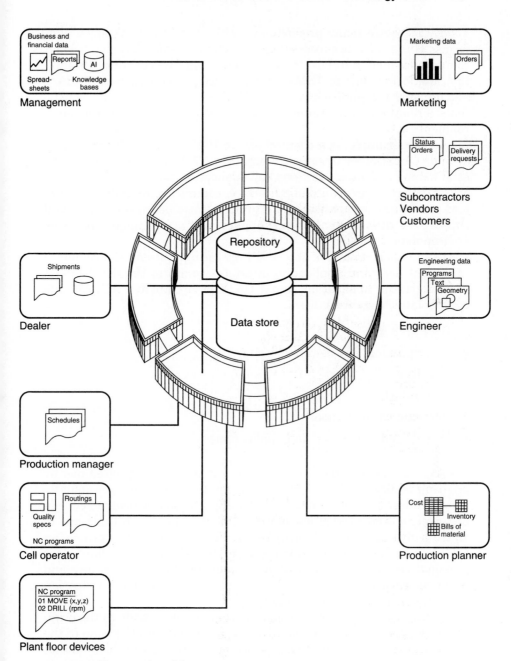

Figure 4.17   Presentation of data.

Today, there are many products available that utilize standard platforms. These include processors, operating systems, and enablers for communications, database management, and presentation. In addition, platforms such as IBM's *systems application architecture* (SAA) and *advance interactive executive* (AIX) help make application consistency a reality across strategic IBM and UNIX operating system environments.

SAA, for example, is a comprehensive IBM blueprint for consistency and compatibility of software products. SAA begins the process by establishing definitions for four key application aspects: common user access, common programming interface, common communication support, and common applications. Through these definitions, SAA will support the development of new applications across major operating environments.

AIX is IBM's version of the UNIX operating system. AIX combines consistent user and application interfaces to aid in the development of a integrated application across UNIX environments. AIX consists of six related system enablers:

1. Base systems

2. Programming

3. Interface

4. User interface

5. Communication support

6. Distributed processing and applications

4.7.6.2 Data integration.   Integration requirements are often met by creating bridges between individual applications. Bridges usually copy a collection of data between two applications. A bridge between engineering and production planning allows these two functions to share a bill of material. Another bridge permits an engineering CAD/CAM application to download an NC program to a plant floor personal computer. Or a bridge between production planning and plant operations may be used to provide a copy of the production schedule to the plant floor system.

However, a problem with bridges is that changes made to the original set of data are not immediately incorporated into the copy of the data. This results in out-of-date information. Another problem is that bridges become difficult to maintain when more than two applications must work together.

As enterprises begin to integrate their operations, it will be imperative that the latest information is shared among multiple applications and across business functions. For example, engineering, marketing, cost accounting, production planning, and plant operations may all need access to inventory status information. At other times, the enterprise's various business functions may need information about product specifications, order status, operating cost, and more.

A CIM architecture must be able to simplify and accelerate this integration. It must provide the facilities to integrate data across the various applications of the business functions—facilities such as data query, data communication, controlled access and editing, and consistent data definitions.

4.7.6.3 Installed base investment. Today's industrial enterprises have made considerable investments in their installed bases, including systems, data, and even training. In the United States alone, manufacturers spend billions of dollars per annum on information systems hardware, software, and integration services for production planning, engineering, and plant operations. CIM with sensory technology must help protect this investment by permitting the integration of existing systems, applications, and data.

4.7.6.4 Heterogeneous systems. In today's heterogeneous environment, data are located on different systems and in different formats. Applications have different needs, which are answered by processors, communications, and displays utilizing different technologies and architectures.

In an enterprise model, production planning may automate its operations on a single mainframe using an interactive database. Engineering may store drawings in an engineering database, then design and analyze products on a network of graphics workstations. Plant operations and sensors and control systems may be automated with personal computers and specialized machine controllers connected by both standard and proprietary networks. The data needed to operate the enterprise are scattered across all these diverse systems.

The heterogeneous environment is also characterized by an installed system base provided by multiple computer system suppliers, software vendors, and systems integrators. A CIM architecture must allow the integration of these varied system solutions and operating platforms.

4.7.6.5 Industry standards and open interfaces. As integration technologies mature, there will be the need to support an expanding set of industry standards. Today these standards include communication protocols such as MAP, token ring, and Ethernet; data exchange formats such as the *initial graphics exchange specifications* (IGES) for engineering drawings; data access methods such as SQL; and programming interfaces such as *programmer's hierarchical interactive graphics standard* (PHIGS). A CIM architecture must be able to accommodate these and other evolving standards. One framework for accomplishing this has already been established, the open systems architecture for CIM (CIM-OSA). CIM-OSA is being defined in the *Esprit* program by a consortium of European manufacturers, universities, and information system suppliers, including IBM. Data exchange formats are also being extended to accommodate product definition in the *product definition exchange specification* (PDES). In addition, a CIM architecture must be able to support well-established solutions, such as IBM's SNA, which have become de facto standards.

In this competitive marketplace, manufacturers must also be able to extend operations as needed and support these new technologies and standards as they become available. These needs may include adding storage to a mainframe system, replacing engineering work-stations, installing a new machine tool, upgrading an operating system, and utilizing new software development tools. A CIM architecture with open interfaces will allow enterprises to extend the integration implementation over time to meet changing business needs.

4.7.6.6   Reduced application support cost.   A CIM architecture incorporating sensors must also yield application solutions at a lower cost than traditional stand-alone computing. This includes reducing the time and labor required to develop integrated applications and data. It also means reducing the time and effort required to keep applications up to speed with the changes in the enterprise's systems environment, technology, and integration needs.

4.7.6.7   Customizable solutions.   Every enterprise has its own business objectives, shared data, system resources and applications. For example, one enterprise may choose to address CIM requirements by reducing the cycle time of product development. It does this by focusing on the data shared between the product development and process development functions in accelerating the product release process.

Another enterprise's aim is to reduce the cycle time required to deliver an order to a customer. It addresses this by exchanging data between production planning and operations functions and automating the order servicing and production processes. It must also be able to customize operations to individual and changing needs over time.

4.7.6.8   Phased implementation.   Implementing enterprise-wide integration will take place in many small steps instead of through a single installation. This is because integration technology is still evolving, implementation priorities are different, installed bases mature at different rates, new users must be trained, and lessons will be learned in pilot installations.

As enterprises begin their implementation efforts in phases, they will be integrating past, present, and future systems and applications. A CIM architecture must be able to support the integration of this diverse installed base.

4.7.6.9 Selectable functions. Most enterprises will want to weigh the benefits of integration against the impact this change will bring to each application and set of users. For example, an emphasis on product quality may require that production management gain greater insight into the quality of individual plant operations activities by implementing advanced sensors and control systems developed for particular applications. When an existing shop floor control application adequately manages schedules and support shop personnel with operating instructions, some additional information, such as that on quality, may be added, but rewriting the existing application may not be justified.

However, the plant manager may plan to develop a new production monitoring application to operate at each workstation. This application will make use of various sensors, share data with shop floor control, and utilize software building blocks for communications, database management, and presentation.

As is evident, the existing application requires only a data sharing capability, while the new application can benefit from both data sharing and the architecture building blocks. A CIM architecture with selectable functions will provide more options that can support the variety of needs within an enterprise.

4.7.6.10 Improved business process. Obviously, an enterprise will not implement integration on the basis of its technical merits alone. A CIM architecture must provide the necessary business benefits to justify change and investment.

The integration of information systems must support interaction between business functions and the automation of business processes. This is a key function if corporate goals, such as improved responsiveness to customer demands and reduced operating cost, are to be met. A CIM architecture must provide the means by which an entire enterprise can reduce the cycle times of business processes required for order processing, custom offerings, and new products. It must also reduce the impact of changes in business objectives and those business processes.

Further Reading

Bucker, D. W., "10 Principles to JIT Advancement," *Manufacturing Systems,* March 1988, p. 55.

Campbell, J., *The RS-232 Solution,* Sybex, Inc., Alameda, Calif., 1984.

Clark, K. E., "Cell Control, The Missing Link to Factory Integration," International Industry Conference Proceedings, Toronto, May 1989, pp. 641–646.

Datapro Research Corporation, "How U.S. Manufacturing Can Thrive," *Management and Planning Industry Briefs,* March 1987, pp. 39–51.

Groover, M. P., and E. W. Zimmer, Jr., *CAD/CAM: Computer Aided Design and Manufacturing,* Prentice Hall, Englewood Cliffs, N.J., 1984.

IBM Corp., *Introducing Advanced Manufacturing Applications,* IBM, Atlanta, Ga., 1985.

"APICS: Tweaks and Distributed Systems Are the Main Focus—MRP No Longer Looks to the Future for Finite Capacity Scheduling," *Managing Automation,* January 1992, pp. 31–36.

Manufacturing Studies Board, National Research Council, *Toward a New Era in U.S. Manufacturing: The Need for a National Vision,* National Academy Press, Washington, D.C., 1986.

Orlicky, J., *Material Requirements Planning,* McGraw-Hill, New York, 1985.

Schonberger, R. J., "Frugal Manufacturing," *Harvard Business Review,* September–October 1987, pp. 95–100.

Skinner, W., *Manufacturing: The Formidable Competitive Weapon,* Wiley, New York, 1985.

"Support for The Manufacturing Floor," *Manufacturing Engineering,* March 1989, pp. 29–30.

Vollmann, T. E., W. L. Berry, and D. C. Whyback, *Manufacturing Planning and Control Systems,* Richard D. Irwin, Homewood, Ill., 1984.

5

Advanced
Sensor Technology
in Precision Manufacturing
Applications

5.1 Identification of Manufactured Components

In an automated manufacturing operation one should be able to monitor the identification of moving parts. The most common means of automatic identification is *bar-code technology*. There are other approaches that offer advantages under certain conditions.

5.1.1 Bar-code identification systems

The *universal product code* (UPC) used in retail stores is a standard 12-digit code. Five of the digits represent the manufacturer and five the item being scanned. The first digit identifies the type of number system being decoded (a standard supermarket item, for example) and the second is a parity digit to determine the correctness of the reading. The first six digits are represented by code in an alternating pattern of light and dark bars. Figure 5.1 shows two encodings of the binary string 100111000. In both cases the minimum printed width is the same. The delta code requires nine such widths (the number of bits), while the width code requires 13 such widths (if a wide element is twice the width of the narrow element). Different bar widths allow for many character combinations. The remaining six digits are formed by dark alternating with light bars reversing the sequence of the first six digits. This allows backward scanning detection (Fig. 5.2).

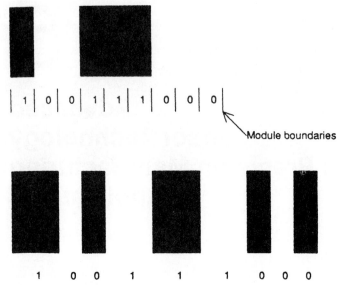

Figure 5.1   Encoding of the binary string 10011100 by delta code (top) and width code (bottom).

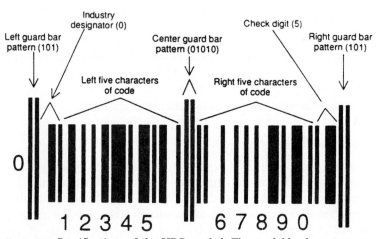

Figure 5.2   Specifications of the UPC symbol. The readable characters are normally printed in OCR-B font.

A bar-code reader can handle several different bar-code standards, decoding the stripes without knowing in advance the particular standard. The military standard, code 1189, specifies the type of coding to be used by the Department of Defense, which is a modification of code 39. Code 39 consists of 44 characters, including the letters A through Z. Because of its alphanumeric capabilities, code 39 is very effective for manufacturing applications. Code 39 is structured as follows: three of nine bars (light and dark) form wide characters; the rest are narrow.

Bar-code labels are simple to produce. Code 39, for example, can be generated by a personal computer. Such labels are ideal for inventory identifications and other types of fixed-information gathering. Bar codes are not necessarily placed on labels. Tools, for example, have had the code etched on their surfaces to allow for tool tracking. Techniques have been developed for molding bar codes onto rubber tires. Holographic scanners allow reading around corners so that parts need not be oriented perpendicular to the reader as they feed down a processing line.

A difficulty with bar coding has been the fact that it cannot be read if the bars become obscured by dirt, grease, or other substances. Infrared scanners are used to read codes that are coated with black substances to prevent secrecy violations through reproduction of the codes. One way to generally offset the problem of a dirty environment is to use magnetic-stripe-encoded information.

5.1.2   Transponders

While bar-code labels and magnetic stripes are very effective on the shop floor, shop circumstances may require more information to be gathered about a product than can be realistically handled with encoded media. For instance, with automobiles being assembled to order in many plants, significant amounts of information are necessary to indicate the options for a particular assembly. Radio-frequency (RF) devices are used in many cases. An RF device, often called a transponder, is fixed to the chassis of a car during assembly. It contains a chip that can store a great amount of information. A radio signal at specific assembly stations causes the transponder to emit information which can be understood by a local receiver. The transponder can be coated with grease and still function. Its potential in any assembly operation is readily apparent. There are several advanced transponders that have read/write capability, thus, supporting local decision making.

5.1.3   Electromagnetic identification of manufactured components

There are many other possible electronic schemes to identify manufactured parts in motion. Information can be coded on a magnetic stripe in much the same way that bars represent information on a bar code label, since the light and dark bars are just a form of binary coding.

Operator identification data are often coded on magnetic stripes that are imprinted on the operators' badges. Magnetic stripe information can be fed into a computer. Such information might include the following: (1) the task is complete, (2) x number of units have been produced, (3) the unit part numbers, (4) the operator's identification number, etc. This same scanning station can also be set up using barcode information; however, with magnetic striping, the information can be read even if the stripe becomes coated with dirt or grease. A disadvantage of magnetic striping is that the reader has to contact the stripe in order to recall the information.

5.1.4   Surface acoustic waves

A process similar to RF identification is *surface acoustic waves* (SAW). With this process, part identification is triggered by a radar-type signal which can be transmitted over greater distances than in RF systems.

5.1.5   Optical character recognition

Another form of automatic identification is *optical character recognition* (OCR). Alphanumeric characters form the information, which the OCR reader can "read." In mail processing centers, high-speed sorting by the U.S. Postal Service is accomplished using OCR. The potential application to manufacturing information determination is obvious.

There are also many other means for part identification, such as vision systems and voice recognition systems. *Vision systems* utilize TV cameras to read alphanumeric data and transmit the information to a digital converter. OCR data can be read with such devices, as can conventionally typed characters. *Voice recognition systems* have potential where an individual's arms and hands are utilized in some function that is not conducive to reporting information. Such an application might be the inspection of parts by an operator who has to make physical measurements on the same parts.

In *laser scanning applications,* a laser beam scans and identifies objects at a constant speed. The object being scanned interrupts the beam for a time proportional to its diameter or thickness. Resolutions of less than 1 mm are possible.

In *linear array applications,* parallel light beams are emitted from one side of the object to be measured to a photooptical diode array on the opposite side. Diameters are measured by the number of array elements that are blocked. Resolutions of 5 mm or greater are possible.

In *TV camera applications,* a TV camera is used in the digitizing of the image of an object and the result is compared to the stored image. Dimensions can be measured, part orientation can be determined, and feature presence can be checked. Some exploratory work is being accomplished with cameras that can fit in a tool changer mechanism. The camera can be brought to the part like a tool and verify part characteristics.

5.2 Digital Encoder Sensors

Digital encoder sensors provide directly an output in a digital form and thus require only simple signal conditioning. They are also less susceptible to electromagnetic interference, and are therefore useful for information processing and display in measurement and control systems. Their ability to rapidly scan a series of patterns provides additional manufacturing automation opportunities when light and dark patterns are placed in concentric rings in a disk. Figure 5.3 illustrates a portion of such a disk that can be rigidly attached to a shaft or an object and housed in an assembly containing optical sensors for each ring (Fig. 5.4). The assembly, called an *optical encoder,* automatically detects rotation of a shaft or an object. The shaft rotation information can be fed back into a computer or controller mechanism for controlling velocity or position of the shaft. Such a device has application in robots and numerical control machine tools and for precision measurements of strip advancement to generate a closed-loop feedback actuation for displacement compensation.

There are two classes of digital encoder sensors:

1. Encoder sensors yielding at the output a digital version of the applied analog input. This class of encoder sensors includes position encoders.

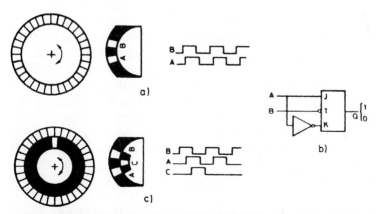

Figure 5.3 Detection of movement direction in directional encoders: (*a*) by means of two outputs with 90° phase shift; (*b*) output electronic circuit; (*c*) additional marker for absolute positioning.

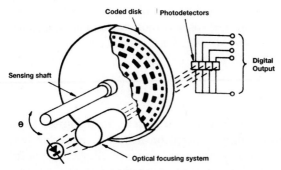

Coded disk   Photodetectors

Sensing shaft

Digital
Output

θ

Optical focusing system

Figure 5.4 Principle of absolute position encoders for
linear and rotary movements.

2. Encoder sensors that rely on some physical oscillatory phenomenon that is transduced by a conventional modulating sensor. This
class of sensors may require an electronic circuit acting as a digital
counter in order to yield a desired digital output signal.

There are no sensors where the transduction process directly yields
a digital output. The usual process is to convert an analog input
quantity into a digital signal by means of a sensor without the
requirement to convert an analog voltage into its digital equivalent.

5.2.1 Position encoder sensors in manufacturing

Position encoder sensors can be categorized as linear and angular
position encoder sensors. The optical encoder sensor can be either
incremental or absolute. The incremental types transmit a series of
voltages proportional to the angle of rotation of the shaft or object.
The control computer must know the previous position of the shaft or
object in order to calculate the new position. Absolute encoders transmit a pattern of voltages that describes the position of the shaft at
any given time. The innermost ring reaches from dark to light every
180°, next ring every 90°, the next 45°, and so on, depending on the
number of rings on the disk. The resulting bit pattern output by the
encoder reveals the exact angular position of the shaft or object. For
an absolute optical encoder disk that has eight rings and eight LED
sensors, and in turn provides 8-bit outputs, (10010110). Table 5.1
shows how the angular position of the shaft or object can be determined.

The incremental position encoder sensor suffers three major weaknesses:

TABLE 5.1   Absolute Optical Encoder

Encoder ring	Angular displacement, degrees	Observed pattern	Computed value, degrees
1 (innermost)	180	1	180
2	90	0	
3	45	0	
4	22.5	1	22.5
5	11.25	0	
6	5.625	1	5.625
7	2.8125	1	2.8125
8	1.40625	0	
			210.94

1. The information about the position is lost whenever the electric supply fails or the system is disconnected, and when there are strong perturbations.

2. The digital output, to be compatible with the input/output peripherals of a computer, requires an up/down counter.

3. The incremental position encoder does not detect the movement direction unless elements are added to the system (Fig. 5.5).

Physical properties used to define the disk pattern can be magnetic, electrical, or optical. The basic output generated by the physical property is a pulse train. By differentiating the signal, an impulse is obtained for each rising or falling edge, increasing by two the number of counts obtained for a given displacement (Fig. 5.6).

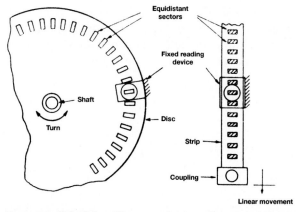

Figure 5.5 Principle of linear and rotary incremental position encoders.

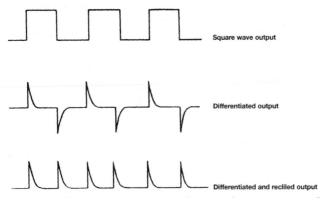

Square wave output

Differentiated output

Differentiated and recliled output

Figure 5.6 Improving output resolution of an incremental encoder by differentiation and rectification.

5.3   Fuzzy Logic for Optoelectronic Color Sensors in Manufacturing

Fuzzy logic will most likely be the wave of the future in practical and economical solutions to control problems in manufacturing. Fuzzy logic is simply a technique that mimics human reasoning. This technology is now being explored throughout various industries. Fuzzy logic color sensors can relay information to microprocessors to determine color variance within an acceptable range of colors. A conventional sensor could not perform this function because it could choose only a specific color and reject all other shades—it uses a very precise set of rules to eliminate environmental interference.

The research and development activities for fuzzy logic technology began in mid-1990. The research has lead to the creation of a fuzzy logic color sensor that can learn a desired color and compare it with observed colors. The sensor can distinguish between acceptable and unacceptable colors for objects on a conveyer belt. The development of new light source technology allows the color sensor to produce more accurate color measurement (Fig. 5.7). Also, the integration of the

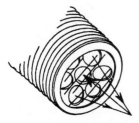

Figure 5.7 Integration of fuzzy logic sensor in a sensing module.

fuzzy logic sensor with a microprocessor enables the data to be collected and interpreted accurately.

5.3.1   Sensing principle

The sensor is designed with a broad-spectrum solid-state light source utilizing a light-emitting diode cluster. The LED-based light source provides stable, long-lasting, high-speed target illumination capabilities. The LED cluster is made up of three representative hues of *red, green, and blue* which provide a triple bell-shaped spectral power distribution for the light source (Fig. 5.8). The light incident on the target is reflected with varying intensities, depending on the particular target color under analysis.

The reflected light is received by a semiconductor receiver in the center of the LED cluster. The amount of light reflected back onto the receiver is transduced to a voltage and converted to digital format immediately via an analog-to-digital converter. The internal processing of the converted red, green, and blue (RGB) values offers variable sample size and averaging to compensate for signal noise (Fig. 5.8).

Ambient light is sampled between every component pulse and immediately subtracted from the sampled signal so that the effects of factory ambient light are suppressed. Thus, hooding the sensor is not totally necessary. In an area of a very bright or high-frequency lighting, it may be beneficial to provide some degree of hooding to at least limit the brightness. The sensor's electronics also employs temperature compensation circuitry to stabilize readings over temperature ranges.

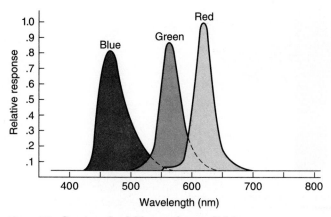

Figure 5.8   Spectra of red, blue, and green light.

5.3.2 Color theory

Color science defines color in a space, with coordinates of *hue, satura-tion,* and *intensity* (HSI). These three general components uniquely define any color within HSI color space. Hue is related to the reflected wavelength of a color when a white light is shined on it. Intensity (lightness) measures the degree of whiteness, or gray scale, of a given color. Saturation is a measure of the vividness of a given hue. The term *chromaticity* primarily includes elements of the hue and satura-tion components. Researchers depict color in space using hue as the angle of a vector, saturation as the length of it, and intensity as a plus or minus height from a center point (Fig. 5.9).

The concepts of hue, saturation, and intensity can be further clari-fied by a simplified pictorial presentation. Consider Fig. 5.10, where a color is depicted at a molecular level. Color is created when light interacts with pigment molecules. Color is generated by the way pig-ment molecules return (bend) incoming light. For example, a red pig-ment causes a measurable hue component of the color. The relative density of the pigment molecules leads to the formation of the satura-tion component. There are some molecules present which return almost all wavelengths, and appear white as a result, leading to the intensity (lightness) component.

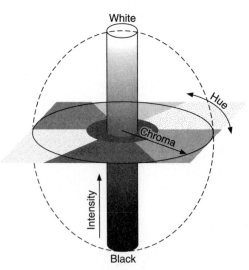

Figure 5.9 Coordinates of hue, saturation, and intensity of color in space.

R-G-B

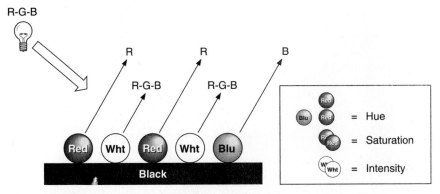

Figure 5.10 Model of color interpretation.

5.3.3 Units of color measurement

If color description depends on measuring the interaction of a target color with a given white light source, it is clear that in order to have the system of measurement standardized, both the light source and means of detection must be well-defined. One very popular set of standardization rules has been set up by the Commission International de l'Eclairage (CIE), a color standardization organization. From color theory, it is known that the response to a color stimulus (its determination), depends on the spectral power distribution of the light source (illuminant), times the spectral reflectance of the target (color) surface, times the spectral response of the detector (observer) (Fig. 5.11).

With this principle in mind, the CIE presented a detailed description of the standard light source and a standard observer (photodetector). The result of the study was the popular CIE diagram, which creates a two-dimensional mapping of a color space (Fig. 5.12).

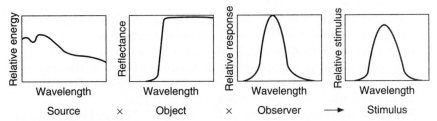

Figure 5.11 Stimulus response to a color (detector/determination) = illuminant × target × observer.

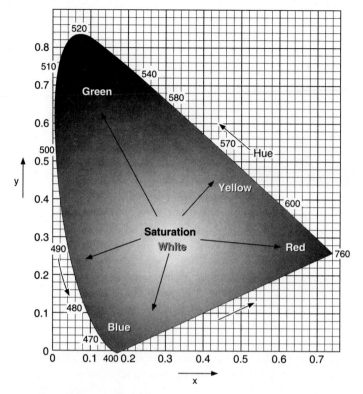

Figure 5.12 Two-dimensional mapping of color in a space.

Further manipulation of the CIE observations has lead to another color coordinate system, the so-called L.a.b numbers for describing a color. The L.a.b. numbering system is fairly prevalent in industrial applications. The machines that measure color according to this theory are referred to as *color spectrophotometers* or *color meters*. These machines are typically expensive, bulky, and not well-suited for distributed on-line color sensing.

The fuzzy color sensor does not offer CIE-based color measurement; however, it is a very high resolution color comparator. The sensor learns a color with its own standard light source (trio-stimulus LEDs) and its own observer (semiconductor photoreceiver). It thereby sets up its own unique color space with the three dimensional coordinates being the red, blue, and green readings (Fig. 5.12).

Theoretically there would be 256^3, or 16,777,216, unique positions in its color space for defining color (Fig. 5.13). In reality, the actual

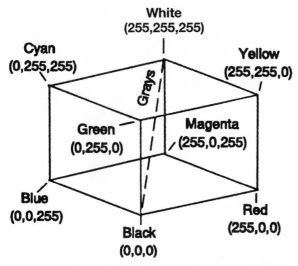

Figure 5.13 Three-dimensional coordinates of color in a space.

number of colors that a sensor can reliably distinguish is much less because of optical noise and practical limitations of the design.

The device compares the RGB colors it observes to the internally learned standard. Each time a standard color is relearned, it essentially recalibrates the sensor.

5.3.4 Color comparators and true color measuring instruments

If the learned color is initially defined by a color spectrometer, the fuzzy logic color sensor (comparator) can be installed in conjunction with it. The color sensor can learn the same area that the spectrometer has read, thereby equating its learned standard to the absolute color. By using the color sensor in this manner, a temporary correlation to an absolute color standard can be established (Fig. 5.14). This permits the relative drift from the standard to be monitored. The advantage is that more color sensing can be economically distributed across the target area. When a significant relative deviation is detected, an alert signal can flag the absolute color sensor to take a reading in the suspect area. If enough storage space is available in a central processing computer, lookup tables can be constructed to relate the serially communicated color sensor readings to a standard color coordinate system like the CIE system.

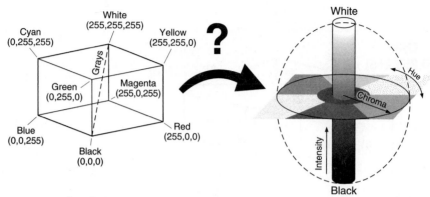

Figure 5.14   Correlation to an absolute color standard.

5.3.5   Color sensor algorithms

There are two internal software algorithms, or sets of rules, for analyzing fuzzy logic color sensor data:

1. The absolute algorithm: compares color on the basis of absolute voltages

2. The relative algorithm: compares color on the basis of relative percentages of each RGB component voltage

The choice of algorithm depends on sensing distance variation and the type of color distinction one needs. If the outputs vary excessively with distance when the absolute algorithm is used, a relative (ratio) algorithm must be considered. While a relative algorithm does not retain the lightness information, it greatly reduces unwanted distance-related variations. The relative algorithm shows changes in chromaticity (hue and chroma) which exist in most color differences. If the sensing distances can be held constant, the absolute algorithm works well at detecting subtle changes in density (shades) of a single color.

5.3.6   Design considerations in fuzzy logic color sensors

The design of fuzzy logic color sensors aims to achieve maximum color sensing ability, while maintaining the expected simplicity of operation and durability of typical discrete industrial sensors. There are several other key goals that must be considered in system design such as high speed, small size, configurability, high repeatability, and long light source life.

The choice of a solid-state light source satisfies the majority of the criteria for a good industrialized color sensor design. The fuzzy logic color sensor utilizes two sets of three different LED photodiodes as its illumination source. The three LED colors, red, green, and blue, were chosen essentially for their coverage of the visible light spectrum (Fig. 5.15).

The light from each LED is sequentially pulsed onto the target and its reflected energy is collected by a silicon photoreceiver chip in the LED cluster. Ambient light compensation circuitry is continually refreshed between each LED pulse, so the reported signals are almost entirely due to the LED light pulses. The LED sources offer a very fast (microsecond response) and stable (low spectral drift, steady power) source of a given wavelength band, without resorting to filters. Emerging blue LEDs, in combination with the more common red and green LEDs, have made it possible to use three solid-state spectra to define a hue. The choice of the specific LED bandwidth (or spectral distribution) is made so as to obtain the best color distinction through broader coverage of the illumination spectrum.

5.3.7 Fuzzy logic controller flowchart

The entire sensor operation is illustrated in Fig. 5.16. An internal microcontroller governs the device operations. It directs signals in and out of the sensor head, to maintain local and remote communications, and provides color discrimination algorithms to produce the appropriate signal output at the control pin. As the device proceeds out of reset, it checks the locally (or remotely) set configuration dipswitches, which in turn define the path, or operating menu, to pro-

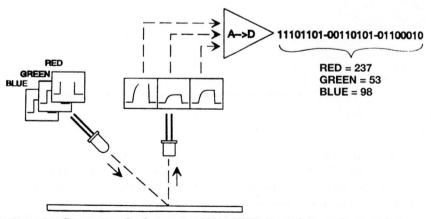

Figure 5.15 Conversion of red, green, and blue LED outputs from analog to digital.

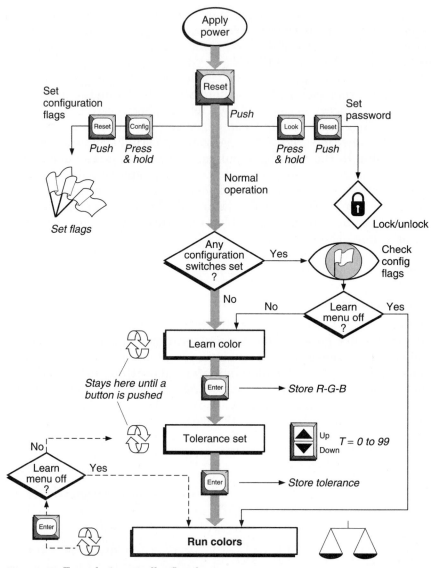

Figure 5.16 Fuzzy-logic controller flowchart.

ceed through. There is permanent storage of learned or remotely loaded values of RGB readings, tolerance, number of readings to average, and the white-card calibration value, so these settings can be available at reset or at power-up.

One or more of three optional menus can be selected before entering into the main operation of learning and running colors. The three

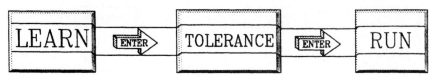

Figure 5.17 Simple path of operation.

alternative menus are (1) white-card gain set, (2) number of reads to average set, and (3) the observed stored reading menus.

If none of the alternative menus is activated, the sensor will proceed directly to the primary modes, which are learn, tolerance set, and run modes (Fig. 5.17). By pressing and holding the appropriate buttons while pushing the reset button, two other programming menus can be entered. These are the set configuration flags (dip-switch) menu and the set password menu.

5.4 Sensors Detecting Faults in Dynamic Machine Parts (Bearings)

A system consisting of analog and digital signal processing equipment, computers, and computer programs would detect faults in ball bearings in turbomachines and predict the remaining operating time until failure. The system would operate in real time, extracting the diagnostic and prognostic information from vibrations sensed by accelerometers, strain gauges, and acoustical sensors, and from the speed of the machine as measured by a tachometer.

The vibrations that one seeks to identify are those caused by impact that occurs when pits in balls make contact with races and pits in races make contact with balls. These vibrations have patterns that are unique to bearings and repeat at known rates that are related to ball-rotation, ball-pass, and cage-rotation frequencies. These vibrations have a wide spectrum that extends up to hundreds of kilohertz, where the noise component is relatively low.

The system in Fig. 5.18 would accept input from one of two sensors. Each input signal would be amplified, bandpass-filtered, and digitized. The digitized signal would be processed in two channels: one to compute the keratosis of the distribution of the amplitudes, the other to calculate the frequency content of the envelope of the signal. The *keratosis* is the fourth statistical moment and is known, from theory and experiment, to be indicative of vibrations caused by impact on faults. The keratosis would be calculated as a moving average for each consecutive digitized sample of the signal by using a number of samples specified by the technician. The trend of a keratosis moving

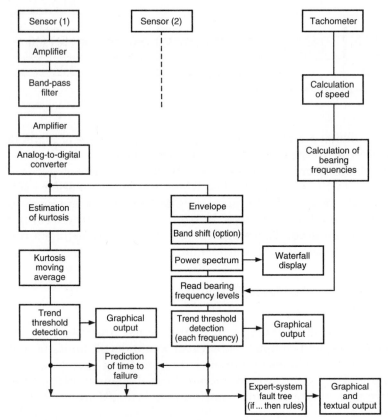

Figure 5.18 Data flow for an automatic bearing fault detection system.

average would be computed several times per second, and the changes in the keratosis value that are deemed to be statistically significant would be reported.

In the other signal processing channel, the amplitude envelope of the filtered, digitized signal would be calculated by squaring the signal. Optionally, the high-frequency sample data would be shifted to a lower frequency band to simplify processing by use of a Fourier transformation. This transformation would then be applied to compute the power spectrum.

The output of the tachometer would be processed in parallel with the spectral calculations so that the frequency bins of the power spectrum could be normalized on the basis of the speed of rotation of the machine. The power spectrum would be averaged with a selected number of previous spectra and presented graphically as a "waterfall" display; this is similar to a sonar display with which technicians can detect a discrete frequency before an automatic system can.

The bearing frequencies would be calculated from the measured speed and the known parameters of the bearings, with allowances for slip. The power spectrum levels would be read for each bearing frequency; a moving average of the amplitude at each bearing frequency and harmonic would be maintained, and trends representing statistically significant increases would be identified by threshold detection and indicated graphically.

By using algorithms based partly on analyses of data from prior tests, the results of both keratosis and power spectrum calculations would be processed onto predictions of the remaining operating time until failure. All the results would then be processed by an expert system. The final output would be a graphical display and text that would describe the condition of the bearings.

5.5   Sensors for Vibration Measurement of a Structure

An advanced sensor was developed to gauge structure excitations and measurements that yield data for design of robust stabilizing control systems (Fig. 5.19).

An automated method for characterizing the dynamic properties of a large flexible structure estimates model parameters that can be used by a robust control system to stabilize the structure and mini-

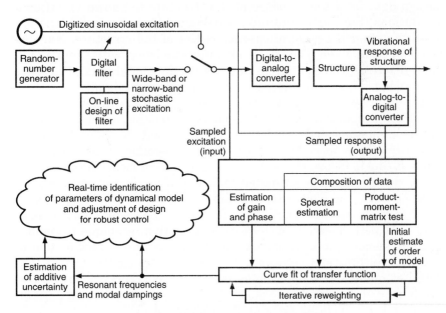

Figure 5.19 Automated characterization of vibrations of a structure.

mize undesired motions. Although it was developed for the control of large, flexible structures in outer space, the method is also applicable to terrestrial structures in which vibrations are important—especially aircraft, buildings, bridges, cranes, and drill rigs.

The method was developed for use under the following practical constraints:

1. The structure cannot be characterized in advance with enough accuracy for purposes of control.

2. The dynamics of the structure can change in service.

3. The numbers, types, placements, and frequency responses of sensors that measure the motions and actuators that control them are limited.

4. Time available during service for characterization of the dynamics is limited.

5. The dynamics are dominated by a resonant mode at low frequency.

6. In-service measurements of the dynamics are supervised by a digital computer and are taken at a low rate of sampling, consistent with the low characteristic frequencies of the control system.

7. The system must operate under little or no human supervision.

The method is based on extracting the desired model and control-design data from the response of the structure to known vibrational excitations (Fig. 5.19). Initially, wideband stochastic excitations are used to obtain the general characteristics of the structure. Narrow-band stochastic and piece-wise-constant (consistent with sample-and-hold discretizations) approximations to sinusoidal excitations are used to investigate specific frequency bands in more detail.

The relationships between the responses and excitations are first computed nonparametrically—by spectral estimation in the case of stochastic excitations and by estimation of gains and phases in the case of approximately sinusoidal excitations. In anticipation of the parametric curve fitting to follow, the order of a mathematical model of the dynamics of the structure is estimated by use of a *product moment matrix* (PMM). Next, the parameters of this model are identified by a least-squares fit of transfer-function coefficients to the nonparametric data. The fit is performed by an iterative reweighting technique to remove high-frequency emphasis and assure minimum-variance estimation of the transfer-function coefficient. The order of the model starts at the PMM estimate and is determined more precisely thereafter by successively adjusting a number of modes in the fit at each iteration until an adequately small output-error profile is observed.

In the analysis of the output error, the additive uncertainty is estimated to characterize the quality of the parametric estimate of the transfer function and for later use in the analysis and design of robust control. It can be shown that if the additive uncertainty is smaller than a certain calculable quantity, then a conceptual control system could stabilize the model structure and could also stabilize the real structure. This criterion can be incorporated into an iterative design procedure. In this procedure, each controller in a sequence of controllers for the model structure would be designed to perform better than the previous one did, until the condition for robust capability was violated. Once the violation occurred, one could accept the penultimate design (if its performances were satisfactory) or continue the design process by increasing a robustness weighting (if available). In principle, convergence of this iterative process guarantees a control design that provides high performance for the model structure while guaranteeing robustness of stability to all perturbations of the structure within the additive uncertainty.

5.6 Optoelectronic Sensor Tracking Targets on a Structure

The location and exact position of a target can be accurately sensed through optoelectronic sensors for tracking a retroreflective target on a structure. An optoelectronic system simultaneously measures the positions of as many as 50 retroreflective targets within 35° of view with an accuracy of 0.1 mm. The system repeats the measurement 10 times per second. The system provides an unambiguous indication of the distance to each target that is not more than 75 m away from its sensor module. The system is called *spatial high-accuracy position-encoding sensor* (SHAPES).

SHAPES fills current needs in the areas of system identification and control of large flexible structures, such as large space- and ground-based antennas and elements of earth-orbiting observational platforms. It is also well-suited to applications in rendezvous and docking systems. Ground-based applications include boresight determination and precise pointing of 70-m deep-space-network antennas.

SHAPES illuminates the retroreflective targets by means of a set of lasers in its sensor module. In a typical application (Fig. 5.20) a laser diode illuminates each target with 30-ps pulses at a repetition rate of 100 MHz. Light reflected from the target is focused by a lens and passed through a beam splitter to form images on a charge-coupled device (CCD) and on the photocathode of a streak tube. The angular position of the target is determined simply from the position of its reflection on the charge-coupled device.

The measurement of the distance to the target is based on the

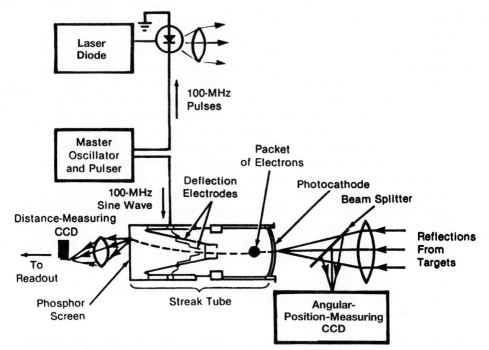

Figure 5.20 Beam splitter diverts reflections from a continuous-wave laser into a CCD camera for measurement of angles of reflection.

round-trip time of the optical pulses. The round-trip distance can be measured in terms of the difference between the phase of the train of return pulses incident on the photocathode and the phase of a reference sine wave that drives the deflection plate of the streak tube. This difference, in turn, manifests itself as a displacement between the swept and unswept positions, at the output end of the streak tube, of the spot of light that represents the reflection from the target. The output of the streak tube is focused on a CCD for measurement and processing of the position of this spot. Three microprocessors control the operation of SHAPES and convert the raw data required from the angular-position and distance-measuring CCDs into position of the target in three dimensions.

5.7 Optoelectronic Feedback Signals for Servomotors through Fiber Optics

In what is believed to be among its first uses to close a digital motor-control loop, fiber-optics transmission provides immunity to noise and rapid transmission of data. An optoelectronic system effects closed-loop control of the shaft angle of four servomotors and could be expanded to control as many as 16. The system includes a full-duplex fiber-optic link (Fig. 5.21) that carries feedforward and feedback digital signals over a distance of many meters, between commercial digital motor-control circuits that execute a PID control algorithm with programmable gain (one such control circuit dedicated to each servomotor) and modules that contain the motor-power switching circuits, digital-to analog buffer circuits for the feedforward control signals, and analog-to-digital buffer circuits for the feedback signals from the shaft-angle encoders (one such module located near, and dedicated to, each servomotor).

Besides being immune to noise, optical fibers are compact and flexible. These features are particularly advantageous in robots, which must often function in electromagnetically noisy environments and in which it would otherwise be necessary to use many stiff, bulky wires (which could interfere with movement) to accommodate the required data rates.

Figure 5.21 shows schematically the fiber-optic link and major subsystems of the control loop of one servomotor. Each digital motor-control circuit is connected to a central control computer, which programs the controller gains and provides the high-level position commands. The other inputs to the motor-control circuit include the sign of the commanded motor current and pulse-width modulation representing the magnitude of the command motor current.

The fiber-optic link includes two optical fibers—one for feedforward, one for feedback. The ends of the fibers are connected to identical bidirectional interface circuit boards, each containing a transmitter and a receiver. The fiber-optic link has a throughput rate of 175 MHz; at this high rate, it functions as though it were a 32-bit parallel link (8 bits for each motor control loop), even though the data are multiplexed into a serial bit stream for transmission. In the receiver, the bit stream is decoded to reconstruct the 8-bit pattern and a programmable logic sequencer expands the 8-bit pattern to 32 bits and checks for errors by using synchronizing bits.

5.8 Acoustooptical/Electronic Sensor for Synthetic-Aperture Radar Utilizing Vision Technology

An acoustooptical sensor operates in conjunction with analog and digital electronic circuits to process frequency-modulated *synthetic-aperture radar* (SAR) return signals in real time. The acoustooptical SAR processor will provide real-time SAR imagery aboard moving aircraft or space SAR platforms. The acoustooptical SAR processor has the

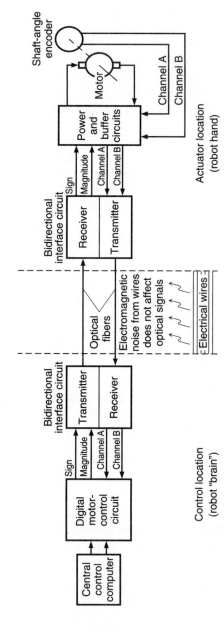

Figure 5.21 Full-duplex fiber-optic transmission link.

potential to replace the present all-electronic SAR processors that are currently so large and heavy and consume so much power that they are restricted to use on the ground in the postprocessing of the SAR in-flight data recorder.

The acoustooptical SAR processor uses the range delay to resolve the range coordinates of a target. The history of the phase of the train of radar pulses as the radar platform flies past a target is used to obtain the azimuth (cross-range) coordinate by processing it coherently over several returns. The range-compression signal processing involves integration in space, while the azimuth-compression signal processing involves integration in time.

Figure 5.22 shows the optical and electronic subsystems that perform the space and time integrations. The radar return signal is het-

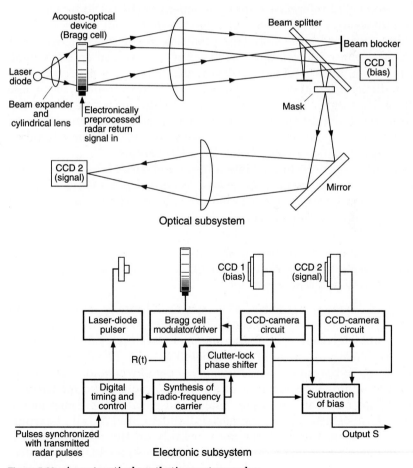

Figure 5.22 Acoustooptical synthetic-aperture radar.

erodyned to the middle frequency of an acoustooptical sensor and added electronically to a reference sinusoid to capture the history of the phase of the return signal interferometrically for compression in azimuth. The resulting signal is applied to the acoustooptical sensor via a piezoelectric transducer. The acoustooptical sensor thus becomes a cell that encodes the evolving SAR return.

Meanwhile, pulses of light a few tens of nanoseconds long are generated by a laser diode in synchronism with the transmitted pulses and are used to sample and process the return signal. Lenses shape that laser light into a plane wave incident upon the acoustooptical sensor. The integration in space is effected at the moment of sampling by the focusing action. The position of the focal point in the cell depends on the range delay of the corresponding target, and light is brought to focus on two CCD imaging arrays at positions that depend on the range.

The sinusoidal reference signal component of the cell interacts with laser radiation to generate a plane wave of light that interferes with the light focused by the cell. This produces interference fringes that encode the phase information in the range-compressed optical signal. These fringes are correlated with a mask that has a predetermined spatial distribution of density and that is placed in front of, or on, one of the CCD arrays. This CCD array is operated in a delay-and-integrate mode to obtain the desired correlation and integration in time for the azimuth compression. The output image is continuously taken from the bottom picture element of the CCD array.

Two CCDs are used to alleviate a large undesired bias of the image that occurs at the output as a result of optical processing. CCD_1 is used to compute this bias, which is then subtracted from the image of CCD_2 to obtain a better image.

5.9   The Use of Optoelectronic/Vision Associative Memory for High-Precision Image Display and Measurement

Storing an image of an object often requires large memory capacity and a high-speed interactive controller. Figure 5.23 shows schematically an optoelectronic associative memory that responds to an input image by displaying one of M remembered images. The decision about which if any of the remembered images to display is made by an optoelectronic analog computation of an inner-product-like measure of resemblance between the input image and each of the remembered images. Unlike associative memories implemented as all-electronic neural networks, this memory does not rely on the precomputation and storage of an outer-product synapse matrix. Instead, the optoelectronic equivalent of this matrix is realized by storing remembered images in two separate spatial light modulators placed in tandem.

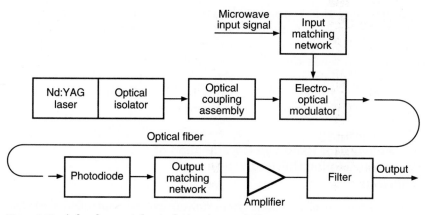

Figure 5.23 A developmental optoelectronic associative memory.

This scheme reduces the required size of the memory by an order of magnitude.

A partial input image is binarized and displayed on a liquid-crystal light valve spatial modulator which reprocesses the image in real time by operating in an edge-enhancement mode. This preprocessing increases the orthogonality (with respect to the inner product) between the input image and each of the remembered images, thereby increasing the ability of the memory to discriminate among different images.

The light from the input image is passed through a polarizing beam splitter, a lens, a binary diffraction grating, and another lens, to focus an array of M replicas of the input image on one face of a liquid-crystal-television spatial light modulator that is displaying the M remembered images. The position of each replica of the input image coincides with that of one of the remembered images. Light from the array of pairs of overlapping input and remembered images is focused by a corresponding array of lenslets onto a corresponding array of photodetectors. The intensity of light falling on each photodetector is proportional to the inner product between the input image and the corresponding remembered image.

The outputs of the photodetectors are processed through operational amplifiers that respond nonlinearly to inner-product level (in effect executing analog threshold functions). The outputs of the amplifiers drive point sources of white light, and an array of lenslets concentrates the light from each source onto the spot occupied by one of M remembered images displayed on another liquid-crystal-television spatial light modulator. The light that passes through this array is reflected by a pivoted ray of mirrors through a lens, which focuses the output image onto a CCD television camera. The output image consists of superpositioned remembered images, the brightest of which are those that represent the greatest inner products (the greatest resemblance to the input image). The television camera feeds the output image to a control computer, which performs a threshold computation, then feeds the images through a cathode-ray tube back to the input liquid-crystal light valve. This completes the associative recall loop. The loop operates iteratively until one (if any) of the remembered images is the sole output image.

5.10   Sensors for Hand-Eye Coordination of Microrobotic Motion Utilizing Vision Technology

The micro motion of a robotic manipulator can be controlled with the help of dual feedback by a new method that reduces position errors by an order of magnitude. The errors—typically of the order of centimeters—are differences between real positions on the one hand and measured and computed positions on the other; these errors arise from several sources in the robotic actuators and sensors and in the kinematic model used in control computations. In comparison with current manufacturing methods of controlling the motion of a robot with visual feedback (the robotic equivalent of hand-eye coordination), the novel method requires neither calibration over the entire work space nor the use of an absolute reference coordinate frame for computing transformations between field of view and robot joint coordinates.

The robotic vision subsystem includes five cameras: three stationary ones that provide wide-angle views of the work space and two mounted on the wrist of an auxiliary robot arm to provide stereoscopic close-up views of the work space near the manipulator (Fig. 5.24). The vision subsystem is assumed to be able to recognize the objects to be avoided and manipulated and to generate data on the coordinates of the objects from sent positions in the field-of-view reference frame.

The new method can be implemented in two steps:

1. The close-up stereoscopic cameras are set initially to view a small region that contains an object of interest. The end effector is commanded to move to a nominal position near the object and within the field of view. Typically, the manipulator stops at a slightly different position, which is measured by the cameras. Then the measured error in position is used to compute a small corrective motion. This procedure is designed to exploit the fact that small errors in relative position can be measured accurately and small relative motions can be commanded accurately.

2. The approximate direct mapping between the visual coordinates and the manipulator joint-angle coordinates can be designed without intermediate transformation to and from absolute coordinates. This

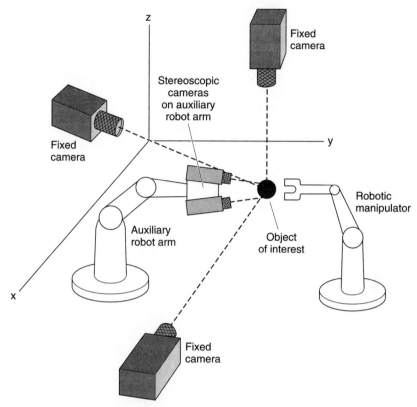

Figure 5.24 Stereoscopic cameras on an auxiliary robot arm.

is, in effect, a calibration, but it requires fewer points than does a conventional calibration in an absolute reference frame over the entire
work space. The calibration is performed by measuring the position of
a target (in field-of-view coordinates) when the target is held rigidly
by the manipulator at various commanded positions (in manipulator
joint-angle coordinates) and when the cameras are placed at various
commanded positions. Interpolations and extrapolations to positions
near the calibration points are thereafter performed by use of the
nonlinear kinematic transformations.

5.11   Force and Optical Sensors Controlling Robotic Gripper for Agriculture and Manufacturing Applications

A robotic gripper operates in several modes to locate, measure, recognize (in a primitive sense), and manipulate objects in an assembly

subsystem of a robotic cell that is intended to handle geranium cuttings in a commercial greenhouse. The basic concept and design of the gripper could be modified for handling other objects, for example, rods or nuts, including sorting the objects according to size. The concept is also applicable to real-time measurement of the size of an expanding or contracting part gripped by a constant force and to measurement of the size of a compliant part as a function of the applied gripping force.

The gripper is mounted on an industrial robot. The robot positions the gripper at a fixed distance above the cutting to be processed. A vision system locates the cutting in the x-y plane lying on a conveyer belt (Fig. 5.25).

The robot uses fiber-optic sensors in the fingertip of the gripper to locate the cutting along the axis. The gripper grasps the cutting under closed-loop digital servo force control. The size (that is, the diameter of the stem) of the cutting is determined from the finger position feedback, while the cutting is being grasped under force con-

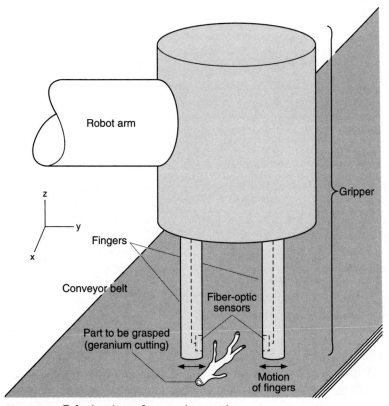

Figure 5.25 Robotic gripper for geranium cutting.

trol. The robot transports the cutting to a scale for weighing, to a trimming station, and finally to a potting station. In this manner, cuttings are sorted according to weight, length, and diameter.

The control subsystem includes a 32-bit minicomputer that processes the vision information and collects system grating data. The minicomputer communicates with a robot controller and the gripper node control. The gripper node control communicates with the scale. The robot controller communicates with the gripper node controller via discrete input/output triggering of each of the gripper functions.

The gripper subsystem incudes a PC/AT-compatible industrial computer; a gripper mechanism, actuated by two dc servomotors, with an integrated load cell; discrete input/output components; and two fiber-optic analog-output distance sensors. The computer includes a discrete input/output circuit card, an 8-channel A/D converter circuit card, a motor-control circuit card with A/D components, two serial ports, and a 286 processor with coprocessor.

A geranium cutting comprises a main stem and several petioles (tiny stem leaves). The individual outputs from the fiber-optic sensors can be processed into an indication of whether a stem or a petiole is coming into view as the gripper encounters the cutting. Consequently the gripper can be commanded to grasp a stem but not a petiole. The axial centerline of a stem can also be recognized from the outputs of the five optic sensors. Upon recognition of a centerline, the gripper signals the robot, and the robot commands the gripper to close.

The motor-controller circuit card supplies the command signals to the amplifier that drives the gripper motors. This card can be operated as a position control with digital position feedback or as a force control with analog force feedback from the load cell mounted in the gripper. A microprocessor is located on the motor control card. Buffered command programs are downloaded from the computer to this card for independent execution by the card.

Prior to a controlled force closure, the motor-control card controls the gripper in position-servo mode until a specified force threshold is sensed, indicating contact with the cutting. Thereafter, the position-servo loop is opened, and the command signal to the amplifier is calculated as the difference between the force set point and the force feedback from the load cell. This distance is multiplied by a programmable gain value, then pulse-width-modulated with a programmable duty cycle of typically 200 percent. This technique provides integral stability to the force-control loop. The force-control loop is bidirectional in the sense that, if the cutting expands between the fingertips, the fingers are made to separate and, if the cutting contracts, the fingers are made to approach each other.

5.12 Ultrasonic Stress Sensor Measuring Dynamic Changes in Materials

An *ultrasonic dynamic vector stress sensor* (UDVSS) has recently been developed to measure the changes in dynamic directional stress that occur in materials or structures at the location touched by the device when the material or structure is subjected to cyclic load. A strain gauge device previously used for the measurement of such a stress measured strain in itself, not in the part being stressed, and thus provided a secondary measurement. Other techniques, such as those that involve thermoelasticity and shearography, have been expensive and placed demands on the measured material. The optical measurement of stress required the application of a phase coat to the object under test. The laser diffraction method required notching or sharp marking of the specimen.

A UDVSS is the first simple portable device able to determine stress directly in the specimen itself rather than in a bonded gauge attached to the specimen. As illustrated in Fig. 5.26, a typical material testing machine applies cyclic stress to a specimen. The UDVSS includes a probe, which is placed in contact with the specimen; an electronic system connected to the probe; and a source of a reference signal. The probe assembly includes a probe handle that holds the probe, a transducer mount that contains active ultrasonic driver and

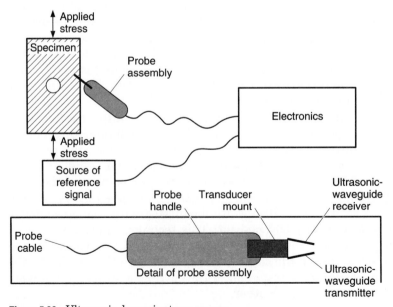

Figure 5.26 Ultrasonic dynamic stress sensor.

receiver, and ultrasonic waveguide transmitter and ultrasonic waveguide receiver that convert the electrical signals to mechanical motion and the inverse, and a cable that connects the probe of the electronics. When in contact with the specimen, the ultrasonic waveguide transmitter causes acoustic waves to travel across the specimen to the ultrasonic waveguide receiver, wherein the wave is converted to an electrical signal.

The operation of the UDVSS is based on the physical phenomenon that the propagation of sound in the specimen changes when the stress in the specimen changes. A pulse phase-locked loop reacts to a change in propagation of sound and therefore in stress by changing its operational frequency. The component of that signal represents that change in voltage needed to keep the system at quadrature to follow the system change in stress. That signal provides the information on changing stress.

The UDVSS can be moved around on the specimen to map out the stress field, and by rotating the probe, one can determine the direction of a stress. In addition, the probe is easily calibrated. The UDVSS should find wide acceptance among manufacturers of aerospace and automotive structures for stress testing and evaluation of designs.

5.13   Predictive Monitoring Sensors Serving CIM Strategy

Computer-integrated manufacturing technology can be well-served by a predictive monitoring system that would prevent a large number of sensors from overwhelming the electronic data monitoring system or a human operator. The essence of the method is to select only a few of the many sensors in the system for monitoring at a given time and to set alarm levels of the selected sensor outputs to reflect the limit of expected normal operation at the given time. The method is intended for use in a highly instrumented system that includes many interfacing components and subsystems, for example, an advanced aircraft, an environmental chamber, a chemical processing plant, or a machining work cell.

Several considerations motivate the expanding effort in implementing the concept of predictive monitoring. Typically, the timely detection of anomalous behavior of a system and the ability of the operator or electronic monitor to react quickly are necessary for the continuous safe operation of the system.

In the absence of a sensor-planning method, an operator may be overwhelmed with alarm data resulting from interactions among sensors rather than data directly resulting from anomalous behavior of the system. In addition, much raw sensor data presented to the operator may by irrelevant to an anomalous condition. The operator is thus presented with a great deal of unfocused sensor information, from which it may be impossible to form a global picture of events and conditions in the system. The predictive monitoring method would be implemented in a computer system running artificial intelligence software, tentatively named *PREMON*. The predictive monitoring system would include three modules: (1) a causal simulator, (2) a sensor planner, and (3) a sensor interpreter (Fig. 5.27).

The word *event* in Fig. 5.27 denotes a discontinuous change in the value of a given quantity (sensor output) at a given time. The inputs to the causal simulator would include a causal mathematical model of the system to be monitored, a set of events that describe the initial state of the system, and perhaps some future scheduled events. The outputs of the causal simulator would include a set of predicted events and a graph of causal dependency among events.

The sensor planner would use the causal dependency graph generated by the causal simulator to determine which few of all the predicted events are important enough that they should be verified. In many cases, the most important events would be taken to be those that either caused, or are caused by, the greatest number of other events. This notion of causal importance would serve as the basis for the election of those sensors, the outputs of which should be used to verify the expected behavior of the system.

The sensor interpreter would compare the actual outputs of the

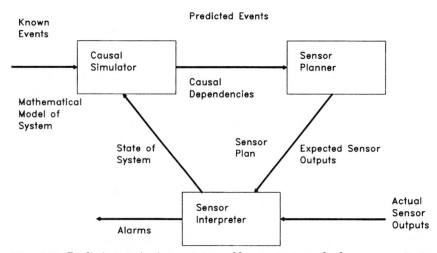

Figure 5.27 Predictive monitoring system would concentrate on the few sensor outputs that are most important at the moment.

selected sensors with the values of those outputs predicted by the causal simulator. Alarms would be raised where the discrepancies between predicted and actual values were significant.

5.14 Reflective Strip Imaging Camera Sensor—Measuring a 180°-Wide Angle

A proposed camera sensor would image a thin striplike portion of a field of view 180° wide. For example, it could be oriented to look at the horizon in an easterly direction from north to south or it could be rotated about a horizonal north/south axis to make a "pushbroom" scan of the entire sky proceeding from the easterly toward the westerly half of the horizon. Potential uses for the camera sensor include surveillance of clouds, coarse mapping of terrain, measurements of the bidirectional reflectance distribution functions of aerosols, imaging spectrometry, oceanography, and exploration of the planets.

The imaging optics would be a segment of concave hemispherical reflecting surfaces placed slightly off center (Fig. 5.28). Like other reflecting optics, it would be achromatic. The unique optical configu-

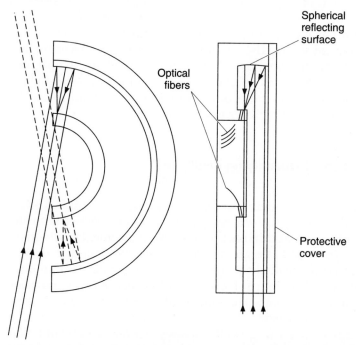

Figure 5.28 A thin segment of a hemispherical concave reflector would form an image from a 180° strip field of view onto optical fibers.

ration would practically eliminate geometric distortion of the image. The optical structure could be fabricated and athermalized fairly easily in that it could be machined out of one or a few pieces of metal, and the spherical reflecting surface could be finished by diamond turning. In comparison, a camera sensor with a fish-eye lens, which provides a nearly hemispherical field of view, exhibits distortion, chromatism, and poor athermalization. The image would be formed on a thin semicircular strip at half the radius of the sphere. A coherent bundle of optical fibers would collect the light from this strip and transfer the image to a linear or rectangular array of photodetectors or to the entrance slit of an image spectrograph. Provided that the input ends of the fibers were properly aimed, the cones of acceptance of the fibers would act as aperture stops; typically, the resulting width of the effective aperture of the camera sensor would be about one-third the focal length (f/3).

The camera sensor would operate at wavelengths from 500 to 1100 nm. The angular resolution would be about 0.5°. In the case of an effective aperture of f/3, the camera would provide an unvignetted view over the middle 161° of the strip, with up to 50 percent vignetting in the outermost 9.5° on each end.

The decentration of the spherical reflecting surface is necessary to make room for the optical fibers and the structure that would support them. On the other hand, the decentration distance must not exceed the amount beyond which the coma that results from decentration would become unacceptably large. In the case of an effective aperture of f/3, the coma would be only slightly in excess of the spherical aberration if the decentration were limited to about f/6. This would be enough to accommodate the fibers and supporting structure.

5.15   Optical Sensor Quantifying Acidity of Solutions

With environmental concerns increasing, a method for taking effective measurements of acidity will minimize waste and reduce both the cost and the environmental impact of processing chemicals. Scientists at Los Alamos National Laboratory (LANL) have developed an optical sensor that measures acid concentration at a higher level than does any current device. The optical high-acidity sensor, reduces the wear generated from acidity measurements and makes possible the economic recycling of acid waste.

The high-acidity sensor (Fig. 5.29) consists of a flow cell (about 87.5 mm across) in which two fused silica lenses are tightly mounted across from one another. Fiber-optic cables connect the two lenses to a spectrophotometer. One lens is coated with a sensing material con-

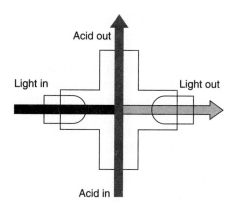

Acid out

Light in

Light out

Acid in

Figure 5.29 Optical high-acidity sensor.

sisting of a polymer that is chemically bound to the lens and an indicator that is physically entrapped with the polymer. Acidic solutions flow vertically up through the flow cell. The light from a fixed source in the spectrophotometer is collimated by the coated lens, passes through the acidic solution, and then is focused by the second lens.

The amount of light absorbed by the indicator depends on the acidity of the solution. The absorption spectrum of the indicator reflects the concentration of hydrogen ions in the solution, which is how acidity is measured.

The LANL sensor's sensitivity range—acid concentrations of 4 to 12 molar—is unmatched by any other acidity measurement technique in current manufacturing. Present techniques are unsuitable for measuring solutions with a negative pH or for measuring chemical processes. With the LANL sensor, high acid concentrations can be measured in 3 min or less, which is 50 times faster than measuring with manual sampling and titration. The sensing material can function in highly acidic solutions for 4 to 6 months with calibration less than once a week, and needs to be replaced only once or twice a year. A sensor using similar principles has recently been developed from measuring lower acidity concentrations.

The sensor is selective in its measurements because the small pores of the polymer allow only hydrogen ions to pass through the indicator. Metal ions do not form a complex with the indicator under acidic solutions. The sensor is reusable, that is, chemically reversible. In short, no comparable product for measuring high acidity is reversible or as sensitive, fast, accurate, and selective as the LANL optical high-acidity sensor.

The prime use for the sensor is monitoring highly acidic chemical processes and waste solutions. High-acidity processes are common in preparing metals from ore-mining operations, treating fuel elements from nuclear power plants, manufacturing bulk acid, and metal finishing including passivation and electroplating.

For highly acidic processor applications, the LANL sensor will save companies thousands of dollars by improving efficiency and decreasing the time devoted to acid measurements. The sensor can be used on manufacturing lines, allowing control and waste management adjustment to be made before acidity fluctuations become a problem. The sensor will improve efficiency at least 25 percent by eliminating the need to reprocess material processed incorrectly on the first try because its true acidity was not known and controlled. Higher efficiency will mean lower cost and minimal waste products generated; the sensor itself generates no caustic waste. Finally, the sensor's waste monitoring capabilities will help ensure that any discharged waste is environmentally benign.

For the 500 acidity measurements done at LANL in 1991, the sensor saved $99,500, a 99.5 percent savings in labor costs in the first year. And, because the sensor generates no waste, 20 L a year of caustic waste was avoided, a 100 percent reduction.

5.16   Sensors for Biomedical Technology

In recent years, advanced imaging and other computer-related technology have greatly expanded the horizons of basic biological and biochemical research. Currently, such drivers as the growing needs for environmental information and increased understanding of genetic systems have provided impetus to biotechnology development. This is a relatively new specialty in the marketplace; nevertheless, the intensity of worldwide competition is escalating. Collaborative research and development projects among the U.S. government, industry, and academia constitute a major thrust for rapid deployment of research and development. The results can place the nation in a position of world leadership in biotechnology.

5.16.1   Sensor for detecting minute quantities of biological materials

A new device based on laser-excited fluorescence provides unparalleled detection of biological materials for which only minuscule samples may be available. This device, invented at the Ames Labora-tories, received a 1991 R&D 100 award.

The Ames Microfluor detector was developed to meet a need for an improved detection technique, driven by important new studies of the human genome, abuse substances, toxins, DNA adduct formation, and amino acids, all of which may be available only in minute amounts. Although powerful and efficient methods have been developed for separating biological mixtures in small volume (i.e., capillary electrophoresis), equally powerful techniques for subsequent detection and identification of these mixtures have been lacking.

The Microfluor detector combines very high sensitivities with the ability to analyze very small volumes. The instrument design is based on the principle that many important biomaterials are fluorescent, while many other biomaterials, such as peptides and oligonucleotides, can be made to fluoresce by adding a fluorescent tag.

When a sample-filled capillary tube is inserted into the Microfluor detector and is irradiated by a laser beam, the sample will fluoresce. The detector detects, monitors, and quantifies the contents by sensing the intensity of the fluorescent light emitted. The signal is proportional to the concentration of the materials. The proportionality constant is characteristic of the material itself.

Analyses can be performed with sample sizes 50 times smaller than those required by other methods, and concentrations as low as 10^{-11} molar (1 part per trillion) can be measured. Often, the critical components in a sample are present at these minute concentrations. These two features make the Microfluor detector uniquely compatible with capillary electrophoresis. In addition, the Ames-developed detector is distinct from other laser-excited detectors in that it is not seriously affected by stray light from the laser itself; it also allows simple alignment and operation in full room light. The Microfluor detector has already been used to determine the extent of adduct formation and base modification in DNA so that the effects of carcinogens on living cells can be studied. Future uses of the sensor will include DNA sequencing and protein sequencing. With direct or indirect fluorescence detection, researchers are using this technique to study chemical contents of individual living cells. This capability may allow pharmaceutical products to be tested on single cells rather than a whole organism, with improved speed and safety.

5.16.2 Sensors for early detection and treatment of lung tumors

A quick, accurate method for early detection of lung cancer would raise chances of patient survival from less than 50 percent to 80 or 90 percent. Until now, small cancer cells deep in the lung have been impossible to detect before they form tumors large enough to show up in x-rays. Researchers at Los Alamos National Laboratory in collaboration with other institutions, including the Johns Hopkins school of medicine and St. Mary's Hospital in Grand Junction, Colorado, are developing methods for finding and treating lung cancer in its earliest stages.

A detection sensor involves a porphyrin, one of an unusual family of chemicals that is found naturally in the body and concentrates in cancer cells. The chemical is added to a sample of sputum coughed up from the lung. When exposed to ultraviolet or laser light, cells in the porphyrin-treated sputum glow a bright red. When the sample is viewed under the microscope, the amount and intensity of fluorescence in the cells determines the presence of cancer.

The first clinical test of a detection technique using porphyrin was done by LANL and St. Mary's Hospital in 1988. Four different porphyrins were tested on sputum samples from two former miners, one known to have lung cancer and one with no detectable cancer. One of the porphyrins was concentrated in certain cells only in the sputum of the miner with lung cancer. Other tests concluded that these were cancer cells. Later, a blind study of sputum samples from 12 patients, 8 of whom had lung cancer in various stages of development, identified all the cancer patients as well as a ninth originally thought to be free of cancer. Further tests showed that this patient also had lung cancer.

Identifying the ninth patient prompted a new study, in which the procedure was evaluated for its ability to detect precancerous cells in the lung. In this study, historical sputum samples obtained from Johns Hopkins were treated with the porphyrin. Precancerous conditions that chest x-rays had not detected were identified in samples from patients who later developed the disease. Although further testing is needed, researchers now feel confident that the technique and the detection sensor can detect precancerous conditions 3 to 4 years before onset of the disease.

Working in collaboration with industry, researchers expect to develop an instrument in several years for rapid automated computerized screening of larger populations of smokers, miners, and other people at risk of developing lung cancer. Because lung cancer is the leading cancer killer in both men and women, a successful screening procedure would dramatically lower the nation's mortality rate from the disease. A similar procedure has potential for the analysis of Pap smears, now done by technicians who screen slides one at a time and can misread them.

In addition to a successful screening program, LANL hopes to develop an effective treatment for early lung cancer. The National Cancer Institute will help investigate and implement both diagnostic and therapeutic programs. Researchers are performing basic cell studies on animals to investigate the effect of porphyrin on lung cancer cells. They are also exploring the use of the LANL-designed porphyrin attached to radioactive copper to kill early cancer cells in the lung in a single search-and-destroy mission. The porphyrin not only seeks out cancer cells but also has a molecular structure, similar to a hollow golf ball, that can take certain metals into its center. A small amount of radioactive copper 67 placed inside the porphyrin should destroy tumors the size of pinhead, as well as function as a tracer.

5.16.3 Ultrasensitive sensor for single-molecule detection

Enhancing the sensitivity of research instruments has long been a goal of physicists, biologists, and chemists. With the single-molecule detector, researchers have achieved the ultimate in this pursuit: detection of a single fluorescent molecule in a liquid. The new instrument—a thousand times more sensitive than existing commercial detectors—brings new capabilities to areas of science and technology that affect lives in many ways, from DNA sequencing, biochemical analysis, and virus studies to identifying environmental pollutants.

For some time, scientists have observed individual molecules trapped in a vacuum, where they are isolated and relatively easy to find. However, many important biological and chemical processes occur in a liquid environment, where many billions of other molecules surround the molecules of interest. Developing a single-molecule detector that operates in such an environment presented a difficult challenge.

The observation technique involves attaching fluorescent dye molecules to the molecules of interest and then exciting the dye molecule by passing them in solution to a rapidly pulsed laser beam and detecting the subsequent faint light, or photons, they emit. The fluorescent lifetimes of the molecules are much shorter than the time the molecules spend in the laser beam; therefore each molecule is reexcited many times and yields many fluorescent photons. The signature of the passing molecule is the burst of photons that occurs when the molecule is passing the laser beam (Fig. 5.30).

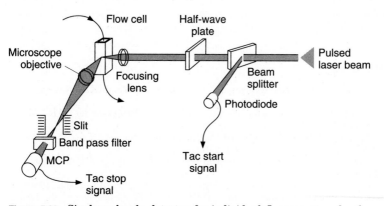

Figure 5.30 Single-molecule detector for individual fluorescent molecules in solution.

A lens, or microscope objective, and a slit are arranged to image the photons from a small region around the laser beam waist onto a microchannel plate photomultiplier (MCPP) that counts individual photons. The intense excitation light from the laser is blocked from reaching the MCPP by a bandpass spectral filter, which is centered near the peak fluorescent wavelength. The excitation light consists of extremely short pulses, each about 70 trillionth of a second. The dye molecule does not emit light until a billionth of a second after excitation, so the flash of laser light fades before the feeble molecular glow occurs. For reliable identification of individual molecules, the technique maximizes the number of the detected photons and minimizes the number of background photons. Although some background remains, the technique registers over 85 percent of the fluorescent molecules.

Developed to aid in sequencing chromosomes for the Human Genome Project, the sensor detector's high activity would allow DNA sequencing rates hundreds of times faster than those obtainable with present techniques. It would also eliminate the need for radioactive materials, gels, and electrophoresis solutions, which often create disposal problems, and it is expected to help make DNA sequencing a routine diagnostic and clinical tool. One eventual benefit of the DNA research may be rapid screening for any of 3500 known genetic diseases such as diabetes, cystic fibrosis, and Alzheimer's disease.

The ultrasensitive detector can be used to find and quantify minute amounts of chemicals, enzymes, and viruses in the blood and monitor the dispersal of extremely low concentrations of environmental pollutants. The device may also be useful in studying the interaction of viruses and their binding sites. It may be possible to develop a procedure for rapidly evaluating the efficiency of vaccines; such a procedure could quickly expedite the search for an effective vaccine against the AIDS virus.

Further Reading

Barks, R. E., "Optical High Acidity Sensor," Los Alamos National Laboratory, June 1991.

Bicknel, T. J., and W. H. Farr, "Acousto Optical/Electronic Processor for SAR," *NASA Tech Briefs,* **16,** May 1992.

Bonnert, R., "Design of High Performance Digital Tachometer with Digital Microcontroller," *IEEE Trans. Instrum.,* **38,** 1104–1108 (1989).

Buser, R. A., and N. F. Rooij, "Resonant Silicon Structures," *Sensors and Actuators,* **17,** 145–154 (1989).

D'Amico A., and E. Verona, "SAW Sensors," *Sensors and Actuators,* **17,** 66–66 (1989).

Fleming Dias, J., "Physics Sensors Using SAW Devices," *Hewlett-Packard J.,* 18–20, December 1981.

Gast, T., "Sensors with Oscillating Elements," *J. Phys. E. Sci. Instrum.,* **18,** 783–789 (1985).

Hayman, J. S., and M. Frogatt, "Ultrasonic Dynamic Vector Stress Sensor (UDVSS)," Langley Research Center, 1992.

Hewlett-Packard, "Design and Operational Considerations for the HELDS-5000 Incremental Shaft Encoder," Application Note 1011, Palo Alto, 1981.

Higbie, B. N., "Automatic Detection of Faults in Turbomachinery Bearings," NASA Tech Briefs, **16,** May 1992.

Higbie, N., Technology Integration and Development Group Inc., *Technical Report,* May 1992.

Huner, B., P. Klinkhachorn, and E. B. Everton, "Hybrid Clock Oscillator Modules as Deposition Monitors," *Rev. Sci. Instrum.,* **59,** 983–986 (1988).

Ito, H., "Balanced Absorption Quartz Hygrometer," *IEEE Trans. Ultrasonics, Ferroelectrics, and Frequency Control,* **34,** 136–141 (1987).

Lokshin, A. M., "Hand/Eye Coordination for Microrobotic Motion—Utilizing Vision Technology," Caltech, 1991.

Los Alamos National Laboratory, Research Team, "Single Molecule Detection," January 1991.

Montgomery, J. L., "Force and Optical Sensors Controlling Robotic Gripper for Agriculture," Martin Marietta Corp., 1992.

Noble, M. N., D. N. Mark, and R. Blue, "Tracking Retroreflective Targets on a Structure," *NASA Tech Briefs,* **16,** May 1992.

"Optical Sensor to Quantify Highly Acidic Solution—Sensing High Acidity without Generating Waste," LANL, *Technology '91,* 57–58 (1991).

Tien-Hsin Chao, "Experimental Optoelectronic Associative Memory," *NASA Tech Briefs,* **16,** May 1992.

"Ultrasensitive Detection for Medical and Environmental Analysis—Single Molecule Detector," LANL, *Technology '91,* 80–81 (1991).

Vaughan, A. H., "Reflective Strip Imaging Camera Sensor—Measuring a 180° Wide Angle," Caltech, NASA's Jet Propulsion Laboratory, 1992.

Williams, D. E., "Laser Sensor Detecting Microfluor," Ames Laboratory, July 1991.

6

Industrial Sensors and Control

6.0 Introduction

Current manufacturing strategy defines manufacturing systems in terms of sensors, actuators, effectors, controllers, and control loops. Sensors provide a means for gathering information on manufacturing operations and processes being performed. In many instances, sensors are used to transform a physical stimulus into an electrical signal that may be analyzed by the manufacturing system and used for making decisions about the operations being conducted. Actuators convert an electrical signal into a mechanical motion. An actuator acts on the product and equipment through an effector. Effectors serve as the "hand" that achieves the desired mechanical action. Controllers are computers of some type that receive information from sensors and from internal programming, and use this information to operate the manufacturing equipment (to the extent available, depending on the degree of automation and control). Controllers provide electronic commands that convert an electrical signal into a mechanical action. Sensors, actuators, effectors, and controllers are linked to a control loop.

In limited-capability control loops, little information is gathered, little decision making can take place, and limited action results. In other settings, "smart" manufacturing equipment with a wide range of sensor types can apply numerous actuators and effectors to achieve a wide range of automated actions.

The purpose of sensors is to inspect work in progress, to monitor the work-in-progress interface with the manufacturing equipment, and to allow self-monitoring of manufacturing by the manufacturing system's own computer. The purpose of the actuator and effector is to transform the work in progress according to the defined processes of the manufacturing system. The function of the controller is to allow for varying degrees of manual, semiautomated, or fully automated control over the processes. In a fully automated case, such as in computer-integrated manufacturing, the controller is completely adaptive and functions in a closed-loop manner to produce automatic system operation. In other cases, human activity is involved in the control loop.

In order to understand the ways in which the physical properties of a manufacturing system affect the functional parameters associated with the manufacturing system, and in order to determine the types of physical manufacturing system properties that are necessary to implement the various desired functional parameters, it is necessary to understand the technologies that are available for manufacturing systems that use automation and integration to varying degrees.

The least automated equipment makes use of detailed operator control over all equipment functions. Further, each action performed by the equipment is individually directed by the operator. Manual equipment thus makes the maximum use of human capability and adaptability. Visual observations can be enhanced by the use of microscopes and cameras, and the actions that are undertaken can be enhanced by the use of simple effectors. The linkages between the sensory information (from microscopes or through cameras) and the resulting actions are obtained by placing the operator in the loop.

This type of system is clearly limited by the types of sensors used and their relationship to the human operator, the types of the effectors that can be used in conjunction with the human operator, and the capabilities of the operator. The manufacturing equipment that is designed for a manual strategy must be matched to human capabilities. The human–manufacturing equipment interface is extremely important in many manufacturing applications. Unfortunately, equipment design is often not optimized as a sensor-operator-actuator/effector control loop.

A manufacturing system may be semiautomated, with some portion of the control loop replaced by a computer. This approach will serve the new demands on manufacturing system design requirements. Specifically, sensors now must provide continuous input data for both the operator and computer. The appropriate types of data must be provided in a timely manner to each of these control loops. Semiautomated manufacturing systems must have the capability for a limited degree of self-monitoring and control associated with the computer portion of the decision making loop. An obvious difficulty in designing such equipment is to manage the computer- and operator-controlled activities in an optimum manner. The computer must be able to recognize when it needs operator support, and the operator must be able to recognize which functions may appropriately be left to computer control. A continuing machine-operator interaction is part of normal operations.

Another manufacturing concept involves fully automated manufacturing systems. The processing within the manufacturing system itself is fully computer-controlled. Closed-loop operations must exist between sensors and actuators/effectors in the manufacturing system. The manufacturing system must be able to monitor its own performance and decision making for all required operations. For effective automated operation, the mean time between operator interventions must be large when compared with the times between manufacturing setups.

$$\text{MTOI} = (\Sigma^n_i\, \tau_1 + \tau_2 + \tau_3 + \tau_4 + \ldots + \tau_4)/n \qquad (6.1)$$

where τ = setup time
i = initial setup
n = number of setups

The processes in use must rarely fail; the operator will intervene only when such failures occur. In such a setting, the operator's function is to ensure the adequate flow of work in progress and respond to system failure.

Several types of work cells are designed according to the concept of total manufacturing integration. The most sophisticated cell design involves fully automated processing and materials handling. Computers control the feeding of work in progress, the performance of the manufacturing process, and the removal of the work in progress. Manufacturing systems of this type provide the opportunity for the most advanced automated and integrated operations. The manufacturing system must be modified to achieve closed-loop operations for all of these functions.

Most manufacturing systems in use today are not very resourceful. They do not make use of external sensors that enable them to monitor their own performance. Rather, they depend on internal conditioning sensors to feed back (to the control system) information regarding manipulator positions and actions. To be effective, this type of manufacturing system must have a rigid structure and be able to determine its own position based on internal data (largely independent of the load that is applied). This leads to large, heavy, and rigid structures.

The more intelligent manufacturing systems use sensors that enable them to observe work in progress and a control loop that allows corrective action to be taken. Thus, such manufacturing systems do not have to be as rigid because they can adapt.

The evolution toward more intelligent and adaptive manufacturing systems has been slow, partly because the required technologies have evolved only in recent years and partly because it is difficult to design work cells that effectively use the adaptive capabilities. Enterprises are not sure whether such features are cost-effective and wonder how to integrate smart manufacturing systems into the overall strategy.

The emphasis must be on the building-block elements that are necessary for many types of processing. If the most advanced sensors are combined with the most advanced manufacturing systems, concepts, and state-of-the-art controllers and control loops, very sophisticated manufacturing systems can result. On the other hand, much more rudimentary sensors, effectors, and controllers can produce simple types of actions.

In many instances today sensors are analog (they involve a continuously changing output property), and control loops make use of digital computers. Therefore, an analog-to-digital converter between the preprocessor and the digital control loop is often required.

The sensor may operate either passively or actively. In the passive case, the physical stimulus is available in the environment and does not have to be provided. For an active case, the particular physical stimulus must be provided. Machine vision and color identification sensors are an active means of sensing, because visible light must be used to illuminate the object before a physical stimulus can be received by the sensor. Laser sensors are also active-type sensors. Passive sensors include infrared devices (the physical stimulus being generated from infrared radiation that is associated with the temperature of a body) and sensors to measure pressure, flow, temperature, displacement, proximity, humidity, and other physical parameters.

6.1   Sensors in Manufacturing

Many types of sensors have been developed during the past several years, especially those for industrial process control, military uses, medicine, automotive applications, and avionics. Several types of sensors are already being manufactured by commercial companies.

Process control sensors in manufacturing will play a significant role in improving productivity, qualitatively and quantitatively, throughout the coming decades. The main parameters to be measured and controlled in industrial plants are temperature, displacement, force, pressure, fluid level, and flow. In addition, detectors for leakage of explosives or combustible gases and oils are important for accident prevention.

Optical-fiber sensors may be conveniently divided into two groups: (1) intrinsic sensors and (2) extrinsic sensors.

Although intrinsic sensors have, in many cases, an advantage of higher sensitivity, almost all sensors used in process control at present belong to the extrinsic type. Extrinsic-type sensors employ light sources such as LEDs, which have higher reliability, longer life, and lower cost than semiconductor lasers. They also are compatible with multimode fibers, which provide higher efficiency when coupled to light sources and are less sensitive to external mechanical and thermal disturbances.

As described in Chap. 1, objects can be detected by interrupting the sensor beam. Optical-fiber interrupters are sensors for which the principal function is the detection of moving objects. They may be classified into two types: reflection and transmission.

In the reflection-type sensor, the light beam emitted from the fiber is reflected back into the same fiber if the object is situated in front of the sensor.

In the transmission-type sensor, the emitted light from the input fiber is interrupted by the object, resulting in no received light in the output fiber located at the opposite side. Typical obstacle interrupters employ low-cost large-core plastic fibers because of the short transmission distance. The minimum detectable size of the object is typically limited to 1 mm by the fiber core diameter and the optical beam. The operating temperature range of commercially available sensors is typically −40 to +70°C. Optical-fiber sensors have been utilized in industry in many ways, such as:

1. Detection of lot number and expiration dates, for example, in the pharmaceutical and food industries

2. Color difference recognition, for example, colored objects on a conveyer

3. Defect detection, for example, missing wire leads in electronic components

4. Counting discrete components, for example, bottles or cans

5. Detecting absence or presence of labels, for example, packaging in the pharmaceutical and food industries

Fiber-optic sensors for monitoring process variables such as temperature, pressure, flow, and liquid level are also classified into two types: (1) normally OFF type in which the shutter is inserted between the fibers in the unactivated state. Thus, this type of sensor provides high and low levels as the light output corresponds to ON and OFF states, respectively, and (2) normally ON type where the shutter is retracted from the gap in the unactivated state.

In both types, the shutter is adjusted so that it does not intercept the light beam completely but allows a small amount of light to be transmitted, even when fully closed. This transmitted light is used to monitor the cable (fiber) continuity for faults and provides an intermediate state. Commercially available sensors employ fibers of 200-μm core diameter. The typical differential attenuation which determines the ON-OFF contrast ratio is about 20 dB. According to manufacturers' specifications, these sensors operate well over the temperature range −40 to +80°C with 2-dB variation in light output.

6.2   Temperature Sensors in Process Control

Temperature is one of the most important parameters to be controlled in almost all industrial plants, since it directly affects material properties and thus product quality. During the past few years, several temperature sensors have been developed for use in electrically or chemically hostile environments. Among these, the practical temperature sensors, which are now commercially available, are classified into two groups: (1) low-temperature sensors with a range of −100 to +400°C using specific sensing materials such as phosphors, semiconductors, and liquid crystals and (2) high-temperature sensors with a range of 500 to 2000°C based on blackbody radiation.

6.2.1   Semiconductor absorption sensors

Many of these sensors can be located up to 1500 m away from the optoelectronic instruments. The operation of semiconductor temperature sensors is based on the temperature-dependent absorption of semiconductor materials. Because the energy and gap of most semiconductors decrease almost linearly with increasing temperature T, the band-edge wavelength $\lambda_g(T)$ corresponding to the fundamental optical absorption shifts toward longer wavelengths at a rate of about 3 Å/°C [for gallium arsenide (GaAs)] with T. As illustrated in Fig. 6.1, when a light-emitting diode with a radiation spectrum covering the

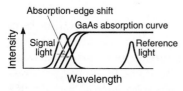

Figure 6.1   Operating principle of optical-fiber thermometer based on temperature-dependent GaAs light absorption.

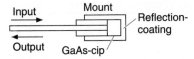

Figure 6.2 Sensing element of the optical-fiber thermometer with GaAs light absorber.

wavelength $\lambda_g(T)$ is used as a light source, the light intensity transmitted through a semiconductor decreases with T.

Figure 6.2 shows the reflection-type sensing element. A polished thin GaAs chip is attached to the fiber end and mounted in a stainless-steel capillary tube of 2-mm diameter. The front face of the GaAs is antireflection-coated, while the back face is gold-coated to return the light into the fiber.

The system configuration of the thermometer is illustrated in Fig. 6.3. In order to reduce the measuring errors caused by variations in parasitic losses, such as optical fiber loss and connector loss, this thermosensor employs two LED sources [one aluminum gallium arsenide (AlGaAs), the other indium gallium arsenide (InGaAs)] with different wavelengths. A pair of optical pulses with different wavelengths $\lambda_s = 0.88$ µm and $\lambda_r = 1.3$ µm are guided from the AlGaAs LED and the InGaAs Led to the sensing element along the fiber. The light of λ_s is intensity-modulated by temperature. On the other hand, GaAs is transparent for the light of λ_r, which is then utilized as a reference light. After detection by a germanium avalanche photodiode (GeAPD), the temperature-dependent signal λ_s is normalized by the reference signal λ_r in a microprocessor.

The performance of the thermometer is summarized in Table 6.1. An accuracy of better than ±2°C is obtained within a range of −20 to +150°C. The principle of operation for this temperature sensor is

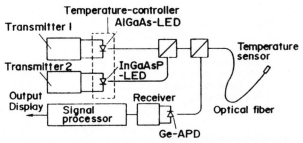

Figure 6.3 System configuration of the optical-fiber thermometer with GaAs light absorber.

TABLE 6.1   **Characteristics of Semiconductor Sensors (Thermometer Performance)**

Property	Semiconductor absorption sensor (Mitsubishi, Japan)	Semiconductor photolumines- cence sensor (ASEA Innovation, Sweden)	Phosphor A sensor (Luxton, U.S.)	Phosphor B sensor (Luxton, U.S.)
Range, °C	−20 to +150	0 to 200	20 to 240	20 to 400
Accuracy, °C	±2.0	±1.0	±2.0	±2.0
Diameter, m	2	From 0.6	0.7	1.6
Time constant, s	0.5	From 0.3	From 0.25	
Fiber type	100-μm silica core	100-μm silica core	400-μm polymer clad	Fiber silica
Fiber length, m	300	500	100	300
Light source	AlGaAs LED	AlGaAs LED	Halogen lamp	Halogen lamp

based on the temperature-dependent direct fluorescent emission from phosphors.

6.2.2   Semiconductor temperature detector using photoluminescence

The sensing element of this semiconductor photoluminescence sensor is a double-heterostructure GaAs epitaxial layer surrounded by two $Al_xGa_{1-x}As$ layers. When the GaAs absorbs the incoming exciting light, the electron-hole pairs are generated in the GaAs layer. The electron-hole pairs combine and reemit the photons with a wavelength determined by temperature. As illustrated in Fig. 6.4, the luminescent wavelength shifts monotonically toward longer wavelengths as the temperature T increases. This is a result of the decrease in the energy

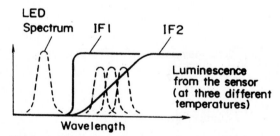

Figure 6.4 Operating principle of optical-fiber thermometer based on temperature-dependent photoluminescence from a GaAs epitaxial film.

gap E_g with T. Therefore, analysis of the luminescent spectrum yields the required temperature information. The double heterostructure of the sensing element provides excellent quantum efficiency for the luminescence because the generated electron-hole pairs are confined between the two potential barriers (Fig. 6.5).

The system is configured as shown in Fig. 6.6. The sensing element is attached to the end of the silica fiber (100-μm core diameter). The excitation light from an LED, with a peak wavelength of about 750

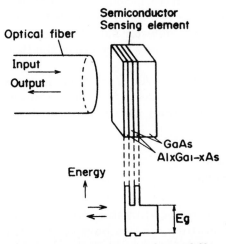

Figure 6.5 Sensing element of optical-fiber thermometer based on temperature-dependent photoluminescence.

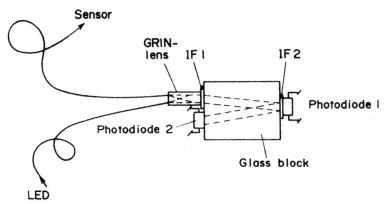

Figure 6.6 Optical system of optical-fiber thermometer based on temperature-dependent photoluminescence.

nm, is coupled into the fiber and guided to a special GRIN lens mounted to a block of glass. A first optical inference filter IF_1, located between the GRIN lens and the glass block, reflects the excitation light which is guided to the sensing element along the fiber. However, this optical filter is transparent to the returned photoluminescent light. The reflectivity of the second interference filter IF_2 changes at about 900 nm. Because the peak wavelength of the luminescence shifts toward longer wavelength with temperature, the ratio between the transmitted and the reflected light intensifies if IF_2 changes. However, the ratio is independent of any variation in the excitation light intensity and parasitic losses. The two lights separated by IF_2 are detected by photodiodes 1 and 2. The detector module is kept at a constant temperature in order to eliminate any influence of the thermal drift of IF_2.

The measuring temperature range is 0 to 200°C, and the accuracy is ±1°C. According to the manufacturer's report, good long-term stability, with a temperature drift of less than 1°C over a period of 9 months, has been obtained.

6.2.3 Temperature detector using point-contact sensors in process manufacturing plant

Electrical sensors are sensitive to microwave radiation and corrosion. The needs for contact-type temperature sensors have lead to the development of point-contact sensors which are immune to microwave radiation, for use in: (1) electric power plants using transformers, generators, surge arresters, cables, and bus bars; (2) industrial plants utilizing microwave processes; and (3) chemical plants utilizing electrolytic processes.

The uses of microwaves include drying powder and wood; curing glues, resins, and plastics; heating processes for food, rubber, and oil; device fabrication in semiconductor manufacturing; and joint welding of plastic packages, for example.

Semiconductor device fabrication is currently receiving strong attention. Most semiconductor device fabrication processes are now performed in vacuum chambers; they include plasma etching and stripping, ion implantation, plasma-assisted chemical vapor deposition, radio-frequency sputtering, and microwave-induced photoresist baking. These processes alter the temperature of the semiconductors being processed. However, the monitoring and controlling of temperature in such hostile environments is difficult with conventional electrical temperature sensors. These problems can be overcome by the contact-type optical-fiber thermometer.

6.2.4 Noncontact sensors—pyrometers

Because they are noncontact sensors, pyrometers do not affect the temperature of the object they are measuring. The operation of the pyrometer is based on the spectral distribution of blackbody radiation, which is illustrated in Fig. 6.7 for several different temperatures. According to the Stefan-Boltzmann law, the rate of the total radiated energy from a blackbody is proportional to the fourth power of absolute temperature and is expressed as:

$$W_t = \sigma \, T^4 \tag{6.2}$$

where σ is the Stefan-Boltzmann constant and has the value of $5.6697 \times 10^{-8} \ \text{W/m}^2 \bullet \text{K}^4$.

The wavelength at which the radiated energy has its highest value is given by Wien's displacement law,

$$\lambda_m T = 2.8978 \times 10^{-3} \ \text{m} \bullet \text{K} \tag{6.3}$$

Thus, the absolute temperature can be measured by analyzing the intensity of the spectrum of the radiated energy from a blackbody. A source of measurement error is the emissivity of the object, which depends on the material and its surface condition. Other causes of error are deviation from the required measurement distance and the presence of any absorbing medium between the object and the detector.

Use of optical fibers as signal transmission lines in pyrometers allows remote sensing over long distances, easy installation, and accurate determination of the position to be measured by observation

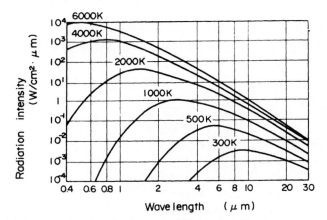

Figure 6.7 Spectral distribution of blackbody radiation.

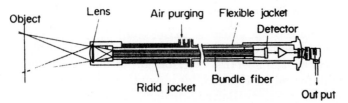

Figure 6.8 Schematic diagram of an optical-fiber pyrometer.

of a focused beam of visible light from the fiber end to the object. The sensing head consists of a flexible bundle with a large number of single fibers and lens optics to pick up the radiated energy (Fig. 6.8).

The use of a single silica fiber instead of a bundle is advantageous for measuring small objects and longer distance transmission of the picked-up radiated light. The lowest measurable temperature is 500°C, because of the unavoidable optical loss in silica fibers at wavelengths longer than 2 μm. Air cooling of the sensing head is usually necessary when the temperature exceeds 1000°C.

Optical-fiber pyrometers are one of the most successful optical-fiber sensors in the field of process control in manufacturing. Typical applications are

1. Casting and rolling lines in steel and other metal plants

2. Electric welding and annealing

3. Furnaces in chemical and metal plants

4. Fusion, epitaxial growth, and sputtering processes in the semiconductor industry

5. Food processing, paper manufacturing, and plastic processing

Figure 6.9 is a block diagram of the typical application of optical-fiber pyrometers for casting lines in a steel plant, where the tempera-

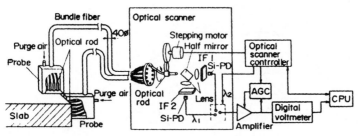

Figure 6.9 Temperature distribution measurement of steel slabs by an optical-fiber pyrometer using two-wavelength method.

ture distribution of the steel slab is measured. The sensing element consists of a linear array of fused-silica optical rods, thermally protected by air-purge cooling. Light radiated from the heated slabs is collected by the optical rods and coupled into a 15-m-long bundle of fibers, which transmits light to the optical processing unit. In this system, each fiber in the bundle carries the signal from a separate lens which provides the temperature information at the designated spot of the slabs. An optical scanner in the processing unit scans the bundle and the selected light signal is analyzed in two wavelength bands by using two optical interference filters.

6.3 Pressure Sensors

If a pressure P acting on a diaphragm compresses a spring until an equilibrium is produced, the pressure can be represented as

$$F \text{ (kg)} = A \text{ (m}^2) \times P \text{ (kg/m}^2) \tag{6.4}$$

In this equation F represents the force of the spring and A represents a surface area of the diaphragm. The movement of the spring is transferred via a system of levers to a pointer whose deflection is a direct indication of the pressure (Fig. 6.10). If the measured value of the pressure has to be transmitted across a long distance, the mechanical movement of the pointer can be connected to a variable electrical resistance (potentiometer). A change in the resistance results in a change in the measured voltage, which can then easily be evaluated by an electronic circuit or further processed. This example illustrates the fact that a physical quantity is often subject to many transformations before it is finally evaluated.

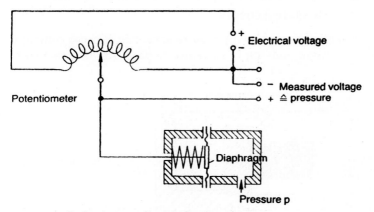

Figure 6.10 Deflection as a direct indication of pressure.

6.3.1 Piezoelectric crystals

Piezoelectric crystals may be utilized to measure pressure. Electrical charges are produced on the opposite surfaces of some crystals when they are mechanically loaded by deflection, pressure, or tension. The electrical charge which is produced in the process is proportional to the effective force. This change in the charge is very small. Therefore, electrical amplifiers are used to make it possible to process the signals (Fig. 6.11).

Pressure in this situation is measured by transforming it into a force. If the force produced by pressure on a diaphragm acts on a piezoelectric crystal, a signal which is proportional to the pressure measured can be produced by using suitable amplifiers.

6.3.2 Strain gauges

Strain gauges can also measure pressure. The electrical resistance of a wire-type conductor is dependent, to certain extent, on its cross-sectional area. The smaller the cross section, i.e., the thinner the wire, the greater the resistance of the wire. A strain gauge is a wire which conducts electricity and stretches as a result of the mechanical influence (tension, pressure, or torsion) and thus changes its resistance in a manner which is detectable. The wire is attached to a carrier which in turn is attached to the object to be measured. Conversely, for linear compression, which enlarges the cross-sectional area of a strain gauge, resistance is reduced. If a strain gauge is attached to a diaphragm (Fig. 6.12), it will follow the movement of the diaphragm. It is either pulled or compressed, depending on the flexure of the diaphragm.

6.4 Fiber-Optic Pressure Sensors

A Y-guide probe can be used as a pressure sensor in process control if a reflective diaphragm, moving in response to pressure, is attached to

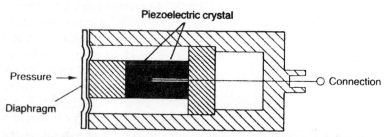

Figure 6.11 Electrical amplifiers are connected to a piezoelectric crystal.

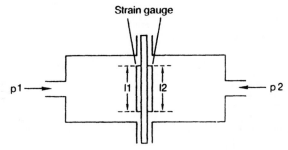

Figure 6.12 Strain gauge for measurement of pressure.

the end of the fiber (Fig. 6.13). This type of pressure sensor has a significant advantage over piezoelectric transducers, since it works as a noncontact sensor and has a high frequency response. The pressure signal is transferred from the sealed diaphragm to the sensing diaphragm, which is attached to the end of the fiber. With a stainless-steel diaphragm about 100 μm thick, hysteresis of less than 0.5 percent and linearity within ±0.5 percent are obtained up to the pressure level of 3×10^5 kg/m^2 (2.94 MPa) in the temperature range of –10 to +60°C.

The material selection and structural design of the diaphragm are important to minimize drift. Optical-fiber pressure sensors are expected to be used under severe environments in process control. For example, process slurries are frequently highly corrosive, and the temperature may be as high as 500°C in coal plants. The conventional metal diaphragm exhibits creep at these high temperatures. In order

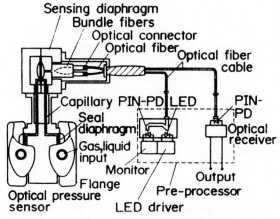

Figure 6.13 Schematic diagram of a fiber-optic pressure sensor using Y-guide probe with a diaphragm attached.

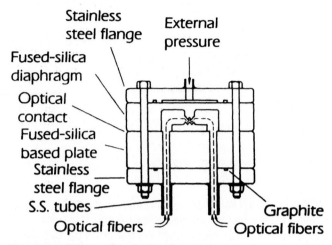

Figure 6.14 Fiber-optic microbend sensor.

to eliminate these problems, an all-fused-silica pressure sensor based on the microbending effect in optical fiber has been developed (Fig. 6.14). This sensor converts the pressure applied to the fused silica diaphragm into an optical intensity modulation in the fiber.

A pressure sensor based on the wavelength filtering method has been developed. The sensor employs a zone plate consisting of a reflective surface, with a series of concentric grooves at predetermined spacing. This zone plate works as a spherical concave mirror whose effective radius of curvature is inversely proportional to the wavelength. At the focal point of the concave mirror, a second fiber is placed which transmits the returned light to two photodiodes with different wavelength sensitivities. When broadband light is emitted from the first fiber to the zone plate, and the zone plate moves back and forth relative to the optical fibers in response to the applied pressure, the wavelength of the light received by the second fiber is varied, causing a change in the ratio of outputs from the two photodiodes. The ratio is then converted into an electrical signal which is relatively unaffected by any variations in parasitic losses.

6.5 Displacement Sensors for Robotic Applications

The operating principle of a displacement sensor using Y-guide probes is illustrated in Fig. 6.15. The most common Y-guide probe is a bifurcated fiber bundle. The light emitted from one bundle is back-reflected by the object to be measured and collected by another bundle (receiv-

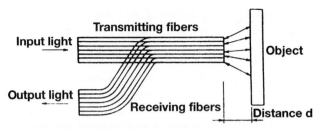

Figure 6.15 Principle of operation of fiber-optic mechanical sensor using a Y-guide probe.

ing fibers). As a result, the returned light at the detector is intensity-modulated to a degree dependent on the distance between the end of the fiber bundle and the object. The sensitivity and the dynamic range are determined by the geometrical arrangement of the array of fiber bundles and by both the number and type of the fibers. Figure 6.16 shows the relative intensity of the returned light as a function of distance for three typical arrangements: random, hemispherical, and concentric circle arrays. The intensities increase with distance and reach a peak at a certain discrete distance. After that, the intensities fall off very slowly. Most sensors use the high-sensitivity regions in these curves. Among the three arrangements, the random array has the highest sensitivity but the narrowest dynamic range. The displacement sensor using the Y-guide probe provides resolution of 0.1 µm, lin-

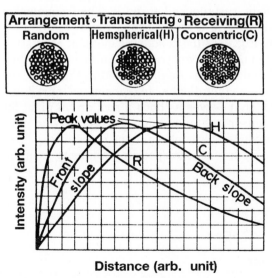

Figure 6.16 Relative intensity of returned light for three fiber-optic arrangements.

earity within 5 percent, and dynamic range of 100 µm displacement. Y-guide probe displacement sensors are well-suited for robotics applications as position sensors and for gauging and surface assessment, since they have high sensitivity to small distances.

One profound problem of this type of displacement sensor is the measuring error arising from the variation in parasitic losses along the optical transmission line. Recalibration is required if the optical path is interrupted, which limits the range of possible applications. In order to overcome this problem, a line-loss-independent displacement sensor with electrical subcarrier phase encoder has been implemented. In this sensor, the light from an LED modulated at 160 MHz is coupled into the fiber bundle and divided into two optical paths. One of the paths is provided with a fixed retroreflector at its end. The light through the other is reflected by the object. The two beams are returned to the two photodiodes separately. Each signal, converted into an electric voltage, is electrically heterodyned into an intermediate frequency at 455 kHz. Then the two signals are fed to a digital phase comparator, the output of which is proportional to the path distance. The resolution of the optical path difference is about 0.3 mm, but improvement of the receiver electronics will provide a higher resolution.

6.6   Process Control Sensors Measuring and Monitoring Liquid Flow

According to the laws of fluid mechanics, an obstruction inserted in a flow stream creates a periodic turbulence behind it. The frequency of shedding the turbulent vortices is directly proportional to the flow velocity. The flow sensor in Fig. 6.17 has a sensing element consisting of a thin metallic obstruction and a downstream metallic bar attached to a multimode fiber-microbend sensor. As illustrated in Fig. 6.18, the vortex pressure produced at the metallic bar is transferred, through a diaphragm at the pipe wall that serves as both a seal and a pivot for the bar, to the microbend sensor located outside the process line pipe. The microbend sensor converts the time-varying mechanical force caused by the vortex shedding into a corresponding intensity modula-

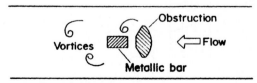

Figure 6.17 Principle of operation of a vortex-shedding flow sensor.

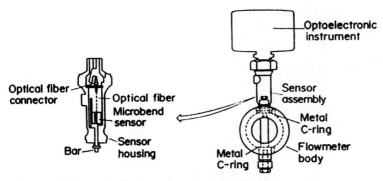

Figure 6.18 Schematic diagram of a vortex-shedding flow sensor.

tion of the light. Therefore, the frequency of the signal converted into the electric voltage at the detector provides the flow-velocity information. This flow sensor has the advantage that the measuring accuracy is essentially independent of any changes in the fluid temperature, viscosity, or density, and in the light source intensity. According to the specifications for typical optical vortex-shedding flow sensors, flow rate can be measured over a Reynolds number range from 5×10^3 to 6000×10^3 at temperatures from -100 to $+600°C$. This range is high compared to that of conventional flow meters. In addition, an accuracy of ±0.4 and ±0.7 percent, respectively, is obtained for liquids and gases with Reynolds numbers above 10,000.

6.6.1 Flow sensor detecting small air bubbles for process control in manufacturing

Another optical-fiber flow sensor employed in manufacturing process control monitors a two-fluid mixture (Fig. 6.19). The sensor can distinguish between moving bubbles and liquid in the flow stream and display the void fraction, namely, the ratio of gas volume to the total volume. The principle of operation is quite simple. The light from the LED is guided by the optical fiber to the sensing element, in which the end portion of the fiber is mounted in a stainless steel needle of 2.8-mm outer diameter. When liquid is in contact with the end of the fiber, light enters the fluid efficiently and very little light is returned. However, when a gas bubble is present, a significant fraction of light is reflected back. With this technique, bubbles as small as 50 µm may be detected with an accuracy of better than 5 pecent and a response time of only 10 µs.

Potential applications of this flow sensor for the control of processes in manufacturing systems are widespread—for example, detection of

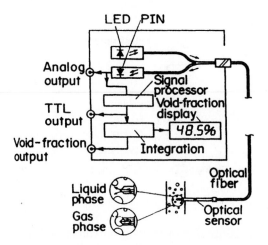

Figure 6.19 Flow sensor for two-phase mixtures.

gas plugs in production wells in the oil industry and detection of fermenters and distillers in the blood-processing and pharmaceutical industries.

An optical-fiber flow sensor for a two-phase mixture based on Y-guide probes is shown in Fig. 6.20. Two Y-guide probes are placed at different points along the flow stream to emit the input light and to pick up the retroreflected light from moving solid particles in the flow. The delay time between the signals of the two probes is determined by the average velocity of the moving particles. Therefore, measurement of the delay time by a conventional correlation technique provides the flow velocity. An accuracy of better than ±1 per-

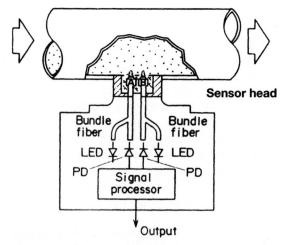

Figure 6.20 Flow sensor using two Y-guided probes based on a correlation technique.

cent and a dynamic range of 20:1 are obtained for flow velocities up to 10 m/s. A potential problem of such flow sensors for two-phase mixtures is poor long-term stability, because the optical fibers are inserted into the process fluid pipes.

6.6.2   Liquid level sensors in manufacturing process control for petroleum and chemical plants

Several optical-fiber liquid level sensors developed in recent years have been based on direct interaction between the light and liquid. The most common method in commercial products employs a prism attached to the ends of two single optical fibers (Fig. 6.21). The input light from an LED is totally internally reflected and returns to the output fiber when the prism is in air. However, when the prism is immersed in liquid, the light refracts into the fluid with low reflection, resulting in negligible returned light. Thus, this device works as a liquid level switch. The sensitivity of the sensor is determined by the contrast ratio, which depends on the refractive index of the liquid. Typical examples of signal output change for liquids with different refractive indices are indicated in Table 6.2.

The output loss stays at a constant value of 33 dB for refractive indices higher than 1.40. The signal output of a well-designed sensor can be switched for a change in liquid level of only 0.1 mm.

Problems to be solved for this sensor are dirt contamination on the prism surface and bubbles in the liquid. Enclosing the sensing ele-

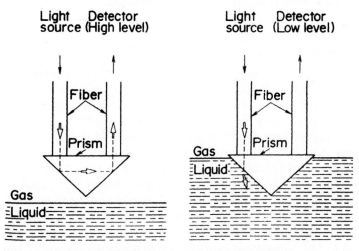

Figure 6.21 Principle of operation of a liquid level sensor with a prism attached to two optical fibers.

TABLE 6.2   Refractive Index versus Output

Refractive index, n	Loss change, dB
1.333	2.1
1.366	4.6
1.380	6.0
1.395	31.0

ment with a fine filter helps keep it clean and simultaneously reduces level fluctuations caused by bubbles. Since optical-fiber liquid level sensors have the advantages of low cost and electrical isolation, their use is widespread in petroleum and chemical plants, where the hazardous environment causes difficulties with conventional sensors. They are used, for example, to monitor storage tanks in a petroleum plant.

Another optical-fiber liquid level sensor, developed for the measurement of boiler-drum water level, employs a triangularly shaped gauge through which red and green light beams pass. The beams are deflected as it fills with water, so that the green light passes through an aperture. In the absence of water, only red light passes through. Optical fibers transmit red or green light from individual gauges to a plant control room located up to 150 m from the boiler drum (Fig. 6.22). The water level in the drum is displayed digitally.

This liquid level sensor operates at temperatures up to 170°C and pressures up to 3200 lb/in^2 gauge. Many sensor units are installed in the boiler drum, and most have been operating for 7 years. This sensor is maintenance-free, fail-safe, and highly reliable.

6.6.3 On-line measuring and monitoring of gas by spectroscopy

An optical spectrometer or optical filtering unit is often required for chemical sensors because the spectral characteristics of absorbed, fluorescent, or reflected light indicate the presence, absence, or precise concentration of a particular chemical species (Fig. 6.23).

Sensing of chemical parameters via fibers is usually done by monitoring changes in a suitably selected optical property—absorbance, reflectance, scattering (turbidity), or luminescence (fluorescence or phosphorescence), depending on the particular device. Changes in parameters such as the refractive index may also be employed for sensing purposes. The change in light intensity due to absorption is determined by the number of absorbing species in the optical path, and is related to the concentration C of the absorbing species by the Beer-Lambert relationship. This law describes an exponential reduc-

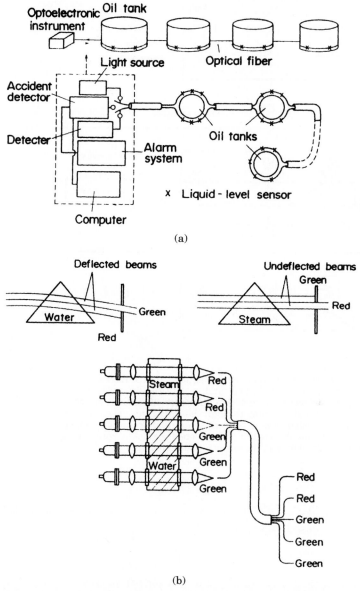

(a)

(b)

Figure 6.22 (*a*) Principle of operation of liquid level sensor for the measurement of boiler-drum water. (*b*) Liquid level sensor for the measurement of boiler-drum water with five-port sensors.

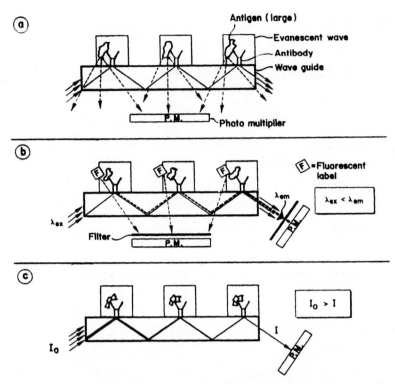

Figure 6.23 (*a*) Light scattering from the waveguide surface increases as the antigen-antibody complexes are formed, and is detected by a side-mounted photodetector. (*b*) Fluorescence is excited in a fluorescent marker, attached to an antigen molecule, by light passing through the waveguide. The fluorescent light can be collected either sideways from the guide, or as light which is retrapped by the guide and directed to a photodetector. (*c*) Absorption of light by antibody-antigen complexes on the surface attenuates light traveling down the waveguide.

tion of light intensity with distance (and also concentration) along the optical path. Expressed logarithmically,

$$A = \log I_0/I = \eta l C$$

where A is the optical absorbance, l is the path length of the light, η is the molar absorptivity, and I_0 and I are the incident and transmitted light, respectively. For absorption measurements via optical fibers, the medium normally must be optically transparent.

An accurate method for the detection of leakage of flammable gases such as methane (CH_4), propane (C_3H_8), and ethylene (C_2H_4) is vital in gas and petrochemical plants in order to avoid serous accidents. The recent introduction of low-loss fiber into spectroscopic measurements of these gases offers many advantages for process control in manufacturing:

1. Long-distance remote sensing

2. On-line measurement and monitoring

3. Low cost

4. High reliability

The most commonly used method at present is to carry the sample back to the measuring laboratory for analysis; alternatively, numerous spectrometers may be used at various points around the factory. The new advances in spectroscopic measurements allow even CH_4 to be observed at a distance of 10 km with a detection sensitivity as low as 5 percent of the lower explosion limit (LEL) concentration. The optical-fiber gas measuring system employs an absorption spectroscopy technique, with the light passing through a gas-detection cell for analysis. The overtone absorption bands of a number of flammable gases are located in the near-infrared range (Fig. 6.24).

The optical gas sensing system can deal with a maximum of 30 detection cells (Fig. 6.25). The species to be measured are CH_4, C_3H_8, and C_2H_4 molecules. Light from a halogen lamp (an infrared light source) is distributed into a bundle of 30 single optical fibers. Each of the distributed beams is transmitted through a 1-km length of fiber to a corresponding gas detection cell. The receiving unit is constructed of three optical switches, a rotating sector with four optical interference

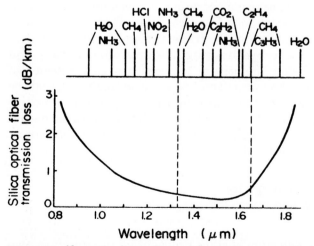

Figure 6.24 Absorption lines of typical flammable gases in the near-infrared and the transmission loss of silica fiber.

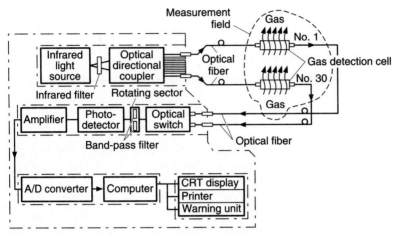

Figure 6.25 Gas detection system with 30 detection cells.

filters, and three Ge photodiodes. Each optical switch can select any 10 returned beams by specifying the number of the cell. The peak transmission wavelength of the optical filter incorporated in the sensor is 1.666 μm for CH_4, 1.690 μm for C_3H_8, 1.625 μm for C_2H_2, and 1.600 μm for a reference beam. After conversion to electrical signals, the signal amplitudes for the three gases are normalized by the reference amplitude. Then the concentration of each gas is obtained from a known absorption-concentration calibration curve stored in a computer.

An intrinsic distributed optical-fiber gas sensor for detecting the leakage of cryogenically stored gases such as CH_4, C_2H_4, and N_2 has also been developed. The sensor's operation is based on the temperature-dependent transmission loss of optical fiber. That is, the optical fiber is specially designed so that the transmission loss increases with decreasing temperature by choosing the appropriate core and cladding materials. Below the critical temperature, in the region of −55°C, most of the light has transferred to the cladding layer, and the light in the core is cut off. By connecting this temperature-sensitive fiber between a light source and a detector and monitoring the output light level, the loss of light resulting from a cryogenic liquid in contact with the fiber can be detected directly.

6.7 Crack Detection Sensors for Commercial, Military, and Space Industry Use

Accurate and precise detection of crack propagation in aircraft components is of vital interest for commercial and military aviation and the space industry. A system has been recently developed to detect cracks and crack propagation in aircraft components. This system uses optical fibers of small diameter (20 to 100 µm) which can be etched to increase their sensitivity. The fibers are placed on perforated adhesive foil to facilitate attachment to the desired component for testing. The fiber is in direct contact with the component (Fig. 6.26). The foil is removed after curing of the adhesive. Alternatively, in glass-fiber-reinforced plastic (GFRP) or carbon-fiber-reinforced plastic (CFRP), materials which are used more and more in aircraft design, the fiber can be easily inserted in the laminate without disturbing the normal fabrication process. For these applications, bare single fiber or prefabricated tape with integrated bundles of fibers is used. The system initially has been developed for fatigue testing of aircraft components such as frames, stringers, and rivets. In monitoring mode, the system is configured to automatically interrupt the fatigue test. The system has also been applied to the inspection of the steel rotor blades of a 2-MW wind turbine. A surveillance system has been developed for the centralized inspection of all critical components of the Airbus commercial jetliner during its lifetime. This fiber nervous system is

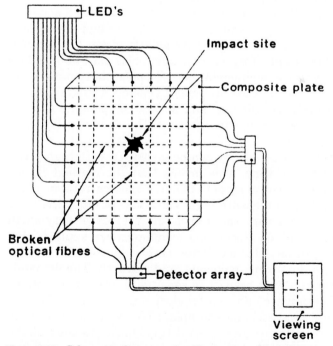

Figure 6.26 Schematic diagram of a fiber-optic system showing the location of impact damage in a composite structure.

designed for in-flight monitoring and currently is accessible to flight and maintenance personnel.

An optical-fiber mesh has been tested for a damage assessment system for a GFRP submarine sonar dome. Two sets of orthogonally oriented fibers are nested in the laminate during the fabrication process. When the fibers of the mesh are properly connected to LEDs and the detectors, the system can be configured to visualize the location of a damaged area.

As an alternative, a video camera and image processing are applied to determine the position of the damaged area. The fiber end faces at the detection side of the mesh are bundled and imaged into the camera tube. Two images are subtracted: the initial image before the occurrence of damage and the subsequent image. If fibers are broken, their location is highlighted as a result of this image subtraction.

6.8   Control of Input/Output Speed of Continuous Web Fabrication Using Laser Doppler Velocity Sensor

A laser Doppler velocimeter (LDV) can be configured to measure any desired component velocity, perpendicular or parallel to the direction of the optical axis. An LDV system has been constructed with a semiconductor laser and optical fibers and couplers to conduct the optical power. Frequency modulation of the semiconductor laser (or, alternatively, an external fiber-optic frequency modulator) is used to introduce an offset frequency. Some commercial laser Doppler velocimeters are available with optical-fiber leads and small sensing heads. However, these commercial systems still use bulk optical components such as acoustooptic modulators or rotating gratings to introduce the offset frequency.

With an LDV system the velocity can be measured with high precision in a short period of time. This means that the method can be applied for real-time measurements to monitor and control the velocity of objects as well as to measure their vibration. Because the laser light can be focused to a very small spot, the velocity of very small objects can be measured, or if scanning techniques are applied, high spatial resolution can be achieved. This method is used for various applications in manufacturing, medicine, and research. The demands on system performance with respect to sensitivity, measuring range, and temporal resolution are different for each of these applications.

In manufacturing processes, for example, LDV systems are used to control continuous roll milling of metal (Fig. 6.27), to control the rolling speed of paper and films, and to monitor fluid velocity and turbulence in mixing processes. Another industrial application is vibra-

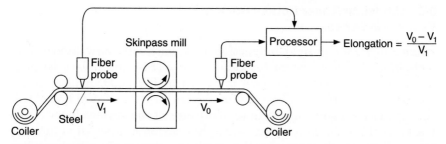

Figure 6.27 Fiber-optic laser Doppler velocimeter at a rolling mill controls pressure by measuring input speeds.

tion analysis. With a noncontact vibrometer, vibration of machines, machine tools, and other structures can be analyzed without disturbing the vibrational behavior of the structure.

Interestingly, the LDV system proved useful in the measurement of arterial blood velocity (Fig. 6.28), thereby providing valuable medical information. Another application in medical research is the study of motion of the tympanic membrane in the ear.

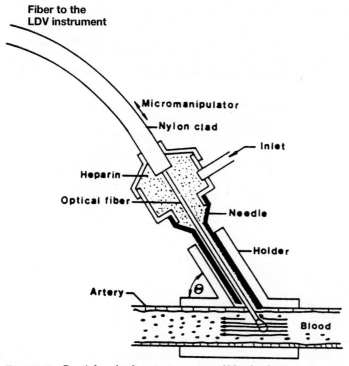

Figure 6.28 Special probe for measurement of blood velocity.

6.9   Ultrasonic/Laser Nondestructive Evaluation Sensor

Ultrasonic/laser optical inspection is a relatively new noncontact technique. A laser system for generating ultrasound pulses without distortion of the object surface is shown in Fig. 6.29. A laser pulse incident on a surface will be partly absorbed by the material and will thus generate a sudden rise in temperature in the surface layer of the material. This thermal shock causes expansion of a small volume at the surface, which generates thermoelastic strains. Bulk optical systems have been used previously to generate the laser pulse energy. However, the omnidirectionality of bulk sources is completely different from other well-known sources, and is regarded as a serious handicap to laser generation.

To control the beamwidth and beam direction of the optically generated ultrasonic waves, a fiber phased array has been developed. In this way the generated ultrasonic beam can be focused and directed to a particular inspection point below the surface of an object (Fig. 6.29). This system has been optimized for the detection of fatigue cracks at a rivet holes in aircraft structures.

The combination of laser-generated ultrasound and an optical-fiber interferometer for the detection of the resultant surface displacement has led to a technique which is useful for a wide variety of inspection tasks in manufacturing, including areas which are difficult to access and objects at high temperature, as well as more routine inspection and quality control in various industrial environments. Such a sys-

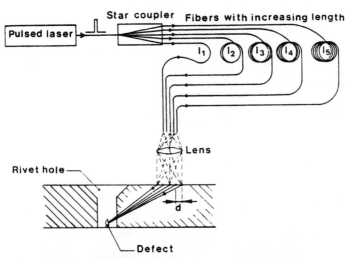

Figure 6.29 Setup for beam steering of laser-generated ultrasound by fiber-optic phased array.

tem can be applied to the measurement of thickness, velocity, flaws, defects, and grain size in a production process.

6.10 Process Control Sensor for Acceleration

The principle of operation of the process control acceleration sensor is illustrated in Fig. 6.30. The sensor element, consisting of a small cantilever and a photoluminescent material, is attached to the end of a single multimode fiber. The input light of wavelength λ_s is transmitted along the fiber from a near-infrared LED source to the sensor element. The sensor element returns light at two different wavelengths, one of which serves as a signal light and the other as a reference light, into the same fiber. The signal light at wavelength λ_s is generated by reflection from a small cantilever. Since the relative angle of the reflected light is changed by the acceleration, the returned light is intensity-modulated. The reference light of wavelength λ_r is generated by photoluminescence of a neodymium-doped glass element placed close to the sensor end of the fiber.

The optoelectronic detector module has two optical filters to separate the signals λ_s and λ_r and two photodiodes to convert the signal and the reference light into separate analog voltages. The signal processing for compensation is then merely a matter of electrical division. A measuring range of 0.1 to 700 m/s² and a resolution of 0.1 m/s² is obtained over the frequency range of 5 to 800 Hz.

6.11 An Endoscope as Image Transmission Sensor

An imaging cable consists of numerous optical fibers, typically 3000 to 100,000, each of which has a diameter of 10 µm and constitutes a

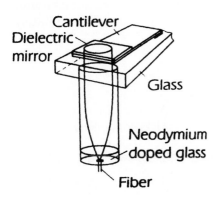

Figure 6.30 Cantilever-type acceleration sensor.

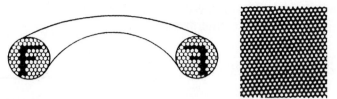

Figure 6.31 Image transmission through an image fiber.

picture element (pixel). The principle of image transmission through the fibers is shown in Fig. 6.31. The optical fibers are aligned regularly and identically at both ends of the fibers. When an image is projected on one end of the image fiber, it is split into multiple picture elements. The image is then transmitted as a group of light dots with different intensities and colors, and the original picture is reduced at the far end. The image fibers developed for industrial use are made of silica glass with low transmission loss over a wide wavelength band from visible to near infrared, and can therefore transmit images over distances in excess of 100 m without significant color changes. The basic structure of the practical optical-fiber image sensing system (endoscope) is illustrated in Fig. 6.32. It consists of the image fiber, an objective lens to project the image on one end, an eyepiece to magnify the received image on the other end, a fiber protection tube, and additional fibers for illumination of the object.

Many examples have been reported of the application of image fibers in process control. Image fibers are widely employed to observe the interior of blast furnaces and the burner flames of boilers, thereby facilitating supervisory control. Image fibers can operate at temperatures up to 1000°C, when provided with a cooling attachment for the objective lens and its associated equipment. Another important application of the image fiber bundles is observation, control, and inspection of nuclear power plants and their facilities. Conventional image fibers cannot be used within an ionizing radiation environment because ordinary glass becomes colored when exposed to radiation,

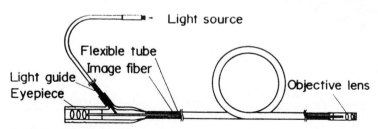

Figure 6.32 Basic structure of fiber scope.

causing increasing light transmission loss. A high-purity silica core fiber is well-known as a radiation-resistant fiber for nuclear applications.

The endoscope has demonstrated its vital importance in medical and biochemical fields such as:

1. Angioplasty
2. Laser surgery
3. Gastroscopy
4. Cystoscopy
5. Bronchoscopy
6. Cardioscopy

6.12   Sensor Network Architecture in Manufacturing

In fiber-optic sensor networks, the common technological base with communication is exploited by combining the signal generating ability of sensors and the signal transmitting capability of fiber optics. This combination needs to be realized by a suitable network topology in various manufacturing implementations. The basic topologies for sensor networking are illustrated in Fig. 6.33. The basic network topologies are classified into seven categories:

1. Linear array network with access-coupled reflective sensors (Fig. 6.33a).
2. Ring network with in-line transmissive sensors (Fig. 6.33b).
3. Star network with reflective sensors (Fig. 6.33c).
4. Star network with reflective sensors; one or more sensors can be replaced by a separate star network, in order to obtain a tree network (Fig. 6.33d).
5. Star network that can also be operated with transmissive sensors (Fig. 6.33e).
6. Ladder network with two star couplers. A star coupler is replaced by several access couplers, the number required being equal to the number of sensors (Fig. 6.33f).

Topological modifications, especially of sensor array and ladder network, may be desirable in order to incorporate reference paths of transmissive (dummy sensors) or reflective sensors (splices, open fiber end).

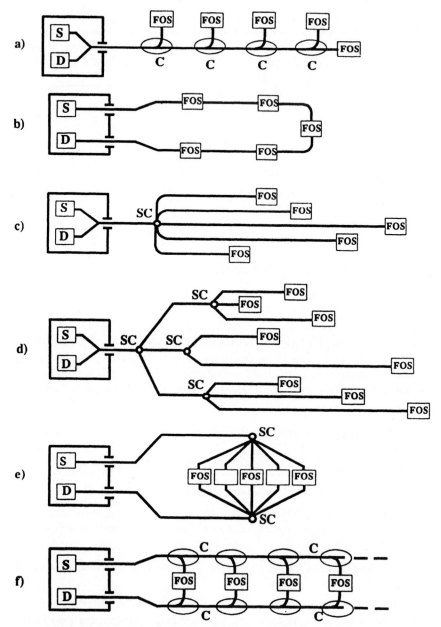

Figure 6.33 Basic network topologies: (*a*) linear array, (*b*) ring, (*c*) reflective star, (*d*) reflective tree, (*e*) transmissive star and (*f*), ladder network.

The transmit and return fibers, or fiber highway, generally share a common single-fiber path in networks using reflective sensors.

When a suitable fiber-optic network topology is required, various criteria must be considered:

1. The sensor type, encoding principle, and topology to be used

2. The proposed multiplexing scheme, required number of sensors, and power budget

3. The allowable cross-communication level

4. The system cost and complexity constraints

5. The reliability (i.e., the effect of component failure on system performance)

6.13   Power Line Fault-Detection System for Power Generation and Distribution Industry

In power distribution lines, faults such as short circuits, ground faults, and lightning strikes on the conductors must be detected in a very short time to prevent damage to equipment and power failure, and to enable quick repair. If the transmission line is divided in sections and a current or magnetic-field sensor is mounted in each section, a faulty section can be determined by detection of a change of the level and phase of the current on the power line. A system was developed as a hybrid optical approach to a fault-locating system that detects the phase and current difference between two current transformers on a composite fiber-optic ground wire (OPGW) wherein, due to induction, current is constantly passing (Fig. 6.34). The signal from

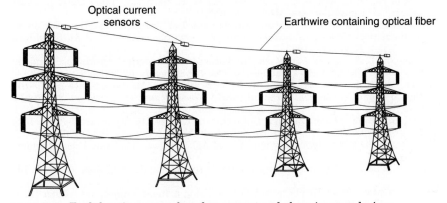

Figure 6.34   Fault locating system based on current and phase in ground wire.

a local electrical sensor, powered by solar cells and batteries, is transmitted over a conventional optical-fiber communication link. By three-wavelength multiplexing, three sensor signals can be transmitted over a single fiber. Seven sensors, three at each side of a substation and one at the substation itself, can monitor one substation on the power line, using one fiber in the OPGW.

Another system uses current transformers to pick up lightning current and thus detect lightning strikes. The signal is transmitted to a central detection point using the OPGW. Every sensor has its own OPGW fiber. This system is on a 273-kV power line in Japan.

The OPGW opens the possibility for using all kinds of sensors along the transmission line. These sensors may not only be for locating faults, but also for monitoring structural integrity. The use of optical time-domain reflectometry (OTDR) combined with passive intrinsic (distributed) sensors along the OPGW has future potential for providing a convenient and powerful monitoring method for power lines.

Further Reading

Bailey Control Systems, Wickliffe, Ohio.

Bartman, R. K., B. R. Youmans, and N. M. Nerheim. "Integrated Optics Implementation of a Fiber Optic Rotation Sensor: Analysis and Development," *Proc. SPIE*, **719**, 122–134.

Berthold, J. W., "Industrial Applications of Optical Fiber Sensors," Fiber Optic and Laser Sensors III, *Proc. SPIE*, **566**, 37–44.

Carrol, R., C. D. Coccoli, D. Cardelli, and G. T. Coate, "The Passive Resonator Fiber Optic Gyro and Comparison to the Interferometer Fiber Gyro," *Proc. SPIE*, **719**, 169–177 (1986).

Chappel, A. (ed.), *Optoelectronics—Theory and Practice*, McGraw-Hill, New York, 1978.

Crane, R. M., A. B. Macander, D. W. Taylor, and J. Gagorik, "Fiber Optics for a Damage Assessment System for Fiber Reinforced Plastic Composite Structures," *Rev. Progress in Quantitative NDE*, 2B, Plenum Press, New York, 1419–1430, 1982.

Doeblin, E. O., *Measurement Systems—Application and Design*, 4th ed., McGraw-Hill, New York, 1990.

Fields, J. N., C. K. Asawa, O. G. Ramer, and M. K. Barnoski, "Fiber Optic Pressure Sensor," *J. Acoust. Soc. Am.*, **67**, p. 816 (1980).

Finkelstein, L., and R. D. Watts, "Fundamental of Transducers—Description by Mathematical Models," *Handbook of Measurement Science*, vol. 2, P. H. Sydenham (ed.), Wiley, New York, 1983.

Friebele, E. L. and M. E. Gingerich, "Radiation-Induced Optical Absorption Bands in Low Loss Optical Fiber Waveguides," *J. Non-Crust. Solids*, **38**(39), 245–250 (1980).

Henze, M., "Fiber Optics Temperature and Vibration Measurements in Hostile Environments," Technical Material, *ASEA Research and Innovation*, CF23-1071E (1987).

Hofer, B., "Fiber Optic Damage Detection in Composite Structure," *Proc. 15th Congress. Int. Council Aeronautical Science*, ICAS-86-4.1.2, 135–143 (1986).

Kapany, N. S., *Fiber Optics, Principles and Applications*, Academic Press, London, 1976.

Lagakos, N., et al., "Multimode Optical Fiber Displacement Sensor," *Appl. Opt.*, **20**, 167 (1981).

Liu, K., "Optical Fiber Displacement Sensor Using a Diode Transceiver," Fiber Optic Sensors II, A. M. Sheggi (ed.), *Proc. SPIE*, **798**, 337–341 (1987).

Mizuno, Y., and T. Nagai, "Lighting Observation System on Aerial Power Transmission Lines by Long Wavelength Optical Transmission," *Applications of Fiber Optics in Electrical Power Systems in Japan,* C.E.R.L. Letterhead, paper 5.

Mori, S., et al., "Development of a Fault-Locating System Using OPGW," *Simitomo Electric Tech. Rev.* (25), 35–47.

Norton, H. N., *Sensors and Analyzer Handbook,* Prentice-Hall, Englewood Cliffs, N.J., 1982.

Neubert, H. K. P., *Instrument Transducers,* 2d ed., Clarendon Press, Oxford, 1975.

Ogeta, K., *Modern Control Engineering,* 2d ed. Prentice-Hall, Englewood Cliffs, N.J., 1990.

Petrie, G. R., K. W. Jones, and R. Joones, "Optical Fiber Sensors in Process Control," *4th Int. Conf. Optical Fiber Sensors, OFS'86,* Informal Workshop at Tsukuba Science City, VIII I–VIII 19, 1986.

Place, J. D., "A Fiber Optic Pressure Transducer Using A Wavelength Modulation Sensor," *Proc. Conf. Fiber Optics '85 (Sira),* London, 1985.

Ramakrishnan, S., L. Unger, and R. Kist, "Line Loss Independent Fiberoptic Displacement Sensor with Electrical Subcarrier Phase Encoding," *5th Int. Conf. Optical Fiber Sensors, OFS '88,* New Orleans, pp. 133–136, 1988.

Sandborn, V. A., *Resistance Temperature Transducers,* Metrology Press, Fort Collins, Colo., 1972.

Scruby, C. B., R. J. Dewhurst, D. A. Hutchins, and S. B. Palmer, "Laser Generation of Ultrasound in Metals," *Research Techniques in Nondestructive Testing,* vol. 15, R. S. Sharpe (ed.), Academic Press, London, 1982.

Tsumanuma, T., et al., "Picture Image Transmission-System by Fiberscope," *Fujikura Technical Review* (15), 1–10 (1986).

Vogel, J. A., and A. J. A. Bruinsma, "Contactless Ultrasonic Inspection with Fiber Optics," Conf. *Proc. 4th European Conf. Non-Destructive Testing,* Pergamon Press, London, 1987.

Yasahura, T., and W. J. Duncan, "An Intelligent Field Instrumentation System Employing Fiber Optic Transmission," *Advances in Instrumentation,* ISA, Wiley, London, 1985.

Sensors in
Flexible Manufacturing Systems

7.0 Introduction

Flexibility has become a key goal in manufacturing, hence the trend toward flexible manufacturing systems. These are designed to produce a variety of products from standard machinery with a minimum of workers. In the ultimate system, raw material in the form of bars, plates, and powder would be used to produce any assembly required without manual intervention in manufacture. Clearly, this is a good breeding ground for robots.

But it should be emphasized that the early FMSs are in fact direct numerical control (DNC) systems for machining. And it must be acknowledged that an NC machine tool is really a special-purpose robot. It manipulates a tool in much the same way as a robot handles a tool or welding gun. Then, with no more than a change in programming, it can produce a wide range of products. Moreover, the controllers for robots and NC machines are almost the same. But for an NC machine to be converted into a self-supporting flexible system, it needs some extra equipment, including a handling device. It then forms an important element in an FMS.

The principle of flexible manufacturing for machining operations is that the NC machining cells are equipped with sensors to monitor tool wear and tool breakage. Such cells are able to operate unmanned so long as they can be loaded by a robot or similar device, since the sensors will detect any fault and shut the operation down if necessary. The requirements can be summarized as:

1. CNC system with sufficient memory to store many different machining programs

2. Automatic handling at the machining tool either by robot or other material handling system

3. Workpieces stored near the machine to allow unmanned operation for several hours. A guided vehicle system may be employed if workpieces are placed at a designated storage and retrieval system away from the machine

4. Various sensors to monitor, locate, and/or diagnose any malfunction

7.1   The Role of Sensors in FMS

The monitoring sensor devices are generally situated at the location of the machining process, measuring workpiece surface textures, cutting-tool vibrations, contact temperature between cutting tool and workpiece, flow rate of cooling fluid, electrical current fluctuations, etc. Data in the normal operating parameters are stored in memory with data on acceptable manufacturing limits. As the tool wears, the tool changer is actuated. If the current rises significantly, along with other critical signals from sensors, a tool failure is indicated. Hence, the machine is stopped. Thus, with the combination of an NC machine, parts storage and retrieval, handling devices, and sensors, the unmanned cell becomes a reality. Since the control system of the NC machine, robot, and unmanned guided vehicles are similar, central computer control can be used effectively.

Systems based on these principles have been developed. In Japan, Fanauc makes numerical controllers, small NC and EDM machines, and robots; Murata makes robot trailers as well as a variety of machinery including automated textile equipment; and Yamazaki makes NC machines. In France, Renault and Citroen use FMS to machine gear boxes for commercial vehicles and prototype engine components respectively, while smaller systems have been set up in many other countries.

The central element in establishing an error-free production environment is the availability of suitable sensors in manufacturing. The following represents a summary of sensing requirements in manufacturing applications:

1. Part identification

2. Part presence or absence

3. Range of object for handling

4. Single-axis displacement of measurement

5. Two-dimensional location measurement

6. Three-dimensional location measurement

7.1.1 Current available sensor technology for FMS

The currently available sensors for manufacturing applications can be classified into four categories:

1. Vision sensors
   a. Photodetector
   b. Linear array
   c. TV camera
   d. Laser triangulation
   e. Laser optical time-domain reflectometry
   f. Optical fiber

2. Tactile sensors
   a. Probe
   b. Strain gauges
   c. Piezoelectric
   d. Carbon material
   e. Discrete arrays
   f. Integrated arrays

3. Acoustic sensors
   a. Ultrasonic detectors and emitters
   b. Ultrasonic arrays
   c. Microphones (voice control)

4. Passive sensors
   a. Infrared
   b. Magnetic proximity
   c. Ionizing radiation
   d. Microwave radar

Integrating vision sensors and robotics manipulators in flexible manufacturing systems presents a serious challenge in production. Locating sensors on the manipulator itself, or near the end effector, provides a satisfactory solution to the position of sensors within the FMS. Locating an image sensor above the work area of a robot may cause the manipulator to obscure its own work area. Measurement of the displacement of the end effector also may suffer distortion, since the destination will be measured in relative terms, not absolute. Placing the sensor on the end effector allows absolute measurement to be taken, reducing considerably the need for calibration of mechanical position and for imaging linearity. Image sensory feedback in this situation can be reduced to the simplicity of range finding in some applications.

Extensive research and development activities were conducted recently to find ways to integrate various sensors close to the gripper jaws of robots. The promise of solid-state arrays for this particular application has not entirely materialized, primarily because of diversion of effort resulting from the commercial incentives associated with the television industry. It might be accurate to predict that, over the next decade, imaging devices manufactured primarily for the television market will be both small and affordable enough to be useful for robotics applications. However, at present, array cameras are expensive and, while smaller than most thermionic tube cameras, are far too large to be installed in the region of a gripper. Most of the early prototype arrays of modest resolution (developed during the mid-1970s) have been abandoned.

Some researchers have attacked the problem of size reduction by using coherent fiber optics to retrieve the image from the gripper array, which imposes a cost penalty on the total system. This approach can, however, exploit a fundamental property of optical fiber in that a bundle of coherent fibers can be subdivided to allow a single high-resolution imaging device to be used to retrieve and combine a number of lower-resolution images from various paths of the work area including the gripper with each subdivided bundle associated with its own optical arrangement.

Linear arrays have been used for parts moving on a conveyer in such a way that mechanical motion is used to generate one axis of a two-dimensional image. The same technique can be applied to a robot manipulator by using the motion of the end effector to generate a two-dimensional image.

Tactile sensing is required in situations involving placement. Both active and passive compliant sensors have been successfully applied in the field. This is not the case for tactile array sensors because they are essentially discrete in design, are inevitably cumbersome, and have very low resolution.

Acoustic sensors, optical sensors, and laser sensors are well developed for effective use in manufacturing applications. Although laser range-finding sensors are well developed, they are significantly underused in FMS, especially in robotic applications. Laser probes placed at the end effector of an industrial robot will form a natural automated inspection system in manufacturing.

Sensing for robot applications does not depend on a relentless pursuit for devices with higher resolution; rather, the fundamental consideration is selecting the optimum resolution for the task to be executed. There is a tendency to assume that the higher the resolution, the greater the application range for the system. However, considerable success has been achieved with a resolution as low as 50×50 picture elements. With serial processing architectures, this resolution will generate sufficient gray-scale data to test the ingenuity of image processing algorithms. Should its processing time fall below 0.5 s, an algorithm can be used for robots associated with handling. However, in welding applications, the image processing time must be faster.

7.2   Robot Control through Vision Sensors

An increasing number of manufacturing processes rely on machine-vision sensors for automation. The tasks for which vision sensors are used vary widely in scope and difficulty. Robotic applications in which vision sensing has been used successfully include inspection, alignment, object identification, and character recognition.

Human vision involves transformation, analysis, and interpretation of images. Machine-vision sensing can be explained in terms of the same functions: image transformation, image analysis, and image interpretation.

7.2.1   Image transformation

Image transformation involves acquiring camera images and converting them to electrical signals that can be used by a vision computer (Fig. 7.1). After a camera image is transformed into an electronic (digitized) image, it can be analyzed to extract useful information in the image such as object edges, alignment, regions, boundaries, colors, and absence or presence of vital components.

Once the image is analyzed, the vision sensing system can interpret what the image represents so that the robot can continue its task. In robot vision execution, design considerations entail cost, speed, accuracy, and reliability.

7.2.2   Robot vision and human vision

Given that robot vision systems typically execute only part of what is normally called *seeing,* some similarities with human vision nevertheless arise. Other similarities arise in the "hardware" of human and robot visual systems.

The similarities in hardware could include an analogy between eyes and video cameras—both have lenses to focus an image on a sensitive

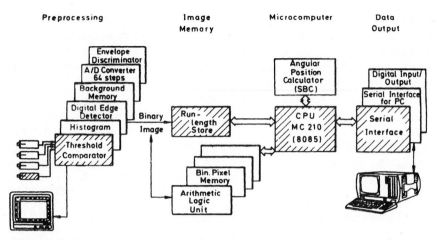

Figure 7.1 Image transformation involves the acquisition and conversion of camera images to electrical signals that can be used by a vision computer.

"retina" that produces a visual signal interpreted elsewhere. In both human and robot vision this signal is passed to a device that can remember important aspects of the image for a time, perform specialized image processing functions to extract important information from the raw visual signal, and analyze the image in a more general way.

Some similarities in performance follow. Human and robot vision work well only where lighting is good. Both can be confused by shadows, glare, and cryptic color patterns. Both combine size and distance judgments, tending to underestimate size when they underestimate distance, for instance.

However, humans and machines have far more differences than similarities. A human retina contains several million receptors, constantly sending visual signals to the brain. Even the more limited video camera gathers over 7 Mbytes of visual information per second. Many of the surprising aspects of machine vision arise from the need to reduce this massive flow of data so that it can be analyzed by a computer system.

Machine-vision systems normally only detect, identify, and locate objects, ignoring many of the other visual functions. However, they perform this restricted set of functions very well, locating and even measuring objects in a field of view more accurately than any human can.

7.2.3 Robot vision and visual tasks

Several standard visual tasks are performed by robot-vision systems. These tasks include recognizing when certain objects are in the field

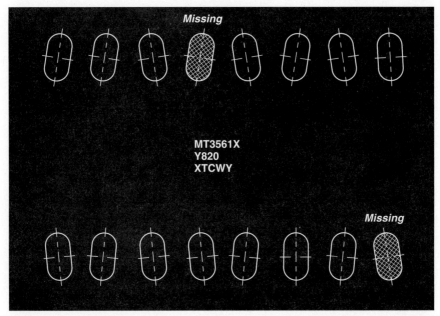

Figure 7.2 Image of a flaw.

of view, determining the location of visible objects, assisting a robot hand with pickup and placement, and inspecting known objects for the presence of certain characteristics (usually specific flaws in manufacture) (Fig. 7.2).

A robot-vision system must exercise some judgment in performing *visual tasks*—those for which the input is a visual image (normally obtained from an ordinary video camera). Which visual tasks are relatively easy for machine vision, and which are hard? The distinction is not so much that some tasks are hard and others easy; rather, it is the detail within a task that distinguishes easy problems from hard ones.

What makes a problem hard? Some of the contributing factors are

1. Objects that vary widely in detail. (Examining stamped or milled product may be easy, while molded or sculpted items may be more difficult. Natural objects are by far the hardest with which to deal.)

2. Lighting variations, including reflections and shadows, as well as fluctuations in brightness (as found in natural sunlight). These variations may go unnoticed by human inspectors, but they can make otherwise easy problems difficult or impossible for robot vision.

3. In general, ignoring "unimportant" variations in an image while responding to "significant" ones is very hard. (Most hard problems can be placed in this category.)

7.2.4 Robot visual sensing tasks

Robots will work in unpleasant locations. Health hazards are of no concern to them. Special safety equipment is not required for a robot spraying paint, welding, or handling radioactive materials or chemicals. All this adds up to reduced production costs. As the day progresses, the tired worker has a tendency to pay less attention to details, and the quality of the finished product may suffer. This is especially noticeable in automobiles where spray paint can run or sag and weld joints may not be made perfectly. The panels of the car may not be aligned, and the finished product may not operate properly, with predictable customer dissatisfaction. In pharmaceutical production, too, an operator inspecting and verifying lot number and expiration date may become fatigued and fail to inspect for sensitive information.

Robots, on the other hand, do not tire or change their work habits unless they are programmed do so. They maintain the same level of operation throughout the day. With vision-sensing systems and robots, it is possible for American manufacturers to compete against lower labor costs in foreign countries. The initial investment is the only problem. After the initial investment, the overall operation costs of the production line are reduced or held constant. Small educational robots can be used to retrain humans to operate and maintain robots.

The roles of robots with machine vision can be summarized as follows:

1. *Handling and assembly:* recognizing position/orientation of objects to be handled or assembled, determining presence or absence of parts, and detecting parts not meeting required specifications

2. *Part classification:* identifying objects and recognizing characters

3. *Inspection:* checking for assembly and processing, surface defects, and dimensions

4. *Fabrication:* making investment castings, grinding, deburring, water-jet cutting, assembling wire harness, gluing, sealing, puttying, drilling, fitting, and routing

5. *Welding:* automobiles, furniture, and steel structures

6. *Spray painting:* automobiles, furniture, and other objects

7.2.5 Robots utilizing vision systems to recognize objects

A basic use of vision is recognizing familiar objects. It is easy to see that this ability should be an important one for robot vision. It can be a task in itself, as in counting the number of each kind of bolt in a mixed lot on a conveyer belt. It can also be an adjunct to other tasks, for example, recognizing a particular object before trying to locate it precisely, or before inspecting it for defects.

It is important to note that this task actually has two distinct parts: first, object familiarization, that is learning what an object looks like, then object recognition (Fig. 7.3).

There are many ways of learning to recognize objects. Humans can learn from verbal descriptions of the objects, or they can be shown one or more typical items. A brief description of a *pencil* is enough to help someone identify many unfamiliar items as *pencil*. Shown a few *pencils,* humans can recognize different types of *pencils,* whether they look exactly like the samples or not.

Robot-vision systems are not so powerful, but both these approach-

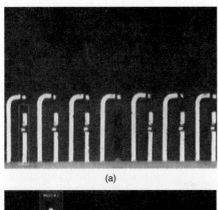

(a)

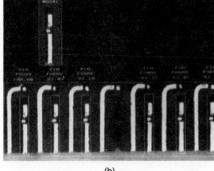

(b)

Figure 7.3 Recognition of a learned object.

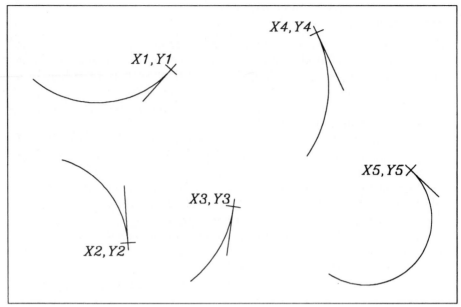

Figure 7.4 Machine guidance.

es to training still apply to them. Robots can be given descriptions of what they are to recognize, perhaps derived from CAD data to guide a machine tool (Fig. 7.4) or they can be shown samples, then be expected to recognize objects more or less like the samples.

Recognizing objects once they have been learned is the second, and more difficult, part of the task. Several basic questions arise in virtually every recognition task. Among them are "What are the choices?" and "What can change?"

Specifying the actual task completely is normally the hardest part of the application. When this has been accomplished, the basic question is, for what particular vision features should the search begin to recognize an object, and, when the features are found, how should they be analyzed to decide which object (if any) has been found?

7.3 Robot Vision Locating Position

Humans use several techniques for gauging distances, especially triangulation on the left- and right-eye views, feedback from the eye's focusing mechanism, and the apparent motion produced by small head movements. This kind of object location may make use of knowledge about the object being located. By knowing how large the object really is, one can judge how far away it is from the size of its retinal image.

Few robot-vision systems use binocular vision, autofocus feedback, or moving cameras to estimate distances. However, with a rigidly mounted camera, it is possible to interpret each visible point as lying on a particular line of sight from the camera. Accurate information about the true distances between visible points on an object allows the robot-vision system to accurately calculate its distance. Similarly, if an object is resting on a platform at an accurately known distance, the robot vision system can interpret distances in the camera image as accurate distances on the object.

However locations are determined, the visual location task for a robot-vision system normally includes calibration as well as object location. Calibration normally is carried out by providing the system with a view that has easily identifiable points at known spatial locations. Once calibrated, the system can then locate objects in its own coordinate system (pixels) and translate the position into work cell coordinates (inches, millimeters, etc.).

7.4   Robot Guidance with Vision System

Another use of machine vision is in robot guidance—helping a robot to handle and place parts and providing it with the visual configuration of an assembly after successive tasks. This can involve a series of identification and location tasks. The camera can be attached to a mobile arm, making the location task seem somewhat more like normal vision. However, the camera typically is mounted on a fixed location to reduce system complexity.

While each image can give the location of certain features with respect to the camera, this information must be combined with information about the current location and orientation of the camera to give an absolute location of the object. However, the ability to move the camera for a second look at an object allows unambiguous location of visible features by triangulation. Recognition is a useful tool for flexible manufacturing systems within a CIM environment. Any of several parts may be presented to a station where a vision system determines the type of part and its exact location. While it may be economical and simpler to send the robot a signal giving the part type when it arrives, the ability to detect what is actually present at the assembly point, not just what is supposed to be there, is of real value for totally flexible manufacturing.

7.4.1   Robot vision performing inspection tasks

Visual inspection can mean any of a wide variety of tasks, many of which can be successfully automated. Successful inspection tasks are those in which a small number of reliable visual cues (features) are to be checked and a relatively simple procedure is used to make the required evaluation from those cues.

The differences between human and robot capabilities are most evident in this kind of task, where the requirement is not simply to distinguish between good parts and anything else—a hard enough task—but usually to distinguish between good parts or parts with harmless blemishes, and bad parts. Nevertheless, when inspection can be done by robot vision, it can be done very predictably.

Many inspection tasks are well suited to robot vision. A robot-vision system can dependably determine the presence or absence of particular items in an assembly (Fig. 7.5), providing accurate information on each of them. It can quickly gauge the approximate area of each item passing before it, as long as the item appears somewhere in its field of view (Fig. 7.6).

Figure 7.5 Presence or absence of items in an assembly.

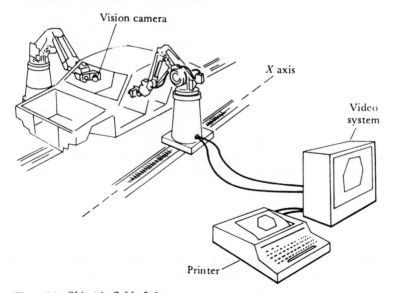

Figure 7.6 Object in field of view.

7.4.2 Components of robot vision

Figure 7.7 is a schematic diagram of the main components of a robot in a typical vision process for manufacturing. A fixed camera surveys a small, carefully lighted area where the objects to be located or inspected are placed. When visual information is needed (as signaled by some external switch or sensors), a digitizer in a robot vision system converts the camera image into a "snapshot": a significant array of integer brightness values (called gray levels). This array is sorted in a large random-access memory (RAM) array in the robot-vision system called an *image buffer* or a *frame buffer*. Once sorted, the image can be displayed on a monitor at any time. More importantly, the image can be analyzed or manipulated by a vision computer, which can be programmed to solve robot vision problems. The vision computer is often connected to a separate general-purpose (host) computer, which can be used to load programs or to perform tasks not directly related to vision.

Once an image is acquired, vision processing operations follow a systematic path. Portions of the image buffer may first be manipulated to suppress information that will not be valuable to the task at hand and to enhance the information that will be useful. Next, the vision program extracts a small number of cues from the image—perhaps allowing the region of interest to be reduced to exclude even more extraneous data.

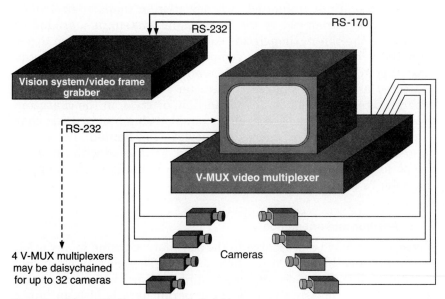

Figure 7.7 Schematic diagram of typical vision process.

At this stage, the vision program calculates, from the selected image region, the cues (features) of direct importance to the task at hand and makes a decision about the presence of a known part or its location in the field of view, or perhaps about the presence of specific defects in the object being inspected. Finally, the robot-vision system activates control lines based on the decisions, and (perhaps) transfers a summary of the conclusion to a data storage device or another computer.

7.5   End Effector Camera Sensor for Edge Detection and Extraction

A considerable amount of development of synchronized dual camera sensors at a strategic location on a robot end effector has been conducted for processing two-dimensional images stored as binary matrices. A large part of this work has been directed toward solving problems of character recognition. While many of these techniques are potentially useful in the present context, it is valuable to note some important differences between the requirements of character recognition and those associated with visual feedback for mechanical assembly.

7.5.1   Shape and size

All objects presented to the assembly machine are assumed to match an exact template of the reference object. The object may have an arbitrary geometric shape, and the number of possible different objects is essentially unlimited. Any deviation in shape or size, allowing for errors introduced by the visual input system, is a ground for rejection of the object (though this does not imply the intention to perform 100 percent inspection of components). The derived description must therefore contain all the shape and size information originally presented as a stored image. A character recognition system has in general to tolerate considerable distortion, or style, in the characters to be recognized, the most extreme example being handwritten characters. The basic set of characters, however, is limited. The closest approach to a template-matching situation is achieved with the use of a type font specially designed for machine reading, such as optical character recognition (Fig. 7.8).

7.5.2   Position and orientation

A component may be presented to the assembly machine in any orientation and any position in the field of view. Though a position- and orientation-invariant description is required in order to recognize the component, the measurement of these parameters is also an impor-

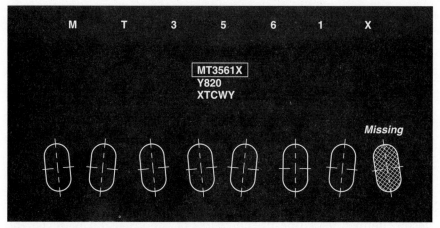

Figure 7.8 Optical character recognition.

tant function of the visual system to enable subsequent manipulation. While a line character may sometimes be skewed or bowed, individual characters are normally presented to the recognition system in a relatively constrained orientation, a measurement of which not required.

7.5.3 Multiple objects

It is a natural requirement that the visual system for an assembly machine should be able to accommodate a number of components randomly positioned in the field of view. The corresponding problem of segmentation in character recognition is eased (for printed characters) by *a priori* knowledge of character size and pitch. Such information has fostered techniques for the segmentation of touching characters. No attempt is made to distinguish between touching objects. Their combined image will be treated by the identification procedures as that of a single, supposedly unknown object.

The essentially unlimited sizes of the set of objects that must be accommodated by the recognition system demands that a detailed description of shapes be extracted for each image. There are, however, a number of basic parameters which may be derived from an arbitrary shape to provide valuable classification and position information. These include:

1. Area

2. Perimeter

3. Minimum enclosing rectangle

4. Center of the area

5. Minimum radius vector (length and direction)

6. Maximum radius vector (length and direction)

7. Holes (number, size, position)

Measurements of area and perimeter provide simple classification criteria which are both position- and orientation-invariant. The dimensionless shape factor area/perimeter has been used as a parameter in object recognition. The coordinates of the minimum enclosing rectangle provide some information about the size and shape of the object, but this information is orientation-dependent. The center of area is a point that may be readily determined for any object, independent of orientation, and is thus of considerable importance for recognition and location purposes. It provides the origin for the radius vector, defined as a line in the center of the area to a point on the edge of an object. The radius vectors of maximum and minimum length are potentially useful parameters for determining both identification and orientation. Holes are common features of engineering components, and the number present in a part is a further suitable parameter. The holes themselves may also be treated as objects, having shape, size, and position relative to the object in which they are found.

The requirements for the establishment of connectivity in the image and the derivation of detailed descriptions of arbitrary geometric shapes are most appropriately met by an edge-following technique. The technique starts with the location of an arbitrary point on the black/white edge of an object in the image (usually by a raster scan). An algorithm is then applied which locates successive connected points on the edge until the complete circumference has been traced and the starting point is reached. If the direction of each edge point relative to the previous point is recorded, a one-dimensional description of the object is built up which contains all the information present in the original shape. Such chains of directions have been extensively studied by Freeman. Measurements of area, perimeter, center of area, and enclosing rectangle may be produced while the edge is being traced, and the resulting edge description is in a form convenient for the calculation of radius vectors.

Edge following establishes connectivity for the object being traced. Continuing the raster scan in search of further objects in the stored image then presents the problem of the rediscovery of an already traced edge.

A computer plot of the contents of the frame with the camera viewing a square and a disk is illustrated in Fig. 7.9, and the result of applying the edge-extracting operation is illustrated in Fig. 7.10. The edge-following procedure may now be applied to the image in the

Figure 7.9 Computer plot of the contents of a frame.

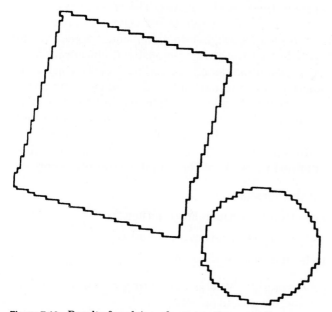

Figure 7.10 Result of applying edge extraction operation.

same way it was to the solid object. The procedure is arranged, however, to reset each edge point as it is traced. The tracing of a complete
object thus removes it from the frame and ensures that it will not be
subsequently retraced.

7.6   End Effector Camera Sensor Detecting Partially Visible Objects

A new method of locating partially visible two-dimensional objects has
been developed. The method is applicable to complex industrial parts
that may contain several occurrences of local features, such as holes
and corners. The matching process utilizes clusters of mutually consistent features to hypothesize objects and uses templates of the objects
to verify these hypotheses. The technique is fast because it concentrates on key features that are automatically selected on the basis on
the detailed analysis of CAD-type models of the objects. The automatic
analysis applies general-purpose routines for building and analyzing
representations of clusters of local features that could be used in procedures to select features for other locational strategies. These routines include algorithms to compute the rotational and mirror symme-
tries of objects in terms of their local features. The class of tasks that
involve the location of the partially visible object ranges from relatively easy tasks, such as locating a single two-dimensional object, to the
extremely difficult task of locating three-dimensional objects jumbled
together in a pallet. In two-dimensional tasks, the uncertainty is in
the location of an object in a plane parallel to the image plane of the
camera sensor. This restriction implies a simple one-to-one correspondence between sizes and orientations in the image, on the one hand,
and sizes and orientations in the plane of the object, on the other.

This class of two-dimensional tasks can be partitioned into four
subclasses that are defined in terms of the complexity of the scene:

1. A portion of one of the objects

2. Two or more objects that may touch one another

3. Two or more objects that may overlap one another

4. One or more objects that may be defective

This list is ordered roughly by the increasing amount of effort
required to recognize and locate the object.

Figure 7.11 illustrates a portion of an aircraft frame member. A
typical task might be to locate the pattern of holes for mounting purposes. Since only one frame member is visible at a time, each feature
appears at most once, which simplifies feature identification. If sever-

Figure 7.11   Portion of an aircraft frame member.

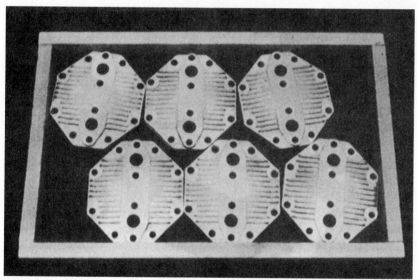

Figure 7.12   Objects touching each other.

al objects can be in view simultaneously and can touch one another, as in Fig. 7.12, the features may appear several times. Boundary features such as corners may not be recognizable, even though they are in the picture, because the objects are in mutual contact. If the objects can lie on one another (Fig. 7.13), even some of the internal holes may

Figure 7.13   Objects lying on top of each other.

be unrecognizable because they are partially or completely occluded. And, finally, if the objects are defective (Fig. 7.14), the features are even less predictable and hence harder to find.

Since global features are not computable from a partial view of an object, recognition systems for these more complex tasks are forced to work with either local features, such as small holes and corners, or extended features, such as a large segment of an object's boundary. Both types of feature, when found, provide constraints on the position and the orientations of their objects. Extended features are in general computationally more expensive to find, but they provide more information because they tend to be less ambiguous and more precisely located.

Given a description of an object in terms of its features, the time required to match this description with a set of observed features appears to increase exponentially with the number of features. The

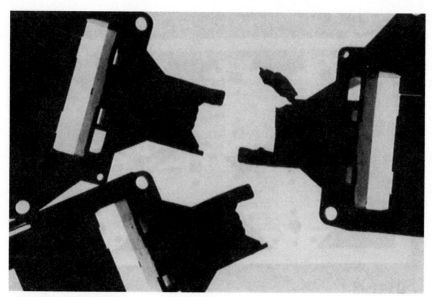

Figure 7.14 Trained image.

multiplicity of features precludes the straightforward application of any simple matching technique. Large numbers of features have been identified by locating a few extended features instead of many local ones. Even though it costs more to locate extended features, the reduction in the combinatorial explosion is often worth it. The other approach is to start by locating just one feature and use it to restrict the search area for nearby features. Concentrating on one feature may be risky, but the reduction in the total number of features to be considered is often worth it. Another approach is to sidestep the problem by hypothesizing massively parallel computers that can perform matching in linear time. Examples of these approaches include graph matching, relaxation, and histogram analysis. The advantage of these applications is that the decision is based on all the available information at hand.

The basic principle of the local-feature-focus (LFF) method is to find one feature of an image, referred to as the *focus feature,* and use it to predict a few nearby features to look for. After finding some nearby features, the program uses a graph-matching technique to identify the largest cluster of image features matching a cluster of object features. Since the list of possible object features has been reduced to those near the focus feature, the graph is relatively small and can be analyzed efficiently.

The key to the LFF method is an automatic feature-selection procedure that chooses the best focus features and the most useful sets of nearby features. This automatic-programming capability makes possible quick and inexpensive application of the LFF method to new objects. As illustrated in Fig. 7.15, the training process, which includes the selection of features, is performed once and the results are used repeatedly.

7.6.1 Run-time phase

The run-time phase of the LFF acquires images of partially visible objects and determines their identities, positions, and orientations. This processing occurs in four steps:

1. Reading task information

2. Locating local features

3. Hypothesizing objects

4. Verifying hypotheses

The procedure (Fig. 7.15) is to input the object model together with the list of focus features and their nearby cofeatures. Then, for each image, the system locates all potentially useful local features, forms

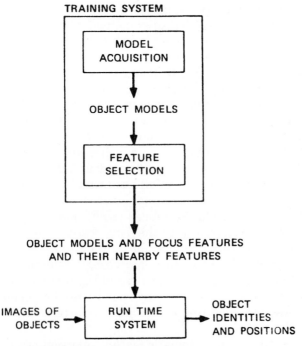

Figure 7.15 Run-time phase procedure.

clusters of them to hypothesize object occurrences, and finally performs template matching to verify these hypostheses.

7.7   Ultrasonic End Effector

An end effector on a welding robot (Fig. 7.16) contains an ultrasonic sensor for inspection of the weld. An ultrasonic sensor detects such flaws as

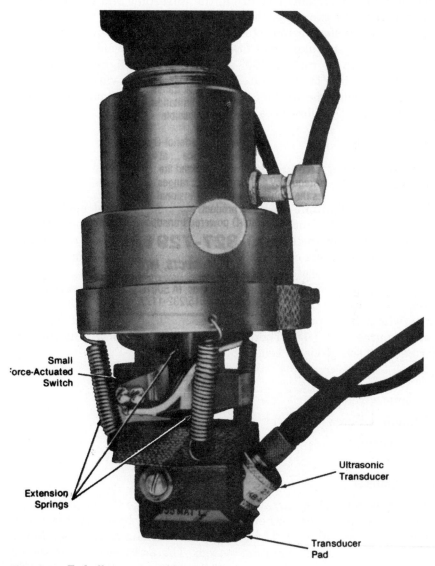

Figure 7.16   End effector on a welding robot.

tungsten inclusions and lack of penetration of weld. The end effector determines the quality of a weld immediately after the weld contact has been made, while the workpiece is still mounted on the weld apparatus; a weld can be reworked in place, if necessary. The delay caused by the paperwork and setup involved in returning the workpiece for rework is thereby avoided.

The ultrasonic end effector can be mounted on any standard gas tungsten arc welding torch. It may also be equipped with a through-the-torch vision system. The size of the ultrasonic end effector is the same as that of a gas cup with cathode.

A set of extension springs stabilizes the sensor and ensures that its elastomeric dry-couplant pad fits squarely in the weldment surface. The sensor can be rotated 360° and locked into alignment with the weld lead. A small force-actuated switch halts downward travel of the robot arm toward the workpiece and sets the force of contact between the sensor and the workpiece.

7.8   End Effector Sound-Vision Recognition Sensor

The sound recognition sensor consists of a source that emits sound waves to an object and a sound receiver that receives the reflected sound waves from the same object (Fig. 7.17). The sound recognition sensor array consists of one sound source and one to as many as 16 receivers fitted intricately on an end effector of a robot.

The sound-vision recognition sensor array measures reflections from some surface of interest on the object, called the *measured surface,* which is perpendicular to the sound waves emitted from the sound source (Fig. 7.18). There are four conditions governing the performance of sound-vision sensors:

1. Standoff

2. Large surfaces

3. Small surfaces

4. Positioning

7.8.1   Standoff

Standoff is how far the array must be located from the measured surface. The standoff, like other measurements, is based on the wavelength of the sound used. Three different wavelengths are used in the sound-vision sensor recognition system. The array standoff d should be one or two wavelengths λ_s from the measured surface for the high-

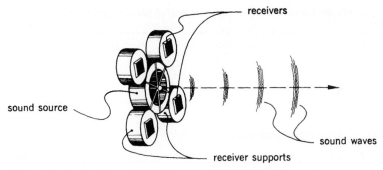

Figure 7.17 End effector sound-vision recognition system.

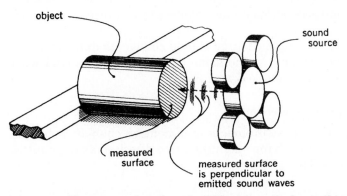

Figure 7.18 The measured surface should be perpendicular to the sound waves emitted from the sound source.

est accuracy. The standoff can be as great as 12 wavelengths, albeit with reduced accuracy (Fig. 7.19).

$$1.5\,\lambda_s \le d \le 12\,\lambda_s \tag{7.1}$$

where λ_s is the sound wavelength and d is the standoff distance. The typical standoff distance in terms of frequency and wavelength is described in Table 7.1.

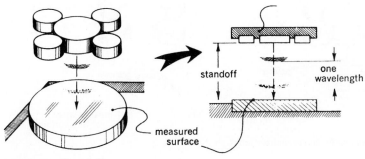

Figure 7.19 Standoff.

TABLE 7.1   Correlation Functions of Typical Standoff *d,* Wavelength λ_s, and Frequency *f*

Frequency, kHz	Wavelength, mm	Standoff distance, mm
20	17	25
40	8	12
80	4	6

7.8.2   Large surface measurements

Large surface measurements achieve more accuracy than those made on small surfaces (Fig. 7.20). Whether a surface is large or small, the accuracy depends on the wavelength of the sound waves selected. A "large" surface must be at least one wavelength distance on each side (Table 7.2).

The large surface being measured should change its dimension perpendicular to the surface, in the same direction as the sound wave emitted from the sound source (Fig. 7.21).

7.8.3   Sensitivity of measurements

The sensitivity of the measurements to dimension changes is 5 or 10 times greater when the change is in the same direction as the emitted sound wave (Fig. 7.22).

7.8.4   Small surfaces

Small surfaces can be measured as long as the robot end effector carrying the sensor array directs the sound wave from the side of the object (Fig. 7.23). Small surfaces are either a small portion of a large object or simply the surface of a small object (Fig. 7.24).

For small surfaces, the sound waves diffract or wrap around the surface rather than diverge from it as in a large surface. Similarly, ocean waves diffract around a rock that is smaller than the distance

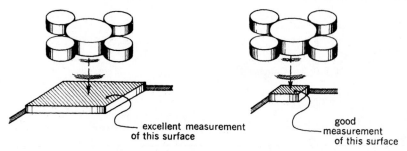

Figure 7.20 Large surface measurements.

TABLE 7.2 **Correlation of Minimum Size of Large Surface, Frequency, and Wavelength**

Frequency f, kHz	20	40	80
Wavelength λ_s, mm	17	8	4
Minimum area of surface, mm²	275	70	20

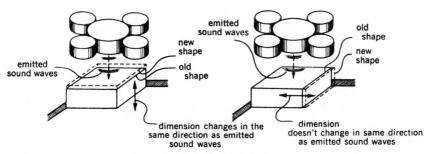

Figure 7.21 The large surface being measured should change its dimension perpendicular to the surface.

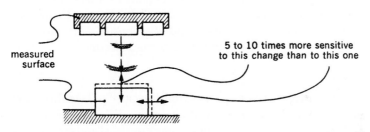

Figure 7.22 Sensitivity to dimension changes.

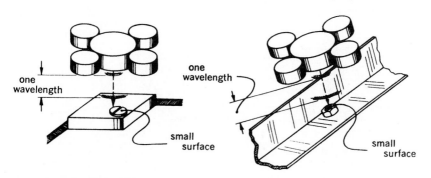

Figure 7.23 Small surfaces.

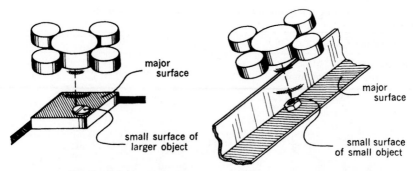

Figure 7.24 Small surfaces are either a small portion of a large object or simply the surface of a small object.

between crests (Fig. 7.25). Because of diffraction, the sound-vision recognition system is sensitive to the volume of the shape change on a small surface (Fig. 7.26).

Volume changes can be positive or negative. Small objects or protrusion from a surface represent positive volume changes, while holes or cavities represent negative volume changes (Fig. 7.27).

Measurement accuracy for small surfaces depends on the change in volume of the surface being measured. Listed in Table 7.3 are the smallest volume change that can be detected and the approximate size of a cube that has that volume for representative acoustic frequencies.

7.8.5 Positioning

The sound-vision sensor array system compares a particular part with a reference or standard part and detects the difference between the two. Objects being inspected—either a part or its standard—must be located relative to the array with at least the same accuracy as the

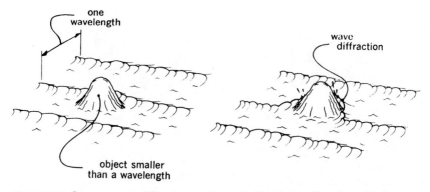

Figure 7.25 Ocean waves diffract around a rock that is smaller than the distance between crests.

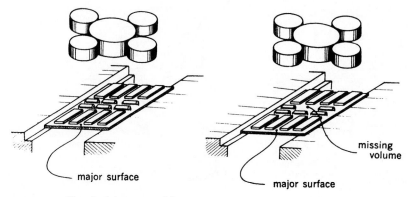

Figure 7.26   Sound-vision recognition system.

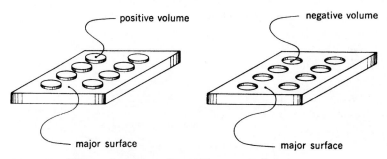

Figure 7.27   Volume changes can be positive or negative.

TABLE 7.3   The Least Measurable Volume

Frequency f, kHz,	20 kHz	40 kHz	80 kHz
Smallest detectable volume change, μm^3	5×10^{-4}	6×10^{-5}	8×10^{-6}
Smallest detectable cube, mm^3	12×10^2	6×10^2	3×10^2

expected measurement. Rectangular parts are usually located against stops; rotational parts are usually located in V-blocks on the face of the end effector (Fig. 7.28).

Rotationally unsymmetric parts must be oriented the same way as the standard in order to be compared. The end effector sensor array system can be used to direct a stepper motor to rotate the part until a match between part and standard is found (Fig. 7.29).

7.9   End Effector Linear Variable-Displacement Transformer Sensor

Sensing capability in a robot can have widely ranging degrees of sophistication in addition to a variety of sensing media. For instance,

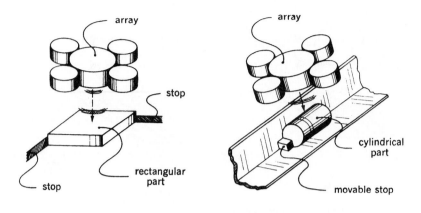

Figure 7.28 Diameter and height measurement.

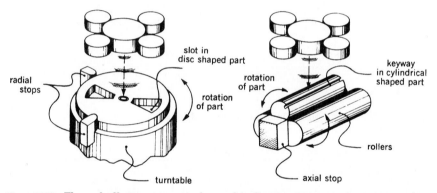

Figure 7.29 The end effector sensor can be used to direct a stepper motor to rotate the part until a match between part and standard is found.

sensing capability can vary from a simple photoelectric cell to a complex, three-dimensional sound-vision system as described in the previous section.

The linear variable-displacement transformer (LVDT) sensor is an electromechanical device that can be attached to a robotic manipulator or can be itself a drive control for a robotic gripper. The LVDT produces an electrical output proportional to the displacement of a separate movable core. It consists of a primary coil and two secondary coils, intricately spaced on a cylindrical form. A free-moving rod-shaped magnetic core inside the coil assembly provides a path for the magnetic flux linking the coils. A cross section of the LVDT sensor and a plot of its operational characteristics are in Figs. 7.30 and 7.31.

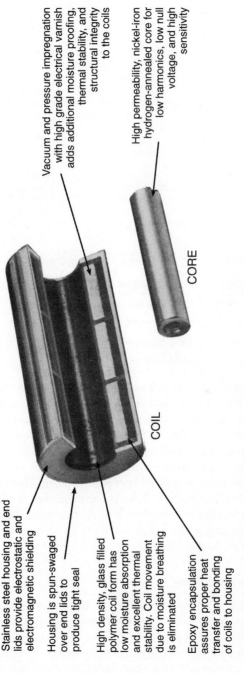

Stainless steel housing and end lids provide electrostatic and electromagnetic shielding

Housing is spun-swaged over end lids to produce tight seal

High density, glass filled polymer coil form has low moisture absorption and excellent thermal stability. Coil movement due to moisture breathing is eliminated

Epoxy encapsulation assures proper heat transfer and bonding of coils to housing

Vacuum and pressure impregnation with high grade electrical varnish adds additional moisture proofing, thermal stability, and structural integrity to the coils

High permeability, nickel-iron hydrogen-annealed core for low harmonics, low null voltage, and high sensitivity

COIL

CORE

Figure 7.30 Cross-section of linear variable displacement transducer sensor.

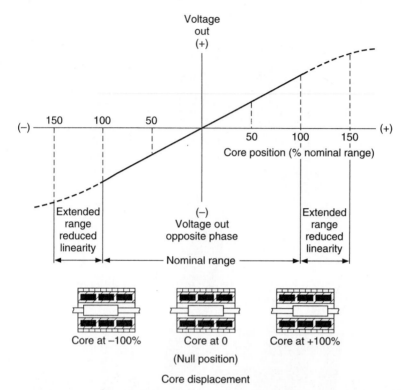

Figure 7.31 Plot of LVDT operational characteristics.

When the primary coil is energized by an external ac source, voltages are induced in the two secondary coils. These are connected series-opposing so that the two voltages are of opposite polarity. Therefore, the net output of the sensor is the difference between these voltages, which is zero when the core is at the center or null position. When the core is moved from the null position, the induced voltage in the coil toward which the core is moved increases, while the induced voltage in the opposite coil decreases. This action induces a differential voltage output that varies linearly with core position. The phase of this output voltage changes abruptly by 180° as the core is moved from one side of the null to the other.

7.9.1   Extreme environments

With increasingly sophisticated technology, more and more instrumentation applications have arisen for sensors capable of operating in such hostile environments as extremely cold temperatures, very high temperatures, and/or intense nuclear radiation. The LVDT has been developed with materials that can tolerate extreme environmental requirements. Although the operating environments vary greatly, these LVDT designs use similar materials of construction and share the same physical configurations.

Currently, these LVDT sensors are built entirely with inorganic materials. The coil form is made of dimensionally stable, fired ceramic wound with ceramic-insulated high-conductivity magnet wire specially formulated for the application. Joints between the windings and lead wires are brazed or welded for mechanical reliability and electrical continuity. Ceramic cements and fillers are chosen to optimize heat transfer and bonding between windings, coil form, and housing. The potted assembly is cured at elevated temperatures, fusing the components together into a solidified structure.

Most inorganic insulations tend to be hygroscopic by nature, so the cured coil assembly is encased in an evacuated stainless steel shell that is hermetically sealed by electron beam (EB) welding. This evacuation and sealing process prevents moisture accumulation and subsequent loss of the insulation's dielectric strength. It also seals out surrounding media from the windings, while permitting the core to move freely.

Electrical connections are made to the windings with nickel conductors mutually insulated from each other by a magnesium oxide filler and sheathed in a length of stainless-steel tubing. This cable assembly can be terminated by a hermetically sealed header for a connector when the application requires it.

The preceding description gives a brief insight into the material and techniques currently used in constructing the sensor for extremely severe environments. However, the state of the art in materials technology is being continually advanced. As new materials and methods of construction are evaluated, tested, and proved to upgrade performance, they will be incorporated into these sensors.

7.9.2 Cryogenic manufacturing applications

An LVDT sensor connected to the gripper of a robot is designed to cover a wide range of cryogenic applications ranging from general scientific research to space vehicle analysis and cryogenic medicine. A significant feature of the LVDT sensor is its ability to withstand repeated temperature cycling from room ambient conditions to the liquefaction temperatures of atmospheric gases such as nitrogen and oxygen. In order to survive such rigorous temperature changes, the sensor is constructed of materials selected for compatible coefficients of expansion while maintaining good electrical and magnetic properties even at −450°F (−270°C). The evacuated and hermetically sealed stainless-steel case prevents damage that could otherwise result from repeated condensation, freezing, and revaporization. Internal magnetic and electrostatic shielding renders the sensor insensitive to external magnetic and electrical influences.

7.9.3 Measurement at high temperatures in manufacturing

The LVDT sensor has been developed for measurements involving very high temperatures. It is capable of operating continuously at 1100°F (600°C) and surviving temperatures as high as 1200°F (650°C) for several hours. Typical uses include position feedback from jet engine controls located close to exhaust gases and measurement of roller position and material thickness in hot strip or slabbing mills. In scientific research it can be used to directly measure dimensional changes in heated test specimens without requiring thermal isolation, which could induce measurement errors. The sensor is the culmination of the development of sophisticated construction techniques coupled with careful selection of materials that can survive sustained operation at high temperatures. Because magnetic properties of a metal vanish above its magnetic transformation temperature (Curie point), the core material must be made from one of the few magnetic materials having Curie temperatures above 1100°F (600°C). Another problem is that, at high temperature, the resistance of windings made of common magnet wire materials increases so much that an LVDT sensor using ordinary conductor materials would become virtually useless. Thus, the winding uses a wire of specially formulated high-conductivity alloy. The sensors are made with careful attention to internal mechanical configuration and with materials having compatible coefficients of expansion to minimize null shifts due to unequal expansion or unsymmetrical construction. Hermetic sealing allows the sensor to be subjected to hostile environments such as fluid pressure up to 2500 psi (175 bars) at 650°F (350°C). Units can be factory calibrated in a special autoclave that permits operation at high temperature while they are hydrostatically pressurized.

7.10 Robot Control through Sensors

In order to pick up an object, a robot must be able to sense the strength of the object being gripped so as not to crush the object. Accordingly, the robot gripper is equipped with sensing devices to regulate the amount of pressure applied to the object being retrieved.

Several industrial sensing devices enable the robot to place objects at desired locations or perform various manufacturing processes:

1. *Transducers:* sensors that convert nonelectrical signals into electrical energy

2. *Contact sensors (limit switches):* switches designed to be turned ON or OFF by an object exerting pressure on a lever or roller that operates the switch

3. *Noncontact sensors:* devices that sense through changes in pressure, temperature, or electromagnetic field

4. *Proximity sensors:* devices that sense the presence of a nearby object by inductance, capacitance, light reflection, or eddy currents

5. *Range sensors:* devices such as laser-interferometric gauges that provide a precise distance measurement

6. *Tactile sensors:* devices that rely on touch to detect the presence of an object; strain gauges can be used as tactile sensors

7. *Displacement sensors:* provide the exact location of a gripper or manipulator. Resistive sensors are often used—usually wire-wound resistors with a slider contact; as force is applied to the slider arm, it changes the circuit resistance

8. *Speed sensors:* devices such as tachometers that detect the motor shaft speed

9. *Torque sensors:* measure the turning effort required to rotate a mass through an angle

10. *Vision sensors:* enable a robot to see an object and generate adjustments suitable for object manipulation; include dissectors, flying-spot scanners, vidicons, orthicons, plumbicons, and charge-coupled devices

7.11 Multisensor-Controlled Robot Assembly

Most assembly tasks are based on experience and are achieved manually. Only when high volume permits are special-purpose machines used. Products manufactured in low-volume batches or with a short design life can be assembled profitably only by general-purpose flexible assembly systems which are adaptable and programmable. A computer-controlled multisensor assembly station responds to these demands for general-purpose assembly. The multisensor feedback provides the information by which a robot arm can adapt easily to different parts and accommodate relative position errors.

The assembly task may be viewed as an intended sequence of elementary operations able to accept an originally disordered and disorganized set of parts and to increase gradually their order and mating degrees to arrive finally at the organization level required by the definition of the assembly. In these terms it may be considered that assembly tasks perform two main functions: (1) ordering of parts and (2) mating of parts in the final assembly.

The ordering function reduces the uncertainty about the assembly parts by supplying information about their parameters, type, and position/orientation. The performance criterion of this function may be expressed as an entropy. The final goal of ordering is to minimize the relative entropy sum of the components of the assembly. In an assembly station the ordering is performed either in a passive way by sensors or in an active way by mechanical means such as containerization, feeding, fixturing, or gripping.

The part mating function imposes successive modifications of the positions of the parts in such a way as to mate them finally in the required assembly pattern. The modification of the position of the parts is carried out by manipulating them by transport systems and robot arms. The part mating function requires *a priori* information about the parts to be manipulated and the present state of the assembly. During this manipulation, the part entropy may increase because of errors in the robot position or by accidental changes in the manipulated parts.

An example of a multisensor system is a testbed for research on sensor control of robotic assembly and inspection, particularly comparisons of passive entropy reduction to active mechanical means. Programs for the system must be easy to to develop and must perform in real time. Commercially available systems and sensors had to be used as much as possible. The system had to be very flexible in interfacing with different subsystems and their protocols. The flexible assembly system (Fig. 7.32) consists of three main parts:

1. Robot equipped with a transport system

2. Vision sensors, force-torque sensors, and ultrasound sensors

3. Control system

The system was built with a variable transport system (VTS) (Fig. 7.33). It is a modular system in which product carriers are transported to the system. Two docking stations which can clamp the product carriers are included in the system. Some of the product carriers have translucent windows to allow backlighting of the vision system. The VTS system has its own controller operated by parallel input/output lines.

The robot is equipped with six degrees of freedom and a payload capacity of 4 kg. The absolute accuracy is 0.2 mm. The robot con-

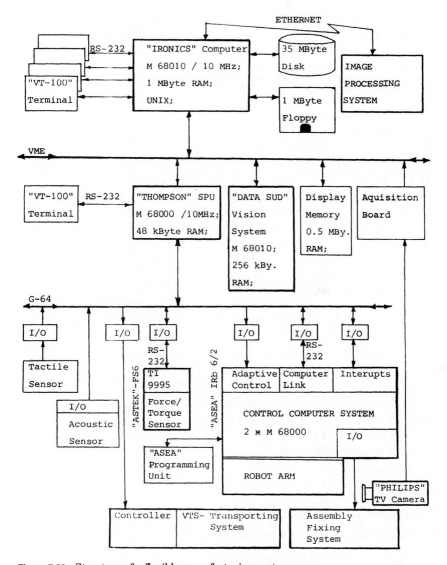

Figure 7.32 Structure of a flexible manufacturing system.

troller consists of two 68000 microprocessors. The standard way of programming the robot is through a teach pendant.

There are three different interfaces between the robot and the control system: (1) computer link at 9600 baud (RS-232), (2) adaptive control, and (3) parallel interrupt lines.

The computer link lets the control computer give commands to the robot such as:

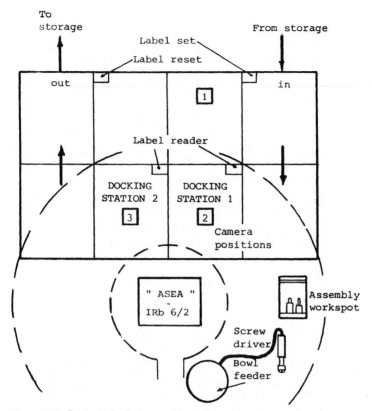

Figure 7.33   Layout of robot assembly and variable transport system.

1. Load programs from disk
2. Start programs
3. Set position registers
4. Read out robot positions
5. Take direct control of robot movement
6. Up- and download robot programs

The adaptive control offers search and contour-following modes. Up to three sensor signals can be used to guide the movements of the robot. The direction in which the robot is to move must be given for each signal. The displacement or velocity is proportional to the signal value generated by the displacement sensor. Both digital and analog signals produced by sensors are accepted. Digital signals may consist of 1 to 8 bits.

The interrupt lines are used to start or to stop a robot movement and to synchronize tasks.

7.11.1 Control computer

The computer used to control the system is microprocessor-based. The main processor unit (MPU) is a VME-bus-compatible single-board computer featuring a 68010 processor, 1-Mbyte dual-ported RAM, a 35-Mbyte Winchester disk, and a 1-Mbyte floppy disk. The system runs under the UNIX V operating system.

Real-time performance is obtained by two special processing units (SPUs). The first SPU realizes the control and interfacing of the robot and the interfacing of the sensors. It links the VME bus to the G64 bus to which the robot, the VTS, the torque-force sensor, and the ultrasound sensor are interfaced. This processor controls the computer link and the adaptive control of the robot. The processor board consists of a 68000 processor and 48-kbit RAM. The second SPU is the vision processing board. This board communicates with the vision boards and the other SPU via the VME bus. It consists of a 68010 processor, 256-kbit RAM, and an NS16081 floating-point processor.

7.11.2 Vision sensor modules

The vision module consists of two Primagraphics VME boards. One board contains the frame image of 768×576 pixels and 128 gray values. A frame-transfer solid-state camera is connected to the system.

Force sensing is available and can measure three force and three torque components. Forces up to 200 N are measured with a resolution of 0.1 N. Torques up to 4 N•m are measured with a resolution of 0.002 N•m. The data rate ranges from 1.8 to 240 Hz. The torque-force sensor is controlled by a Texas Instruments 9995 microprocessor. Besides a high-speed RS-232 link 38400 board, an analog output is available.

An ultrasonic range finder sensor has been interfaced to the G64 bus. A tactile array based on the piezoelectric material PVDF has been developed. A stereo vision system is incorporated to the system.

7.11.3 Software structure

Because an important aim of the testbed is to investigate the use of sensors in an assembly process, it is essential that an optimal environment is present for program development and real-world testing. A user-friendly program environment is provided by the UNIX V operating system. Real-time operation is provided by the special processing units. Programs developed under the UNIX system can be downloaded to the SPUs via the VME bus.

The assembly process consists of a sequence of assembly stages. A succeeding assembly stage is entered when certain entrance conditions are met. This means that the assembly process can be controlled by a state machine. An assembly stage may consist of a number of states. With this approach, a modular and hierarchical control structure is obtained in which certain processes are executed at the lowest level.

The control software for the assembly task consists of three different levels: (1) state machine level, (2) executor level, and (3) driver level.

At the highest level a command interpreter activates certain states of the machine. The state machine activates another (nested) state machine or a process (executor). Executors may run in parallel. For instance, the vision system may analyze an image of a product while the robot is assembling. The synchronization of the processes is obtained from the required entrance conditions for the next state of the state machine. In general, this level requires a moderately fast response and may run under the UNIX system.

At the second level, the executors are activated by the state machine. An executor performs a certain subtask and requires a realtime response. Examples are robot movements and image analysis routines. The executors reside in the SPUs and in the robot controller (robot programs). An executor is the only program that communicates directly with a driver.

The drivers for the different subsystems form the third level and also reside in the SPUs. The drivers take care of the protocol conversion and of the error checking in the communication with the subsystems.

During the assembly task a number of processes (executors) are running, requiring synchronization of data transfer. The data may consist of, for instance, the type, position, and orientation of a detected part by the vision system. The communication between the processes is realized through a shared memory lock in the VME bus.

7.11.4 Vision sensor software

The first step consists of a segmentation of gray-value image into a binary image. Because backlighting is used in the system, contrast is excellent and simple thresholding is sufficient for segmentation. During initialization, a threshold is calculated from the image histogram and used until the next initialization takes place. Before the system starts an assembly operation, the vision system must be calibrated to the robot coordinate system. This is done by moving a ring in the vision field. From the known robot coordinates, the ring position, and the computed ring coordinates in the vision system, the necessary transformation is calculated.

A drawback of connectivity analysis is that objects may not touch. The vision package may be extended with graph-matching techniques to allow recognition of touching and overlapping parts.

7.11.5   Robot programming

The robot is programmed in the teach-in mode with a standard pendant. Different subprograms, which are called from the control computer, are thus realized. The robot has 100 registers in which positions and orientations can be stored by a control computer. In this way position and orientation determined by the vision module can be transferred to the robot controller. The search and contour tracing options of the robot controller are used for adaptive control.

The products arrive in a varying order and in an undefined position. Because the parts (except for bolts) are supplied on a flat surface in their most stable position, i.e., resting on their flat side, they have three degrees of freedom, namely the x and y coordinates and the orientation angle. The different parts are recognized by the vision system on the basis of area, perimeter, and position of the holes in the objects. The position and orientation are found from the center of gravity and the hole positions.

7.11.6   Handling

The assembly starts with the housing, which is to be positioned at a work spot beside the robot where the assembly takes place (Figs. 7.34

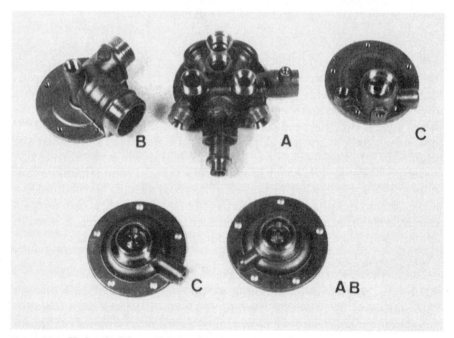

Figure 7.34   Hydraulic lift product family.

Figure 7.35 Exploded view of hydraulic lift assembly. (*a*) Bolts, (*b*) cover, (*c*) O ring, (*d*) diaphragm, and (*e*) housing.

and 7.35). The work spot is not suited to a product-carrier feed, since the supply varies, and some carriers would have to be passed over, until the next suitable part arrives. Each of the three variants has its own work spot because each has a different foundation. After it has been picked up, the product-carrier housing has to be rotated 180° around an axis in the *x-y* plane, so that the flat side is up. Then the diaphragm and the cover are moved and laid down in the correct orientation. For supplying and fastening the bolts, a commercially available automatic screwdriver is used. The sixth axis of the robot adds a considerable compliance in the extreme end of the robot; this is not a disadvantage, since, during placement of the housing, a certain amount of passive compliance is needed for smooth assembly.

7.11.7 Gripper and gripping methods

All the products are manipulated by the same gripper, since gripper alternation is not possible. Handling all the products with a multifunctional gripper requires selection of a common gripping configuration for the various products. The diaphragm is a particular problem;

it does not resemble the other parts in form and composition. The diaphragm has to fit in a groove on the housing and therefore cannot be gripped at its side. Tests showed that the diaphragm could be handled by vacuum from a vacuum line.

Common features of the housing and cover are holes, maximum diameter, and the unprocessed exteriors of the holes. The housing must be positioned on the work spot with high precision in order to obtain a reference for the remaining parts of the assembly. To achieve this precision, the worked surfaces and holes of the housing have to be used, and therefore cannot be used for handling the parts. This leaves the unprocessed exteriors and maximum diameter for handling. The exteriors of the holes were chosen because they allow a smaller gripper and have a larger tangent plane.

As the multifunctional gripper, a pneumatic gripper is used, extended with a suction pad for the diaphragm. The gripper is placed perpendicular to the torque-force (TF) sensor and outside the central axis of the TF sensor. Otherwise, when the housing is rotated 180°, the TF sensor would get in the way as the product is placed on the work spot. The work spots have small dimensional tolerances (about 0.1 mm) relative to the housing. Feedback of information from the TF sensor and the passive compliance of the gripper assist it in placing the housing on the work spot. The gripper can grasp parts in any orientation on the product carrier of the VTS system because the gripper can rotate more than 360° in the x-y plane. The section of the gripper that grasps the scribe on the housing has a friction coating so that a light pneumatic pressure is sufficient to hold the product.

7.11.8 Accuracy

High accuracy in position is not needed to grip the housing or the cover because uncertainty in position is reduced simply by closing the gripper. High accuracy in position for the diaphragm is also unnecessary because the diaphragm can be shifted in the groove by a robot movement. The required accuracy in orientation is about 0.7° for placing the cover. The error in both the robot and the vision system was measured by having the robot present a part to the vision system repeatedly. The maximum error in position was 0.5 mm and in orientation was 0.4° in a field of view of 300 × 400 mm. This amply meets the requirement of automatic assembly.

Further Reading

Ambler, A. P., et al., "A Versatile Computer-Controlled Assembly System," *Proc. of IJCAI-73*, Stanford, California, 298–307, August 1973.

Ballard, D. H., "Generalizing the Hough Transform to Detect Arbitrary Shapes," *Pattern Recognition,* **13**(2): 111–112 (1981).

Barrow, H. G., and R. J. Popplestone, "Relational Descriptions in Picture Processing," *Machine Intelligence* **6,** Edinburgh University Press, 1971.

"Decade of Robotics," 10th anniversary issue of *International Robot,* IFS Publications, 1983.

Duba, R. O., and P. E. Hart, "Use of Hough Transform to Detect Lines and Curves in Pictures," *Communications of AQCM,* **15**(1): 11–15, January 1972.

DuPont-Gateland, C. "Flexible Manufacturing Systems for Gearboxes," *Proc. 1st Int. Conf. on FMS,* Brighton, U.K., 453–563, October 20–22, 1982.

Freeman, H., "Computer Processing of Line-Drawing Images," *Computing Survey,* **6**(1), March 1974.

Gilbert, J. L., and V. Y. Paternoster, "Ultrasonic End Effector for Weld Inspection," *NASA Tech Briefs,* May 1992.

Ingersol Engineers, *The FMS Report,* IFS Publications, 1982.

Karp, R. M., "Reducibility Among Combinatorial Problems," *Complexity of Computer Computations,* Plenum Press, 85–103, 1972.

Morgan, T. K., "Planning for the Introduction of FMS," *Proc. 2d Int. Conf. on FMS,* 349–357, 1983.

Novak, A., "The Concept of Artificial Intelligence in Unmanned Production— Application of Adaptive Control," *Proc. 2d Int. Conf. on FMS,* 669–680, 1983.

Perkins, W. A., "A Model-Based Vision System for Industrial Parts," *IEEE Trans. Computers,* **C-27,** 126–143, February 1978.

Proc. 1st Int. Conf. on Flexible Manufacturing Systems, Brighton, 20–22 October 1982. IFS Publications and North-Holland Publishing Company, 1982.

Proc. 2d Int. Conf. on Flexible Manufacturing Systems, London, 26–28 October 1983, IFS Publications and North-Holland Publishing Company, 1983.

Proc. 4th Int. Conf. on Assembly Automation, Tokyo, 11–13 October 1983.

Ranky, P. G., *The Design and Operation of FMS,* IFS Publications and North-Holland Publishing Company, 1983.

Roe, J., "Touch Trigger Probes on Machine Tools," *Proc. 2d Int. Conf. on FMS,* 411–424, 1983.

Rosenfeld, A., *Picture Processing Computer,* Academic Press, New York, 1969.

Suzuki, T., et al., "Present Status of the Japanese National Project: Flexible Manufacturing System Complex," *Proc. 2d Int. Conf. on FMS,* 19–30, 1983.

Tsuji, S., and F. Matsumoto, "Detection of Ellipse by a Modified Hough Transformation," *IEEE Trans. Computers,* **C-27**(8): 777–781, August 1978.

Tsuji, S., and Nakamura, "Recognition of an Object in Stack of Industrial Parts," *Proc. IJCAI-75,* Tbilisi, Georgia, 811–818, August 1975.

Ullman, J. R., *Pattern Recognition Techniques,* Butterworths, London, 1972.

Woodwark, J. R., and D. Graham, "Automated Assembly and Inspection of Versatile Fixtures," *Proc. 2d Int. Conf. on FMS,* 425–430, 1983.

Zucker, S. W., and R. A. Hummel, "Toward a Low-Level Description of Dot Cluster: Labeling Edge, Interior, and Noise Points," *Computer Graphics and Image Proc.,* **9**(3): 213–233, March 1979.

8.0 Introduction

The signals supplied by the sensors enable the processing subsystems to construct a true picture of the process at a specific point in time. In this context, reference is often made to a *process diagram*. The process diagram is necessary to logically connect individual signals, to delay them, or to store them. The process data processing can be used in the broadest sense for the processing of a conventional control system, even a simple one. The task of the processing function unit is to connect the data supplied by the sensors as input signals, in accordance with the task description, and pass data as output signals on to the actuators. The signals generated are mainly binary signals.

8.1 Single-Board Computer

A single-board computer system is illustrated in Fig. 8.1, with input and output units which are suitable for the solution of control problems. The central processing unit is a microprocessor that coordinates the control of all sequences within the system in accordance with the program stored in the program memory.

A read-only memory (ROM) does not lose its contents even if the line voltage is removed. In the system in Fig. 8.1, the ROM contains a program which allows data to be entered via the keyboard and displayed on the screen. This program is a kind of minioperating system; it is called a *monitor program* or simply a *monitor*.

A RAM loses its contents when the voltage is removed. It is provided to receive programs which are newly developed and intended to solve a specific control task.

The input/output units enable the user to enter data, particularly

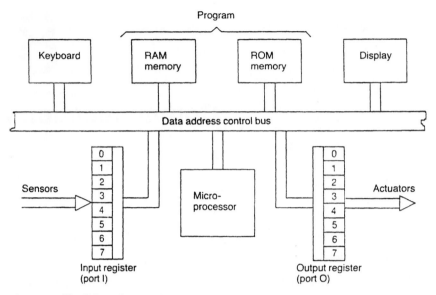

Figure 8.1 Single-board computer.

programs, through the keyboard into the RAM and to monitor the inputs.

Binary signals generated by the sensing process can be read in through the input/output units, generating an actuating process as output response. Registers of this type, which are especially suitable for signal input and output, are also known as *ports*.

The system input/output units are connected by a bus system, through which they can communicate with one another. *Bus* is the designation for a system of lines to which several system components are connected in parallel. However, only two connected units are able to communicate with one another at any time (Fig. 8.1).

8.2   Sensors for Input Control

A microcomputer is used for the dual-level input control of a reservoir (Fig. 8.2). The level of the liquid is kept constant within the specified limits. For this purpose, two sensors are placed at the desired maximum and minimum locations:

Maximum sensing level = maximum limit

Minimum sensing level = minimum limit

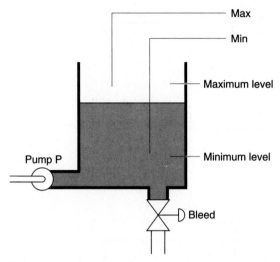

Figure 8.2 Dual-level control.

Maximum and minimum sensors are installed to recognize the upper and lower limits. The motor is switched ON when the minimum level has been reached:

$$Max = 0$$

$$Min = 0$$

and OFF when the maximum level has been reached:

$$Max = 1$$

$$Min = 1$$

If the level is between the two levels

$$Max = 0$$

$$Min = 1$$

the motor maintains current status.

The constant level is disturbed by varying the bleed-off valve. These requirements are shown in a *fact table,* where P represents a current status and PN represents the required pump status (Table 8.1).

The program flowchart in Fig. 8.3 can be used as a programming model. The ports to connect the sensors and actuators are specified in Fig. 8.4.

TABLE 8.1   Fact Table

Max	Min	P	Port	PN
0	0	0	I	1
0	0	1	I	1
0	1	0	I	0
0	1	1	I	1
1	1	0	I	0
1	1	1	I	0

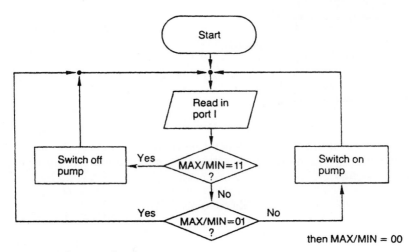

Figure 8.3   Program flowchart.

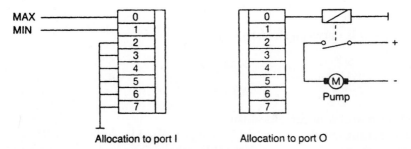

Figure 8.4   Ports to connect sensors and actuators.

The programming model is converted into computer program statements. For this, it is necessary for the statement commands to be formulated in machine code. In turn, this makes it necessary to understand the type of microprocessor being used. The well known Z80 type is used in the example in Table 8.2. The program appears in the abbreviated hexadecimal form.

The hexadecimal counting system is based on the number 16. Like the decimal system, the hexadecimal system uses the numerics 0 to 9 and, in addition, the letters A to F, which represent the numbers 10 to 15. Addresses and commands are entered on a keyboard which comprises keys for the hexadecimal alphanumerics 0 to F and various special keys. An address that has been set and the data it contains can be read from a six-digit numerical display.

Once the program has been developed and tested, there is no further need to use keyboard and display. Thus, these components can now be disconnected from the system. They do not need to be reconnected until program editing becomes necessary.

It always is a good practice to transfer a completed program from a

TABLE 8.2   Program Statements

Address	Command	Description
0001	DB	Port I (address 10) to the actuator
0002	10	
0003	FE	Compare accu = 0000 0000
0004	00	
0005	CA	If accu = 0000 0000
0006	14	jump to address 14
0007	00	
0008	FE	If accu does not = 0000 0000
0009	01	jump to address 00
000A	C2	If accu does not = 0000 0001
000B	00	jump to address 00
000C	00	
000D	3E	Load 000 000 into accu
000E	00	
000F	D3	Bring accu to port O (address 11)
0010	11	
0011	C3	Jump to address 00
0012	00	
0013	00	
0014	3E	Load 0000 0001 into accu
0015	01	
0016	D3	Bring accu to port O (address 11)
0017	11	
0018	C3	Jump to address 00
0019	00	
001A	00	

RAM to a memory whose contents are retained even when the supply voltage is switched off. An electrically programmable ROM (EPROM) could be used. Therefore, the RAM is replaced by an EPROM chip on the PC board.

Sometimes it is necessary to develop extensive programs, thus, a comprehensive knowledge and considerable experience are needed. Programs in machine code, as a hexadecimal sequence of alphanumerics, are very complex, difficult to read and document, and hard to test.

8.3 Microcomputer Interactive Development System

A microcomputer interactive development system consists of the following key elements as illustrated in Fig. 8.5:

1. Software

2. Bulk storage (floppy disk, permanent disk, or other storage)

3. Monitor

4. Keyboard for interaction and data entry

5. Printer

In principle, there are two methods of developing the communication software for an interactive control system:

■ The interactive system is located near and directly connected to the installation workstation so that the development may take place directly at the machine under real conditions.

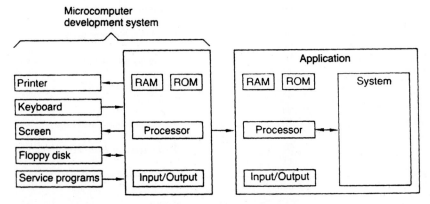

Figure 8.5 Microcomputer interactive development system.

■ The interactive system is located away from the installation work-station. In this case, the installation must be totally or partially simulated, otherwise the system must be connected to the installation from time to time.

When the development of the interactive system has been completed, the system will contain a precise image of the hardware and the software, such as programming, memory size, ports, and number of inputs and outputs. The microcomputer system can then be either built as a printed circuit according to customer specifications or assembled from standard modules (Fig. 8.6).

8.4   Personal Computer as a Single-Board Computer

The personal computer is not a completely new development in the computer evolution. It simply represents a new stage of the development of single-board computers. A single-board computer can play a key role in developing active communication and interfaces for work-stations, since it can be effectively integrated with process computer peripherals by the following steps:

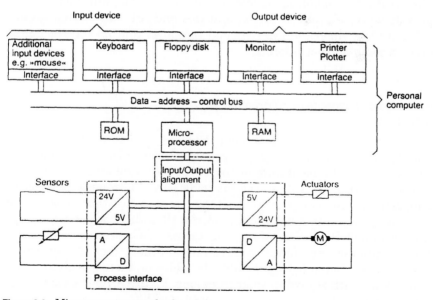

Figure 8.6 Microcomputer standard modules.

1. Inclusion of a single-board computer and power supply into a housing

2. Replacement of the simple switch or hexadecimal input by a keyboard similar to that of a typewriter

3. Replacement of the simple LED or seven-segment display by a screen with 24 lines and 40 or 80 characters per line

4. Addition of external mass storage for programs and data (cassette recorder or floppy disk)

5. Connection to a printer or other peripheral device

6. Integration of an interpreter for high-level programming language

A microcomputer system with this configuration is often referred to as a *personal computer,* which is misleading terminology. Although the microprocessor and single-board computer were initially developed for control purposes, the nomenclature directed interest toward purely commercial applications.

The personal computer (or single-board computer) is similar to any peripheral device that requires interfacing to communicate with a central control unit to perform control functions. The personal computer's input and output signals must also be adapted and made available to peripheral devices. Control technology employs a variety of signal forms and levels, e.g., with both analog and binary signals and voltages that may be as high as 24 V. The microcomputer operates with internal voltage levels of 0 and 5 V. Thus, interfacing must provide voltage-level conversion and D/A and A/D conversion. Figure 8.6 is a block diagram of a personal computer and its interfacing to process computer peripherals.

8.4.1 Role of sensors in programmable logic controllers

The programmable logic controller (PLC) is used in processes where mainly binary signals are to be processed. The PLC first came into the market at the beginning of the 1970s as a replacement for relay circuits. Since then, these systems have been continually developed, supported by the progress in microelectronics. For that reason, the present systems are much more powerful than those which were originally named PLC. Therefore, the terms PLC, process computer, microcomputer, etc., often overlap one another. Technically, these systems are often indistinguishable. In addition, there are often many similarities with regard to their functions. Figure 8.7 shows the functional relations between the components of a PLC.

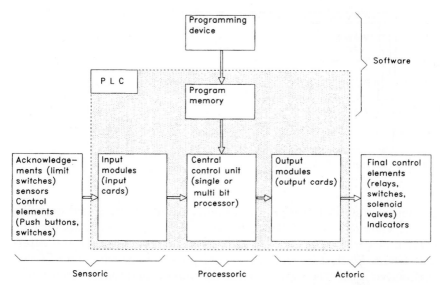

Figure 8.7   Relationships between sensor, processor, and actuator activities in the PLC.

The input signals supplied by the sensors are passed onto the central control unit via an input module. The signals generated by the central control unit are prepared by the output modules and passed on to the actuators. The program is drawn up by using an external programming device and transferred into the program memory (input of the program varies, however, according to the PLC).

The modules are designed to meet the following requirements on the input side:

1. Protection against loss of the signal as a result of overvoltage or incorrect voltage

2. Elimination of momentary interface impulses

3. Possibility of connecting passive sensors (switches, push buttons) and initiators

4. Recognition of error status

It depends on the manufacturer whether all requirements are met or only some of them. Figure 8.8 shows the basic elements of an input module.

The module contains an error voltage recognition facility, which is activated when the input voltage exceeds specified tolerance limits. A trigger circuit with signal delay (added in sequence) ensures that

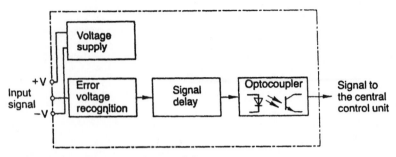

Figure 8.8   Elements of an input module.

momentary interference peaks are suppressed. Positive and negative supply voltages (+V, –V) are provided at the input terminals for connection of various sensors. Optocouplers separate internal and external circuits so that interference is not able to penetrate the PLC via conductive lines. The input module contains visual indicators (mainly LEDs) which indicate whether a 1 or a 0 is present at the input.

The requirements at the output side are:

1. Amplification of the output signals

2. Protection against short circuiting

Figure 8.9 shows the basic elements of an output module. The optocoupler is used to separate the internal from the external circuit to prevent interference via conductive connections. The output signals need to be amplified so that final control elements and actuators which require additional current can be directly connected to the PLC. There are three methods of safeguarding the outputs against short circuiting:

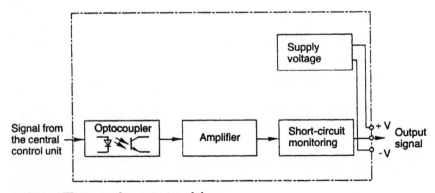

Figure 8.9   Elements of an output module.

1. Protection by a safety fuse

2. Conditional short-circuit-proof outputs (momentary protection)

3. Sustained short-circuit-proof outputs

Signal status of the outputs can be read from indicators.

8.4.2   Central control unit

The first PLCs consisted of a simple single-bit processor. Nowadays, a modern PLC may have at its disposal one or several multibit processors. Thus, in addition to logic comparison operations, a PLC may execute arithmetic, counting, and time functions. By using powerful microprocessors, it is possible to provide PLCs with capabilities which were previously found only in process computers. An example is multitasking, which makes it possible to process a number of unrelated programs simultaneously in the PLC. Of course, the processor itself is still only able to process one command after another. However, because of high processing speeds and complex operating software, it is possible to give the user the illusion of a number of processors working simultaneously.

Almost all PLCs offer important functions like flags, counters, and timers. Flags are single-bit memories in which signal status and program status can be stored temporarily if they are not required until later in the program cycle. Most control tasks require counting and timing functions. Counters and timers help the user in programming such tasks.

With most PLCs it is possible to process simultaneously a number of bits, e.g., 8 bits (rather than just 1 bit such as a switch signal). This is known as *word* processing. Thus, it is possible to connect an A/D converter to the PLC and thus evaluate analog signals supplied by the process. PLCs possess convenient interfaces by means of which they are able to communicate with overriding systems (e.g., control computers) or link up with systems of equal status (e.g., other PLCs).

There are PLC systems available for small, medium, and large control tasks. The required number of inputs and outputs, i.e., the sum of all sensors and actuators, and the necessary program memory capacity generally determine whether a control task is to be designated as large or small. To facilitate matching PLC systems to requirements, they are generally offered as modular systems. The user is then in a position to decide the size of a system to suit the requirements.

Input and output modules are obtainable as compact external modules or as plug-in electronic cards, generally with 8 or 16 inputs/outputs. The number of such modules to be coupled depends on the size of the control task.

A Representative PLC would be composed of plug-in electronic cards in a standard format. The basic housing provides space for system modules and a power supply. Eight card locations might be provided for input/output modules, each containing 16 input/outputs, giving an I/O-connection capacity of 128. If this number were insufficient, a maximum of four housings could be united to form a system, increasing the number of input/outputs to a maximum of 512. With this system, a program memory capacity of 16 kbytes can be installed; i.e., up to 16,000 commands can be programmed. Thus, this controller would be the type used for medium to large control tasks.

8.4.3 Process computer

The process computer is constructed from components similar to those used in a microcomputer system, with the difference that these components are considerably more powerful because of their processing speed and memory capacity. The supporting operating systems are also more powerful, meaning that many control tasks can be processed in parallel. For this reason the process computer is used in places where it is necessary to solve extensive and complicated automatic control tasks. In such cases, it has the responsibility of controlling a complete process (such as a power station) which is made of a large number of subprocesses. Many smaller control systems which operate peripherally are often subordinated to a process computer. In such cases, the process computer assumes the role of coordinating master computer.

8.5 The NC Controller

Numerical control systems are the most widely used control in machine tools. A blank or preworked workpiece is machined by NC machine tools in accordance with specifications given in a technical drawing and other processing data. Two-, three-, or multiple-axis machines may be used, depending on the shape of the workpiece and the processing technology. A two-axis machine tool might be an NC lathe, for example. A milling machine is an example of a three-axis machine tool.

In NC machines, sensors and actuators are connected to the processor by the manufacturer. The operator or programmer needs to know the operational principle and the interplay between the controller and machine.

Figure 8.10 illustrates the location of holes to be drilled in a work-

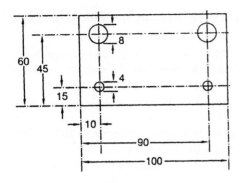

Figure 8.10 Location of holes to be drilled in a workpiece.

piece. The skilled machinist needs to know the material from which the workpiece has been made so that suitable drills and feed values can be selected for machining. A drilling machine is equipped with a work table which is adjustable in two directions. The adjustment is carried out by using a lead screw, and the set value set can be read from a scale. The machine is illustrated in Fig. 8.11.

8.5.1 Manufacturing procedure and control

The manufacturing procedure for a machine tool application is as follows:

1. The table is moved up to the reference point. The reference point values are imparted to the computer. The value zero is read from both scales. The drill is located above the reference point.

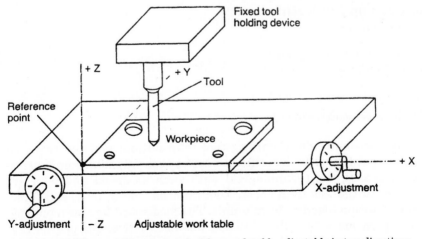

Figure 8.11 Drilling machine equipped with a work table adjustable in two directions.

2. The workpiece is placed on the work table in the precise position required and clamped.

3. The workpiece zero point is determined; a reference edge from which all dimensions are measured is also determined. The drill is positioned at this point. The workpiece zero point can be imagined as the origin of a rectangular system of coordinates: the y axis is in one direction of motion of the table, the x axis is in the other. Similarly, the dimensions in the drawing are referred to this point. Thus, it is easy to move the work table to each point in turn where holes need to be drilled. Modifications can be carried out if required.

4. Machining is commenced; the working sequence should be described in detail.

This description can completely replace the technical drawing. Any operator must be able to carry out the working cycle using this exact description.

8.5.2 Machining program

The machining program is described in terms of sequences:

1. Clamp drill 1
2. Set X to 10 and Y to 15
3. Drill at cutting speed 2
4. Set X to 95
5. Drill at cutting speed 2
6. Unclamp drill 1, clamp drill 2
7. Set Y to 45
8. Drill at cutting speed 1
9. Set X to 10
10. Drill at cutting speed 1
11. Remove workpiece

The positioning data, cutting tool information, and the required cutting speed must be identified to accomplish any machining operation. Usually, the technological data such as speed and feed are included in the tool data sheet. Tool number, speed, feed, and direction of rotation together form a unit. The feed speed has been omitted in the program above; this decision has been left to the operator. If the process is to proceed automatically at a later date, this information will have to be added.

The positioning data is always referenced from the origin of the coordinates. In this case, this is always the point of intersection of the reference lines for the dimensioning of the drawing. The positional information was therefore given in absolute form and can be programmed in absolute dimensions.

The information can also be provided relative to the current position, which is entered as the x value, and a new value added for the next position. This is called *incremental positioning data*. The incremental positioning data may expressed as follows:

1. Set $X + 10$, $Y + 15$

2. Set $X + 80$

3. Set $Y + 30$

4. Set $X - 80$

Choice of the most suitable method often depends on the dimensioning of an engineering drawing. Sometimes incremental and absolute specifications of positional data are mixed. Then, obviously, it is necessary to describe the type of data in question.

In the example described above, the machining of the workpiece takes place only at four different points; only one machining position is approached. There is no rule as to which path should be taken to reach this position. Thus, no changes have been made at the workpiece when the drilling operation occurs, and the table is moved by a zigzag course to the position for the second drilling operation. This is referred to as a *point-to-point* control.

The situation is different if, as shown in Fig. 8.12, a groove is to be milled in a workpiece. The tool to be used is a milling cutter. The worktable might be the same as that used for the drill. The workpiece is to be contacted on the path from point 1 to point 2, and a zigzag course is, of course, impermissible. However, the problem can be easily solved, since the machining parts are always parallel to one of the

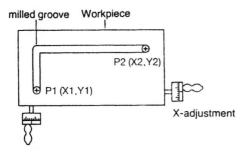

Figure 8.12 Groove to be drilled in workpiece.

two axes. The tool is set to start at position P1. From this point on, only the *y* value is set to Y2. When the tool arrives there, the *x* value is set to the new value X2. This is a simple straight-line control.

Figure 8.13 represents a diagonal groove to be milled from point 1 to point 2. A simple solution can be realized using the zigzag course, but it must be calculated with precision, as illustrated in Fig. 8.14.

Without making tedious calculations, it is clear that the machine arrives at point 2 when the tool is moved by an increment of Δx, then by the same amount Δy and then once more by Δx, etc. If the increments Δx and Δy are small, the zigzag line is no longer noticeable and the deviation from the ideal course remains within the tolerance limits. If the required path is not under 45°, the values for Δx and Δy differ from each other, according to the linear gradient. These values must then be calculated. In an automatic control system, the computer solves this problem. The straight-line control in this example is referred to as an *extended straight-line control*.

A random course from point 1 to point 2 is illustrated in Fig. 8.15. The increments for the *x* direction and the *y* direction (Δx, Δy) must be formed. However, calculations of this path may be complicated if the

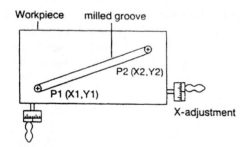

Figure 8.13 Diagonal groove to be milled from point 1 to point 2.

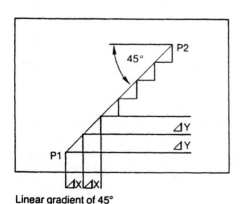

Linear gradient of 45°

Figure 8.14 Zigzag course to mill diagonal groove.

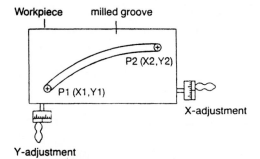

Figure 8.15 Random course from point 1 to point 2.

curve to be traveled cannot be described by simple geometric equations. Accordingly, a contour control method is used in this manufacturing application. Contour controls can generally travel in arcs, even elliptical arcs. Complicated shapes must be transformed into a straight-line path by coordinating points. A programming workstation is necessary for this application.

Figure 8.16 illustrates in diagrammatic form the method of automating a drilling machine. The scales are replaced by incremental or absolute displacement measuring systems. Suitable actuators are used in place of the handwheels. Sensors are precisely placed at the extreme ends of the traveled distance. Today, dc motors, which provide infinite and simple adjustments for all axes and for the main drive, are widely used.

There are two different methods of describing the course of a tool or machine slide, incrementally and absolutely. Accordingly, the subcontrol task, movement of a machine slide, is solved differently for each.

The two types of incremental control system are illustrated in Fig. 8.17. In an incremental control system with incremental displacement encoder, the processor counts the pulses supplied by the linear displacement sensor. A pulse corresponds to the smallest measurable unit of displacement. The distance to be traveled and the direction of travel are passed on to the processor as reference values. From these data the processor forms a control signal for the final control element of the drive motor. If the values do not agree, the motor must continue running. Once they agree, the processor transmits a stop signal to the final control element, indicating that the correct slide position has been reached. When a new reference value is supplied, this process is started up once again. It is important that the slide stop quickly enough. Braking time must be taken into consideration; i.e. before the end position is reached, the speed must be reduced as the remaining gap decreases.

In incremental control with a stepping motor as actuator, there is

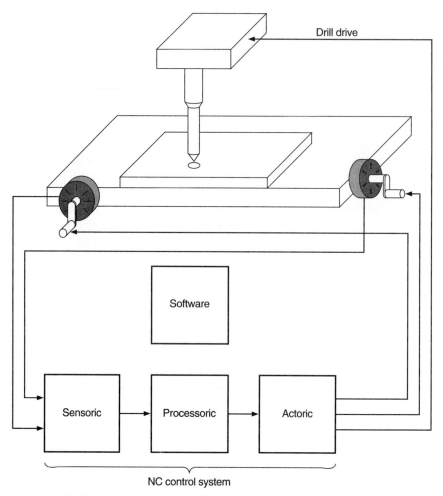

Figure 8.16 Method of automating a drilling machine.

no acknowledgment of the slide position. In place of the impulse from the linear displacement sensor, a pulse is given straight to the processor. The processor now passes on pulses to the stepping motor until a pulse count adds up to the displacement reference value. As the stepping motor covers an accurately defined angle of rotation for each step, each position can be easily approached by the slide. With stepping motors, only very limited speeds are possible, depending on the resolution.

Figure 8.18 illustrates the two types of absolute control system: one with an absolute displacement encoder and the other with an incremental displacement encoder.

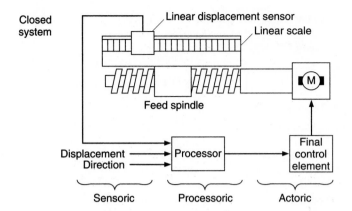

Incremental control system with displacement monitoring

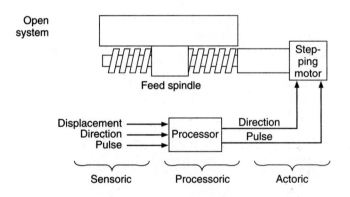

Incremental control system with stepping motor

Figure 8.17 Two types of incremental control system.

8.5.3 Absolute control

In an absolute control system with an absolute displacement encoder, such as an angular encoder or linear encoder, the actual value of absolute position of the slide is known at any given time. The processor receives this value and a reference value. The processor forms the arithmetic difference between the two values. The greater the difference, the greater the deviation of the slide from its reference position. The difference is available as a digital value at the output of the processor. After D/A conversion, the difference is presented as an analog value to a final control element such as a transistor that triggers the drive motor. The greater the deviation from the reference value, the greater the motor speed and the adjustment speed of the slide.

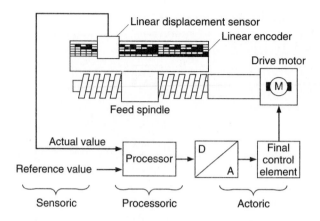

Absolute control system with absolute displacement encoder

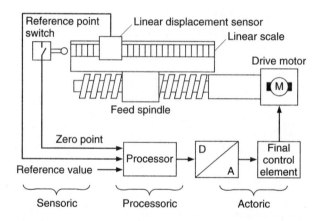

Absolute control system with incremental displacement encoder

Figure 8.18 Two types of absolute control system, with absolute displacement encoder and incremental displacement encoder.

When there is no further deviation, the motor stops. The motor, of course, has both upper and lower speed limits. At lower speeds the torque may become so low that the tool may falter.

In an absolute control system with an incremental displacement encoder, a sensor acts as a limit switch at the end position of the slide. Once the machine is started, the slide is moved to the end position, where an incremental displacement counter in the processor is set to the reference point value. Each additional movement of the slide is recorded by this counter so that, at any given time, it knows the absolute position of the counter, like the absolute system with a displacement encoder. The absolute control system with incremental displacement encoder is the most widely used today.

8.5.4 NC software

As with a PLC system, it is necessary to draw up an accurate model for an NC system before a program is written.

The machine must be told which movements it is to execute. For this purpose, displacement data must be prepared which can be taken from a drawing of the workpiece. In addition, the machine needs to know what tools to use and when, the tool feed speed, and the cutting speed. Certain additional functions such as the addition of lubricants must also be programmed.

All these data are clearly of great importance for conventional machining of workpieces. The data are compiled during the work preparation in a factory to create a work plan for a skilled operator. A conventional work plan is an excellent programming model. However, work plans for manual operation cannot as a rule be converted to NC operations.

To ensure that there are no ambiguities in the description of displacement data there is general agreement about the use of cartesian coordinates for NC (Fig. 8.19). Different layers of the coordinate system apply to different processing machines, as shown in Fig. 8.20.

The commands in an NC instruction set are made up of code letters and figures. The program structure is generally such that sentences containing a number of commands can be constructed. Each sentence contains a sentence number. Table 8.3 shows an NC program for an automatic drilling machine.

Displacement conditions are given the code letter G. The displacement condition G00 is programmed in sentence 1, signifying point-to-point control behavior. This condition thus also applies to all other sentences; it does not have to be repeated. The feed is specified with code letter F, followed by a code number or address where the actual feed speed is located. The cutting speed is given with a code letter S and a code number. The code for the tool is T; the following code number 01 in the program indicates that a drill should be taken from magazine 1. The additional function M09 indicates that a lubricant is to be used. M30 defines the program M.

An NC program for a lathe is described in Table 8.4. A part to be turned is illustrated in Fig. 8.20. It is assumed that the workpiece has already been rough-machined. Therefore, it is only necessary to write a finished program.

Path control behavior is programmed in the beginning sentences with displacement condition G01. The lathe is set in absolute values to

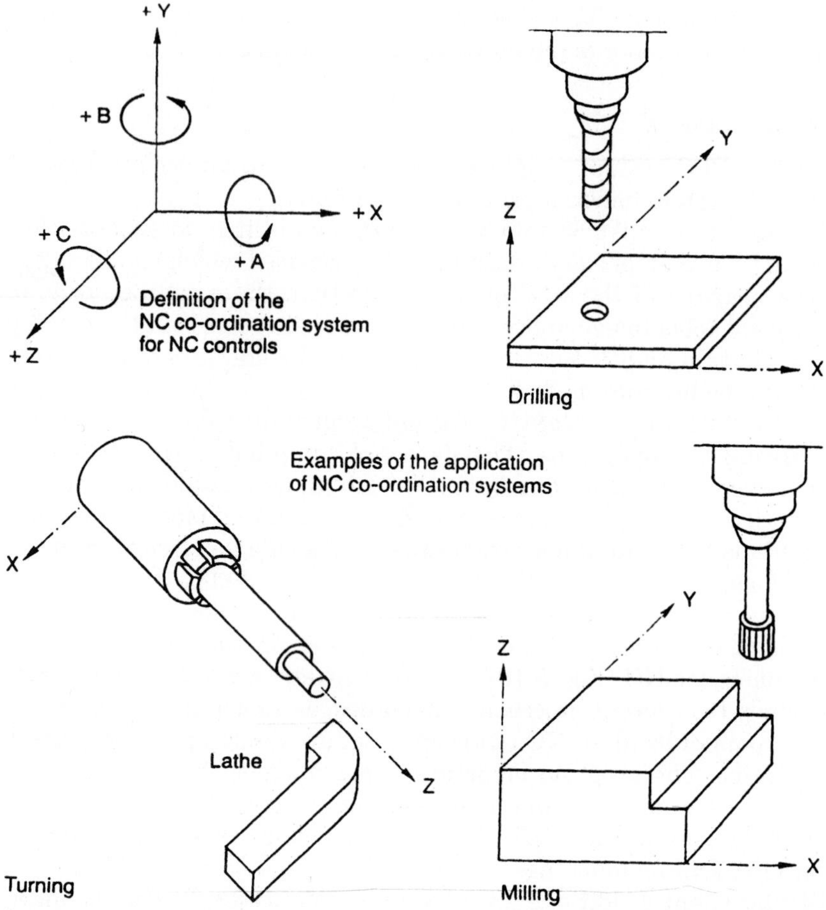

Figure 8.19 Cartesian coordinate system for NC and coordinate system applied to different processing machines.

TABLE 8.3 NC Program—Automatic Drilling Machine

Sentence number	Conditions	Displacement Data x	Displacement Data y	Displacement Data z	Switching data: Feed	Switching data: Tool speed	Switching data: Tool	Switching data: Addition
N001	G00	X10	Y15	Z2	F1000	S55	T01	M09
N002				Z0				
N003		X909		Z2				
N004				Z0				
N005			Y45	Z2	F500	S30	T02	
N006				Z0				
N007		X10		Z2				
N008				Z0				
N009		X0	Y0					M30

TABLE 8.4 **NC Program—Lathe Machining Operation**

Sentence number	Conditions	Displacement data			Switching data			
		x	y	z	Feed	Tool speed	Tool	Addition
N001	G01	X20		Z100	F1500	S60	T03	M09
N002	G91			Z–30				
N003		X + 20						
N004		X + 10		Z–30				
N005		X + 30						
N006				Z–40				M30

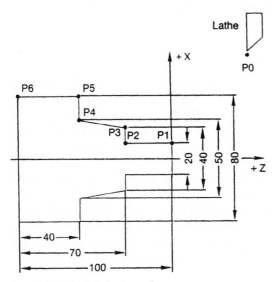

Figure 8.20 Parts to be turned.

coordinates X20 and Z100. The displacement condition G91 of the second sentence determines relative addressing; i.e., the x and z values should be added to the previous values. Where the x values are concerned, it should be noted that diameter rather than radius is specified.

The NC programs shown in these examples must, of course, also be put on a suitable data carrier in machine-readable form. The classic data carrier for NC is punched tape. Each character in the program text is converted into a 7-bit combination of 1s and 0s and stamped into the tape as an appropriate combination of holes and spaces (1 = hole, 0 = no hole).

This process of conversion is called *coding*. The totality of all combinations of holes representing a character set is called a *code*.

Recommendations have been laid down as to which combination of holes represents a specific character. Figure 8.21 shows the N005 sentence statement in Table 8.3 as a punched tape code to ISO standards. Each 7-bit character is accompanied by a test bit. The test bit is a logic 1 (hole punched) when the number of holes is uneven. Thus, the total number of holes in a character is always even. When the punched tape code is checked to ensure it is even-numbered, errors caused by damage or dirt may be detected. This method of checking for errors is called a parity check.

Because punched tape is robust in aggressive industrial environments, it is still preferred by many users. However, modern storage media such as a floppy disks are used more and more. It is also possible to use an economical personal computer as a development and documentation station for NC programs.

8.5.5   Operation of an NC system

Figure 8.22 is a block diagram of a simple NC system. The NC machine is equipped with a punched tape reader into which the program for machining a specific piece is inserted. After starting the program, the processor reads the first sentence. The various displacement and switching data are written into a reference-value memory. Next, the commands are processed. The comparator compares the reference values to the values supplied by the measuring devices (e.g., the current positions of the x- and y-drive units).

If there is a difference between reference values and actual values, the comparator generates suitable control signals for the final control elements of the drives. These are generated only until reference values and actual values agree. The comparator now demands a new

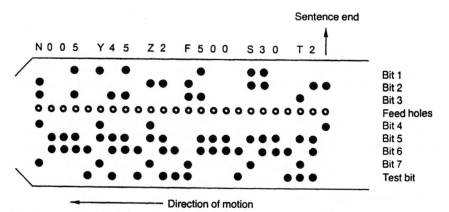

Figure 8.21   N005 sentence statement from Table 8.3 (drilling task) as a punched tape code to ISO standards.

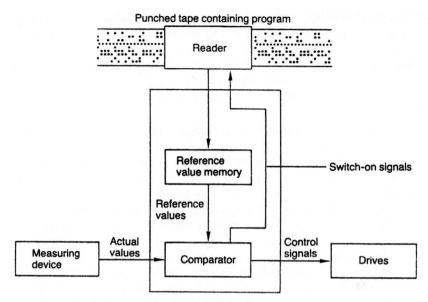

Figure 8.22 Block diagram for a simple NC system.

sentence from the punched tape reader for processing. This sentence is evaluated in the manner described. Reading and processing are repeated until the final sentence has been processed. As an example of reference value/actual value comparison, Fig. 8.23 shows the adjustment of the x axis of an NC machine.

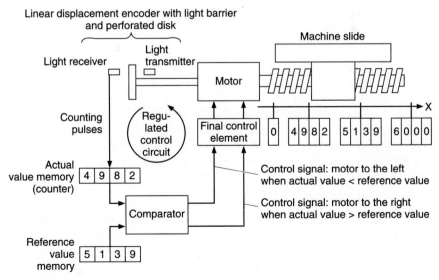

Figure 8.23 Adjustment of the x axis of an NC machine.

If the machine slide is moved from its current position representing the actual value to a new position representing the reference value, a continuous comparison of the reference value to the actual value is carried out. The comparator determines the difference between the values and supplies a control signal to the final control element of the motor in accordance with the arithmetical sign.

The motor runs in a counterclockwise direction when the actual value is smaller than the reference value. This corresponds to a displacement of the machine slide to the right. The motor runs in a clockwise direction when the actual value is greater than the reference value, which corresponds to a displacement of the machine slide to the left.

The number of rotations of the motor, which is proportional to the displacement of the slide, is counted for the actual-value memory by a measuring device which, in the example, is a light transmitter, a perforated disk on the motor shaft, and a light receiver.

The movement of the slide by the motor drive reduces the difference between the reference value and actual value. When the difference is 0, the comparator sets the control signal to 0, and the motor is stopped.

The type of programming in this case is machine-dependent; it corresponds to an assembler language. Application-oriented programming, with which a more general formulation can be written, is also developed for NC systems. An example is the NC programming language Advanced Programming Tool (APT), which can cope with several hundred commands. APT commands are classified into three categories:

1. Geometric commands

2. Motion commands

3. Auxiliary commands

8.5.5.1 Geometric commands. In the NC coordinate system in Fig. 8.24, each point is clearly determined through specification of coordinates x, y, and z. In the APT, each point can be defined symbolically; e.g. the definition for the point $x = 15.7$, $y = 25$, and $z = 13$ is

$$P1 = POINT/15.7,25,13$$

Thus, the variable to the left of the equals sign can be freely selected. Distances from one point to another are defined as follows:

$$S5 = LINE/P2,P3$$

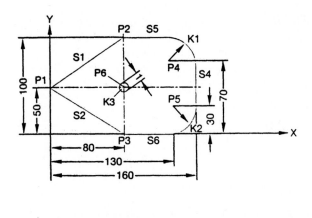

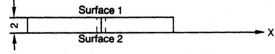

Figure 8.24 NC coordinate system for a workpiece.

P2 and P3 must be defined in advance. Points may also be defined as the intersection of two distances, as for example:

$$P1 = POINT/INTOF \ L6,L8$$

A circle can be defined in the following ways:

$$CIRCLE1 = CIRCLE/CENTER,POINT,RADIUS,5.8$$

$$CIRCLE2 = CIRCLE/P4,P5,P6$$

In the first instance, the circle is defined by specifying the center and the radius, and in the second, by specifying three points on the perimeter. A plane surface is defined by three points located on it, e.g.,

$$SURFACE1 = PLANE/P2,P2,P3$$

Another way of specifying a plane is to indicate a point on a surface and a second surface lying parallel to the first one, e.g.,

$$SURFACE2 = PLANE/P1,PARALLEL,SURFACE1$$

The geometric dimensions of a workpiece given in the technical drawing can be fully described using these geometric commands. The workpiece shown in Fig. 8.24 is to be described using the APT geometric commands:

P1 = POINT/0.0, 50.0, 0.0

P2 = POINT/80.0, 100.0, 0.0

P3 = POINT/80.0, 0.0, 0.0

P4 = POINT/130.0, 70.0, 0.0

P5 = POINT/130.0, 70.0, 0.0

P6 = POINT/80.0, 50.0, 0.0

S1 = LINE/P1, P2

S2 = LINE/P1, P3

K1 = CIRCLE/CENTER, P4, RADIUS, 30.0

K2 = CIRCLE/CENTER, P5, RADIUS, 30.0

K3 = CIRCLE/CENTER, P6, RADIUS, 0.5

S5 = LINE/P2, LEFT, TANTO, K1

S6 = LINE/P3, RIGHT, TANTO, K2

S4 = LINE/LEFT, TANTO, K2, RIGHT, TANTO, K1

SURFACE1 = PLANE/0.0, 0.0, −2

The commands for defining paths S4, S5, and S6 form tangents to cir-
cles K1 and K2 and are for this reason given the additional functions
LEFT, RIGHT, and TANTO. The line S5 starts at P2 and forms a left
(LEFT) tangent to (TANTO) circle K1.

8.5.5.2. Motion commands.   These commands are necessary to
describe the path that the tools must travel. A distinction is made
between point-to-point (PTP) controls and path controls. The commands for point-to-point controls are as follows:

GOTO   Absolute statement of the point to be approached

GODTLA   Relative statement of the point to the approached

Example: To travel from point 2,5,10 to point 8,10,20, the two commands
would have to be used as follows:

Either   GOTO/8,10,20

Or   GODTLA/6,5,10

The drilling operation in this example is programmed as follows:

GOTO/P6

GODTLA/0.0, 0.0, −2.5

GODTLA/0.0, 0.0, 2.5

A path control is programmed with the following commands:

GOFWD   (go forward)

GOLFT   (go left)

GODWN   (go down)

GOBACK   (go back)

GORGT   (go right)

GOUP   (go up)

With these commands it is possible, for example, to program the movement of a cutting tool in accordance with the contour in the example.

GOTO/S1,TO,SURFACE1,TO,S2   Go to start point P1

GOFWD/S1,PAST,S5   Go along line S1 to S5

GOFWD/S5,TANTO,K1   Go along line S5 to K1

GOFWD/K1,TANTO,K2   Go from K1 via S4 to K2

GOFWD/K2,PAST,S6   Go from K2 via S4 to S6

GOFWD/S2,PAST,S1   Go along line S2 to S1

8.5.5.3 Auxiliary commands. Various switching functions are programmed using commands from this group; for example:

MACHIN   Specifies the tool

FEDRAT   Determines the feed speed

COOLNT   Connects coolant

CUTTER   Diameter of the tool

PARTNO   Part number of the workpiece

FINI   Program end

Other application-oriented programming languages for NC programming are EXAPT, ADAPT, and AUTOSPOT. These languages are constructed like APT, and in part build on that language. An NC program written in application-oriented language must, of course, also be translated into machine code and produced in punched tape (Fig. 8.25).

8.5.6 Computer numerical control system

The program for a specific machining task of an NC machine is located on a punched tape. This punched tape must be read separately for every single workpiece to be machined. Reading of punched tape is a mechanical process that subjects the tape to wear, which can lead to errors.

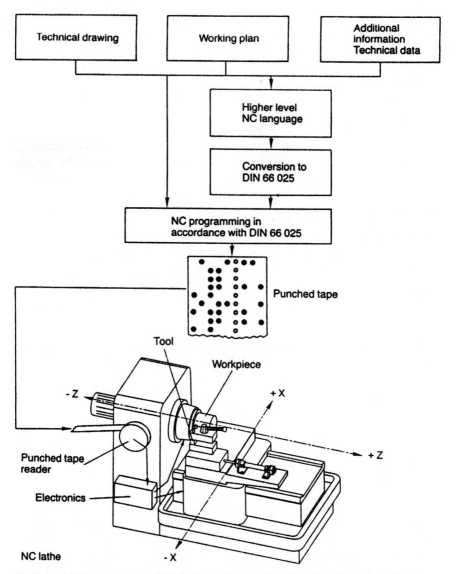

Figure 8.25 NC program translated into machine code and produced in punched tape.

With the development of computer numerical control, it became possible to overcome these disadvantages and, in addition, to create many advantages. A CNC system contains its own computer system based on modern microcomputer technology. A semiconductor memory is used as program memory. The punched tape is in fact also available with the system. However, it is only required when the program is transferred into the semiconductor memory. In addition to this program entry technique, CNC systems offer the possibility of transferring programs directly from external computers via appropriate interfaces.

Moreover, most CNC control systems are equipped with a keyboard and a screen which make it possible to program directly on the machine. This is referred to as *workshop programming*. Keyboard and screen also offer the possibility of editing and improving programs on site. This is a decisive advantage when compared to NC systems, where modifications and adaptations are much more time-consuming.

With microcomputers it is easy to program the frequently complicated arithmetical operations required for machining complicated curves. Cutting and feed speed can be matched to the best advantage. The computer can check and adapt the values continually and, thus, determine the most favorable setting values.

Integrated test and diagnostic systems, which are also made possible by microcomputers, guarantee a high degree of availability for CNC systems. The control system monitors itself and reports specific errors independently to the operator.

The programming languages correspond in part to those discussed for NC systems but are, in fact, frequently manufacturer-specific. The most important components of a CNC system are illustrated in Fig. 8.26.

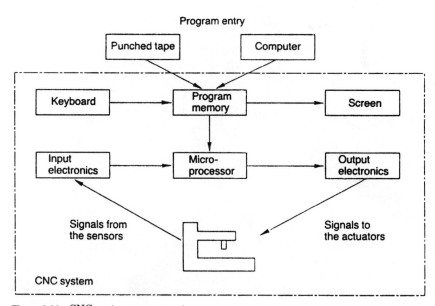

Figure 8.26 CNC system components.

8.6 Industrial Handling

The modern term for handling technology is *industrial handling* or simply *handling*. This concept embraces simple and complex handling and clamping devices.

The best and most versatile handling component is the human hand. Within its physical limitations of weight and size, many possibilities are available: gripping, allocating, arranging, feeding, positioning, clamping, working, drawing, transferring, etc.

Unlike the human hand, a mechanical handling device is able to perform only a very limited number of functions. For this reason, several handling devices have been developed for automatic handling and clamping functions.

It is not necessary to make a full copy of the human hand. A mechanical device is more or less limited to the shape, size, construction material, and characteristics of a specific workpiece to be handled.

Fig. 8.27 shows a device made up of four pneumatic handling units that can insert parts into a lathe for turning.

Cylinder A separates the parts to be turned and assumes the function of allocating parts in the machine-ready position. The part to be turned is brought to the axial center of the machine. At this position the part can be taken over by the gripping cylinder D. Cylinders B and C are used to insert the workpiece into the clamping chuck and remove it again after machining.

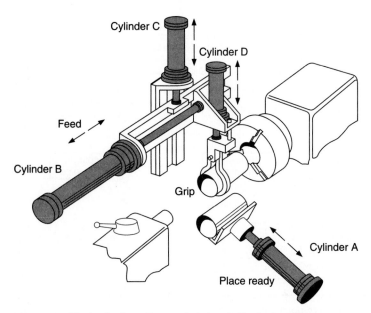

Figure 8.27 Device for inserting parts into a lathe for turning.

The fully turned part may be either dropped or picked up by a handling mechanism that is located opposite cylinder A and directly passed on after being oriented.

A PLC is suitable as a control system for the feed unit. A program has to be developed for this purpose.

First, the subtasks (1) "pass on workpiece to the machine tool" and (2) "take workpiece from the machine tool" are to be analyzed. In order to be able to solve these tasks, signal exchange between the PLC and the machine tool (MT) is necessary. Figure 8.28 shows the necessary interface signals.

The subtasks are

1. The MT informs the PLC that a workpiece has been completed and can be removed. Signal: MT complete.

2. The gripper travels to the MT and grips the workpiece. The PLC requires the MT to release the workpiece. Signal: release workpiece.

3. The MT releases the workpiece and replies with the signal: workpiece released.

4. The gripper can now advance from the MT with the workpiece.

5. The gripper travels with the workpiece into the MT and requires it to clamp. Signal: clamp workpiece.

6. The MT clamps the workpiece and replies with the signal: workpiece clamped.

7. The gripper opens and retracts from the MT.

8. Once the gripper has left the collision area, the PLC reports that the MT can begin machining. Signal: MT start.

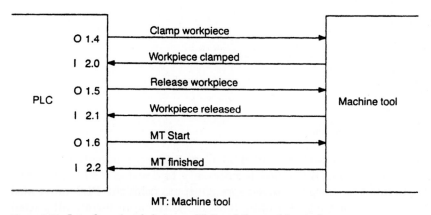

MT: Machine tool

Figure 8.28 Interface signals between PLC and the machine tools.

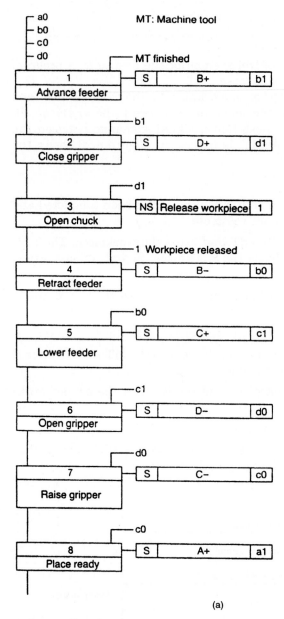

(a)

Figure 8.29 Function chart.

The function chart in Fig. 8.29 illustrates the specified cycle in detail form. Using this chart is necessary to develop a PLC program. Inputs must be assigned to sensors. Outputs from the PLC must also be assigned to actuators. Figure 8.30 shows the appropriate allocation lists. The sequencing program for the feed unit is shown in Fig. 8.31.

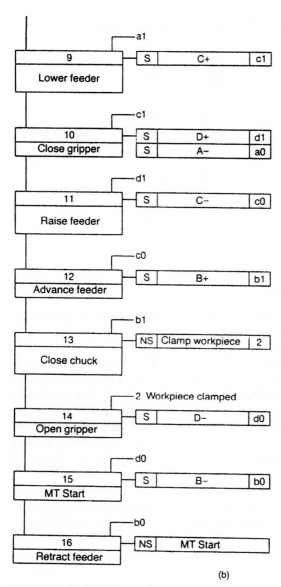

(b)

Figure 8.29 Continued

```
                INSTRUCTION              COMMENT
================================================================
0000 ALLOCATION LISTING MACHINE   V 01
0001                 I  1.0        " LIMIT SWITCH A0
0002                 I  1.1        " LIMIT SWITCH A1
0003                 I  1.2        " B0 HOR. FEED BACK
0006                 I  1.3        " B1 HOR. FEED FORWARD
0007                 I  1.4        " C0 VERT. FEED UP
0008                 I  1.5        " C1 VERT. FEED DOWN
0009                 I  1.6        " D0 GRIPPER OPEN
0010                 I  1.7        " D1 GRIPPER CLOSED
0011
0012                 I  2.0        " MT WORKPIECE CLAMPED
0013                 I  2.1        " MT WORKPIECE RELEASED
0014                 I  2.2        " MT FINISHED
0015
0016                 O  1.0        " CYL.A PREPARE
0017                 O  1.1        " CYL.B HOR. FEED
0018                 O  1.2        " CYL.C VERT. FEED
0019                 O  1.3        " CYL.D GRIP
0020                 O  1.4        " MT CLAMP WORKPIECE
0021                 O  1.5        " MT RELEASE WORKPIECE
0022                 O  1.6        " MT START
```

Figure 8.30 Allocation list.

8.7 Packaging Technology

Arranging and feeding of parts for insertion into boxes, including an intermediate layer, with a feed unit is illustrated in Fig. 8.32. Using this inserting machine, many other functions are carried out either simultaneously with or subsequent to feeding. The various parts which are fed by a conveyer belt are lined up independently and then arranged in rows by actuator A, according to a specific pattern. The parts are transferred by the lifting device in accordance with this pattern.

The horizontal movement of the feed unit is executed by actuator B; the vertical movement, by actuators C and D. In the position shown in the diagram, the parts picked up by the lifting device are inserted into the box by extending actuator D. At the same time, actuator C also extends and picks up a square board which is used as an intermediate layer as two layers of parts are being inserted. As soon as actuator C has retracted, holding the intermediate layer, the device is moved by the indexing actuator B. The end position actuator C located above the box and actuator D located above the collection point are used for the arrangement of parts.

When actuator C has extended, the intermediate layer is inserted and the next layer of parts is taken up by actuator D. When actuator B retracts, the position drawn is reached once again and the second layer is inserted into the box. Then, when actuator B extends once again, an intermediate layer is placed on the second layer and the next parts are picked up.

When actuators C and D retract for the second time, box changeover is triggered. Only when a new empty box is brought into position does a new insertion cycle begin.

```
            INSTRUCTION              COMMENT
===================================================================
0000 PROGRAM MACHINE    0.0 V 01
-------------------------------------------------------------------
0001 STEP 1
0002 IF              I 1.0          " LIMIT SWITCH A0
0003        AND      I 1.2          " B0 HOR. FEED BACK
0004        AND      I 1.4          " C0 VERT. FEED UP
0005        AND      I 1.6          " D0 GRIPPER OPEN
0006        AND      I 2.2          " MT FINISHED
0007 THEN   SET      O 1.1          " CYL.B HOR. FEED
0008        RESET    O 1.6          " MT START
-------------------------------------------------------------------
0009 STEP 2
0010 IF              I 1.3          " B1 HOR. FEED FORMARD
0011 THEN SET        O 1.3          " CYL.D GRIP
-------------------------------------------------------------------
0012 STEP 3
0013 IF              I 1.7          " D1 GRIPPER CLOSED
0014 THEN SET        O 1.5          " MT RELEASE WORKPIECE
-------------------------------------------------------------------
0015 STEP 4
0016 IF              I 2.1          " MT WORKPIECE RELEASED
0017 THEN RESET      O 1.1          " CYL.B HOR. FEED
0018       RESET     O 1.5          " MT RELEASE WORKPIECE
-------------------------------------------------------------------
0019 STEP 5
0020 IF              I 1.2          " B0 HOR. FEED BACK
0021 THEN SET        O 1.2          " CYL.C VERT. FEED
-------------------------------------------------------------------
0022 STEP 6
0023 IF              I 1.5          " C1 VERT. FEED DOWN
0024 THEN RESET      O 1.3          " CYL.D GRIP
-------------------------------------------------------------------
0025 STEP 7
0026 IF              I 1.6          " D0 GRIPPER OPEN
0027 THEN RESET      O 1.2          " CYL.C VERT. FEED
-------------------------------------------------------------------
0028 STEP 8
0029 IF              I 1.4          " C0 VERT. FEED UP
0030 THEN SET        O 1.0          " CYL.A PREPARE
-------------------------------------------------------------------
0031 STEP 9
0032 IF              I 1.1          " LIMIT SWITCH A1
0033 THEN SET        O 1.2          " CYL.C VERT. FEED
-------------------------------------------------------------------
0034 STEP 10
0035 IF              I 1.5          " C1 VERT. FEED DOWN
0036 THEN SET        O 1.3          " CYL.D GRIP
0037       RESET     O 1.0          " CYL.A PREPARE
-------------------------------------------------------------------
0038 STEP 11
0039 IF              I 1.7          " D1 GRIPPER CLOSED
0040 THEN RESET      O 1.2          " CYL.C VERT. FEED
-------------------------------------------------------------------
0041 STEP 12
0042 IF              I 1.4          " C0 VERT. FEED UP
0043 THEN SET        O 1.1          " CYL.B HOR. FEED
-------------------------------------------------------------------
0044 STEP 13
0045 IF              I 1.3          " B1 HOR. FEED FORWARD
0046 THEN SET        O 1.4          " MT CLAMP WORKPIECE
-------------------------------------------------------------------
0047 STEP 14
0048 IF              I 2.0          " MT WORKPIECE CLAMPED
0049 THEN RESET      O 1.3          " CYL.D GRIP
0050       RESET     O 1.4          " MT CLAMP WORKPIECE
-------------------------------------------------------------------
0051 STEP 15
0052 IF              I 1.6          " D0 GRIPPER OPEN
0053 THEN RESET      O 1.1          " CYL.B HOR. FEED
-------------------------------------------------------------------
0054 STEP 16
0055 IF              I 1.2          " B0 HOR. FEED BACK
0056 THEN SET        O 1.6          " MT START
0057       JUMP TO   S 1
0058       PSE
```

Figure 8.31 Sequencing program for the feed unit.

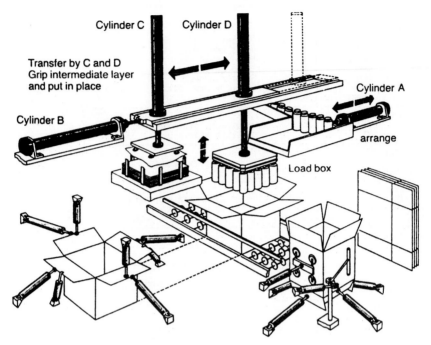

Figure 8.32 Insertion machine.

The feed functions carried out by this inserting machine include arrangement, separation, and insertion cycles.

8.8   Linear Indexing for Manufacturing Applications

Linear indexing is best-suited in applications where strip or rod-shaped materials have to be processed in individual stages throughout their entire length. An example of linear indexing is shown in Fig. 8.33.

A machine stamps holes into a strip of sheet metal at equal intervals. After each stamping process, the metal strip is clamped, pushed forward a specified distance, and clamped again. Pushing forward and clamping are carried out by a pneumatic feed unit controlled by various sensors. This feed unit is a compact handling component specifically intended for strip-indexing tasks.

Figure 8.34 illustrates indexing conveyer belts which are also referred to as *linear indexing systems*. Suitable devices for driving such a belt are either motors or pneumatic actuators with a mechanism such as a ratchet to convert an actuator linear stroke into a rotary movement. A pneumatic device is utilized in this case, as an actuator to drive the belt conveyer.

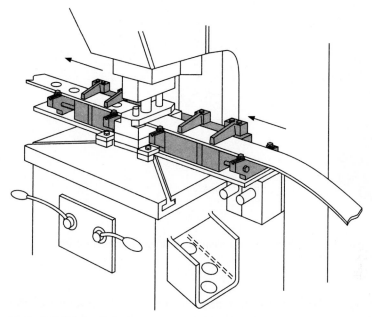

Figure 8.33 Linear indexing.

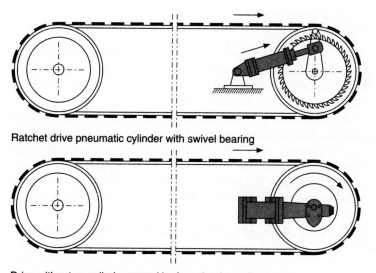

Ratchet drive pneumatic cylinder with swivel bearing

Drive with rotary cylinder moved by freewheel coupling

Figure 8.34 Generating indexing motion.

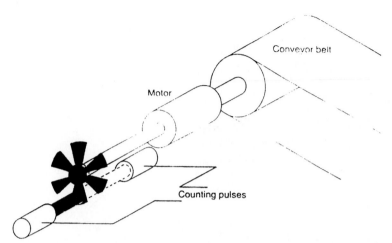

Figure 8.35   Conveyer belt with electric servomotor as actuator and sensor for angular displacement detection.

Figure 8.35 illustrates another application of an indexing driving mechanism of a conveyer belt with an electric servomotor as actuator and a sensor as an angular displacement detection.

8.9   Synchronous Indexing for Manufacturing Applications

Synchronous indexing systems are used when a number of assembly and fabrication processes are required to be carried out on a product. The product components must be fed once and remain on the synchronous indexing table until all processes are complete.

A vertical synchronous rotary indexing system picks up a minimum of two components. The complete product sequence can be carried out in one position without the need for repeated clamping and unclamping procedures. Rearrangement within the cycle becomes unnecessary. Feeding and withdrawing are limited to the last rotary indexing station; the number of manufacturing functions is thus without significance.

As illustrated in Fig. 8.36, each component is individually processed at six stations:

1. Station 1—insert and remove the workpiece

2. Station 2—drill the front face

3. Station 3—drill the top and base

4. Station 4—drill both sides at a right angle to the top and base

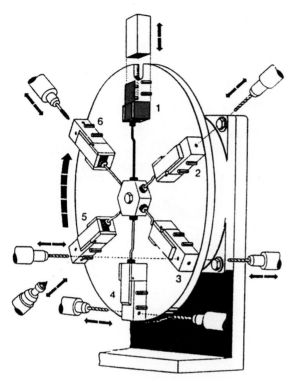

Figure 8.36 Vertical synchronous rotary indexing system.

5. Station 5—countersink the front face drill hole

6. Station 6—cut a thread into the drill hole on the front face

A horizontal synchronous rotary indexing table is illustrated in Fig. 8.37. This system consists of 8 stations. Stations 2 to 7 are the machining stations. This is just an example of the many diverse manufacturing processes which can take place in a synchronous fashion. Station 1 is the feed station; station 8 is the eject station.

The synchronous system can be driven by an electric motor with a special driver and cam configuration, by stepping motors, or by pneumatic rotary indexing units.

8.10 Parallel Data Transmission

Figure 8.38 illustrates the principle of parallel-type data transmission. Each data bit has its own transmission line. The data, e.g., alphanumeric characters, are transmitted as parallel bits from a transmitter such as the terminal to a receiver such as the printer.

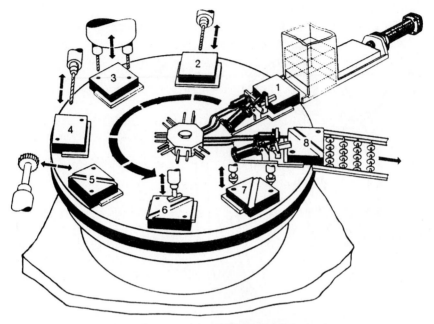

Figure 8.37 Horizontal synchronous rotary indexing table.

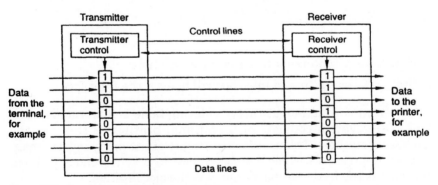

Figure 8.38 Parallel-type data transmission.

Some additional control lines are required to ensure synchronization between the transmitter and the receiver. Via these control lines, the printer informs the terminal whether it is ready to take over a character, and the terminal informs the printer that a character is being sent. The distance between the transmitter and the receiver may amount to a maximum of approximately 8 m, as line capacities may lead to coupling and signal deformation. Transmission speed is hardware-dependent and may theoretically amount to 1 Mbyte/s, i.e., 1 million characters/s—the line, however, can only be 1 m in length.

If several devices are to communicate with one another according to the principle of parallel transmission, precise coordination of the data flow is necessary (Fig. 8.39).

The devices are divided into four groups:

1. Talkers: able only to transmit, such as sensors

2. Listeners: able only to receive, such as printers

3. Talkers/listeners: can transmit and receive, such as floppy disks

4. Controllers: can coordinate data exchange, such as computers

Each device is allocated a specific address. The controller decides between which two devices data exchange is to be carried out. There are two standards describing the technique of transmitting interface functions: standards IEC-625 and IEEE-488. The number of data and control lines is specified in the standards as eight each. Up to 30 devices can be connected to the bus; the complete system of lines may not be longer than 20 m. Maximum transmission speed is between 250 and 500 kbytes/s.

In general, it can be said that relatively fast data transmission, such as several hundred thousand characters per second, can be achieved with parallel interfaces. However, transmission length is limited to several meters.

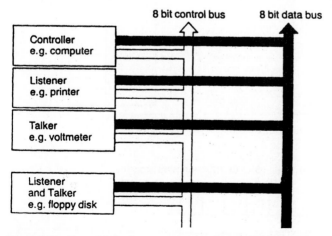

Figure 8.39 Several devices communicating with one another using parallel transmission, with precise coordination of data flow.

8.11 Serial Data Transmission

Figure 8.40 illustrates the principle of serial data transmission. Only one line is available for serial data transmission. This means that an 8-bit piece of data, for example, must be sent bit by bit from the send register via the data line to the receiver register.

A distinction is made between synchronous and asynchronous data transmission. With asynchronous data transmission, data can be sent at any time; pulse recovery is achieved through the start and stop bits.

In the synchronous process a transmitter and receiver have the same time pulse at their disposal. Control lines are necessary for this purpose.

Functional, electrical, and mechanical characteristics of serial data transmission can also be found in several standards. The most widely used standard versions are the V-24 and RS-232 interfaces (these interfaces are almost identical). With standards, it is guaranteed that different manufacturers' devices are able to exchange data. The V-24 is an asynchronous interface.

Synchronization is achieved by ready signals. The transmitter may, for example, pass data on to the receiver only when the receiver has previously signaled its readiness to receive.

The transmission sequence of the data bits determines the data format (Fig. 8.41). Before transmission of the first data bit, a start bit is

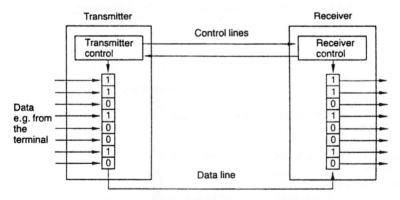

Figure 8.40 Principle of serial transmission.

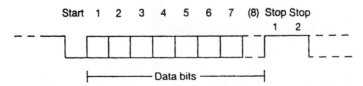

Figure 8.41 Transmission sequence of the data bits determines the data format.

sent at the logic zero level. Then 7 or 8 data bits follow, the one with the lowest value coming first. Then a test bit (parity bit) is transmitted, followed by one or two stop bits.

Transmission speeds are freely selectable within a specified framework. Transmitter and receiver must be set to the same speed, or baud rate. Common rates are 75, 110, 135, 150, 300, 600, 1200, 2400, 4800, 9600, and 19,200 bit/s (baud).

The electrical characteristics determine the signal level (Fig. 8.42). If the voltage of a signal on a data line opposite the signal ground is greater than 3 V and is negative, signal status 1 applies, corresponding to the binary character 1. If the voltage of a signal on a data line opposite the signal ground is greater than 3 V and is positive, signal status 0 applies, corresponding to the binary character 0.

Signal status is undefined in the transmission range +3 to –3 V. Connection of the interface line between two devices is effected with a 24-pin D plug. Figure 8.43 shows a V-24 plug with the most important signal lines.

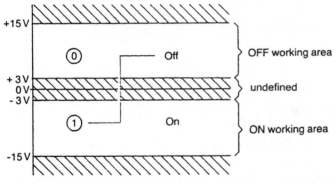

Figure 8.42 Electrical characteristics determine the signal level.

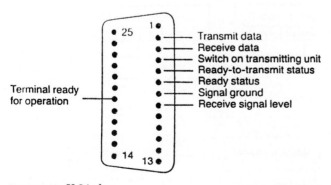

Figure 8.43 V-24 plug.

This is one of the oldest serial interfaces. It is used to drive teleprinters [teletypewriters (TTY)]. Transmission takes place along a pair of send and receive lines. Logic 0 is realized via a 20-mA current, and logic 1, via the lack of this current. The functional characteristics of the V-24 interface may also apply to the 20-mA current loop interface.

8.12   Collection and Generation of Process Signals in Decentralized Manufacturing Systems

Not long ago it was customary to centrally collect all process signals. The processor and control elements were housed in a central control cabinet (Fig. 8.44). Today, with automation systems, extensive control tasks can be solved effectively if these tasks are distributed among several decentralized self-sufficient systems.

A decentralized structure is illustrated in Fig. 8.45. The sensing, actuating, and processing sections of subsystems are connected to a

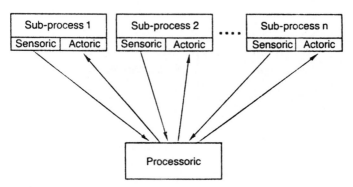

Figure 8.44   Centralized control system.

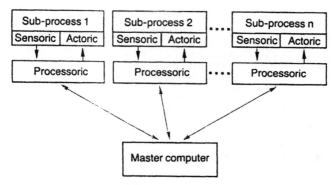

Figure 8.45   Decentralized control system.

master computer which evaluates signals applied by each subprocess and sends back control signals.

These subsystems can be realized in different ways; e.g., a PLC or single-board computer could be used as the control system for the subprocesses. A process computer, personal computer, or mainframe can be used as a master computer.

Figure 8.46 shows a simple control task from chemical process technology as an example of the networking of various systems. In a liq-

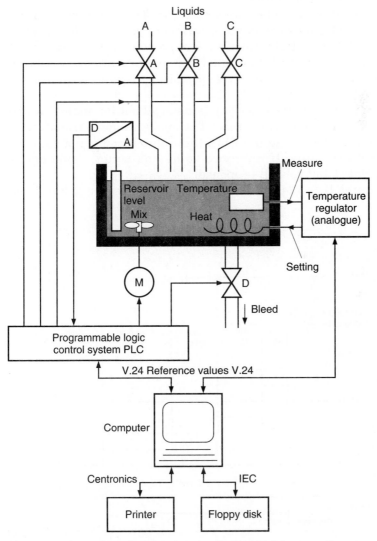

Figure 8.46   Networking of various systems in liquid mixing operation.

uid mixing plant, three different liquids, A, B, and C, are mixed in a specified proportion. Filling is effected by three valves, A, B, and C, and the mixture is extracted via one valve, D. Before the mixture is removed, it is heated to prespecified reference temperature T.

A PLC controls the valves and the mixer. The temperature of the mixer is adjusted to the reference temperature by a two-step controller.

An overriding computer prepares the reference values for the mixture and the mixing temperature and coordinates the PLC and regulator. In addition, the computer stores operational data and downloads the data to the printer as documentation.

Figure 8.47 illustrates the control process. The reference values 1 and 2, which determine the mixture ratio, are passed on by the computer to the PLC. The PLC controls the remainder of the sequence.

First, valve A is opened and liquid A is poured into the reservoir. The liquid level rises until the level specified by reference value 1 has been reached.

The liquid level in the reservoir is determined by analog means using a capacitive measuring probe. The analog value is transformed into 8-bit digital values. In the PLC, this value is compared with the stored reference value. If reference value and actual value agree, valve A is closed and valve B is opened.

Next, liquid B is poured in until reference value 2 has been

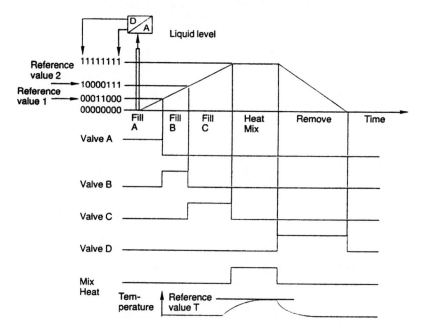

Figure 8.47 Control process in liquid mixing operation.

reached. Then valve B is closed and valve C opened. Liquid C is poured in until the maximum value (reservoir full) is reached. Reference values 1 and 2 determine the mixture ratio A:B:C.

On reaching the maximum level of the liquid, the mixing motor switches ON and sends on a signal to the computer, which, in turn, switches ON the temperature regulator. The regulator adjusts the temperature to the reference value and informs the computer when the reference value has been reached.

Next, the computer requires the PLC to switch ON the mixing motor and to open the bleed valve D. Once the reservoir has been emptied, the cycle is complete.

A new cycle is now started up with the same or modified reference values. Data communication between the PLC regulator and the computer is carried out serially via V-24 interfaces. Computer and printers communicate via a parallel interface, and computer and floppy disk, via a parallel IEC interface.

Further Reading

Boyes, G. S., (ed.), *Synchro and Resolver Conversion,* Memory Devices Ltd., Surrey, U.K., 1980.

Burr-Brown Corp., "Wiring and Noise Considerations," *The Handbook of Personal Computer Instrumentation,* 4th ed., sec 7, Burr-Brown, Tucson, 1989.

Cole, R., *Computer Communications,* Macmillan, London, 1986.

Grant, D., "Applications of the AD537 IC Voltage-to-Frequency Converter," application note, Analog Devices Inc., Norwood, Mass., 1980.

Hagill, A. L., "Displacement Transducers Based on Reactive Sensors in Transformer Ratio Bridge Circuits," *Journal Phys. E: Instrum.,* **16:** 597–606 (1982).

Higham, E. H., *Pneumatic Instrumentation in Instrument Technology: Instrumentation Systems,* Butterworths, London, 1987.

Hilburn, J. L., and D. E. Johnson, *Manual of Active Filter Design,* 2d. ed., McGraw-Hill, New York, 1983.

Huang, S. M., A. L. Scott, and M. S. Green, "Electronic Transducers for Industrial Measurement of Low Value Capacitances," *Journal of Phys. E: Sci. Instrum.,* **21:** 242–250 (1988).

Netzer, Y., "Differential Sensor Uses Two Wires," *Electronic Design News,* March 31, 1982, p. 167.

Oliver, B. M., and J. M. Cage, "Electronic Measurements and Instrumentations," McGraw-Hill, New York, 1971.

Paiva, M. O., "Applications of the 9400 Voltage to Frequency, Frequency to Voltage Converter," application note 10, Teledyne Semiconductor, Mountain View, Calif., 1989.

Tompkins, W. J., and J. G. Webster (eds.), *Interfacing Sensors to IBM PC,* Prentice-Hall, Englewood Cliffs, N.J., 1988.

Economic and Social Interests

This chapter is particularly written for those who possess the talents of innovation but need economic know-how to implement their dreams. It is written for all my students, my peers, and the entrepreneurs who wish to grasp the simple yet practical understanding to build their new manufacturing business with sound financial strength, attempting to make America a better America.

This chapter is dedicated to my students, the hope of tomorrow, whose eyes hold the sparks of inspiration, who see the challenges to our country and rise to them. They can take vision and creativity, mixed with the discipline of science and thereby shape the future. It is dedicated to them in hope that they take the fire of their enthusiasm and turn their ideas into reality. My sincere desire is that students foster their creativity, nurture their confidence, and take the risk to establish their enterprise.

9.0 Financial Planning and Control of Manufacturing Operations

Engineering in the absence of economics is a degenerate science.

The small manufacturing firm, just as surely as the larger manufacturer, must have yardsticks for measuring the results of every activity of its business operation. For this purpose, management should take the following steps:

1. Define its objectives precisely

2. Prepare a written statement of the steps to be taken to accomplish these objectives

3. Refer continually to its statements of objectives to see that its operation is proceeding according to schedule

For some manufacturers, the budget may be simply an informal sales forecast, production schedule, and profit forecast. If only one person is responsible for seeing that activities proceed in accordance with plans, the budget becomes a simple basis for action by a top executive. It helps an executive make on-the-spot decisions that are in line with an overall plan. It charts the course a top manager should take in carrying responsibility for many different functions.

If the business grows to the point where selling, producing, and perhaps financing responsibilities are delegated to different persons, a more specific and formal plan becomes important. The budget must still direct the decisions of the owner-manager and key executive of the firm, but in addition it must coordinate the actions of the different members of the group.

9.1   Developing a Plan

A manufacturer's budgeting procedure must be simple. It must be focused on a limited number of functions that are important to the special circumstances. All individuals who are to have any responsibility for seeing that it works should take part in planning it. The overall budget for most manufacturers should include the following parts:

1. Sales budget
2. Production budget
3. Selling and administrative budget
4. Capital budget
5. Financial budget

The first four of these budgets taken together provide a basis for projecting profits, since the sales budget provides an estimate of revenues and the other three of costs. The financial budget provides a plan for financing the first four plans. Each of the five budgets should be made in advance for periods of 6 months or a year.

9.1.1   Sales budget

The accuracy of any budget or plan for profitable activity depends first on the accuracy of the sales forecast. In many small firms, the volume of sales may be affected substantially by the condition of a single new customer or the loss of a single old one. The sales forecast in these cases has to include a good deal of guesswork.

Nevertheless, a forecast should be made. It will usually be based mostly on past sales figures, but sales tendencies and economic conditions that are likely to increase or decrease sales volume in the coming period should be studied.

A sales forecast or budget that might be suitable for a typical small plant appears in Table 9.1. Forecasted sales are stated in both product units and dollars and are broken down by sales territories.

9.1.2 Production budget

Once the sales forecast has been made, the production budget can be prepared. Enough units of products 1 and 2 must be produced to supply the estimated sales volume and maintain reasonable stock levels. In this example, it has been decided that about a 2-month supply should be kept in stock. If the inventories at the beginning of the period are below that level, production will have to exceed the sales forecast. Table 9.2 shows how the quantity to be produced can be figured.

On the basis of the number of production units required, budgets or standards for material, labor, and overhead costs are prepared. These production costs should be figured in detail so that cash requirements, material purchasing schedule, and labor requirements can be set up. Detailed plans are then made for monthly production levels.

TABLE 9.1   Sales Forecast

| Sales territory | Product 1 | | Product 2 | | Total dollar sales |
	Product units	Dollar sales	Products units	Dollar sales	
A	1850	14,800	2200	48,400	63,200
B	575	4,600	500	11,000	15,600
C	1400	11,200	1000	22,000	33,200
Total	3825	$30,600	3700	$81,400	$112,000

TABLE 9.2   Required Production Based on Sales Forecast

	Product 1	Product 2
Sales forecast	3825	3700
Desired inventories (⅙ annual sales)	637	617
Product units required	4462	4317
Less: beginning inventories	200	400
Total production required	4262	3917

9.1.3 Selling and administrative budget

The sales volume of a small plant usually does not provide a large enough margin over production cost to overcome high selling and administrative costs. These nonmanufacturing overhead items must be watched constantly to see that commitment for fixed costs are kept low.

This can be done best by using a selling and administrative budget. All cost items expected in these areas should be listed and classified according to their fixed and variable tendencies. Account classifications in the accounting system should correspond to the classifications used in the budget.

At the end of each month, the fixed, variable, and total selling and administrative costs budgeted for the expected level of operation should be a adjusted for the level actually reached during that month. Actual costs should be matched against these adjusted figures and any sizable differences investigated. If possible, action should be taken to prevent repetition of any unfavorable variances.

9.1.4 Capital budget

Plans for acquisition of new buildings, machinery, or equipment should be developed annually. A list of amounts to be spent at specified times during the year will provide information for use in a firm's financial budget as well as cost and depreciation figures to be shown on its projected income statement and balance sheet.

9.1.5 Financial budget

Any business owner must be certain that funds are available when needed for plant or equipment replacements or additions. In addition, enough working capital must be available to take care of current needs. This requires a financial plan or budget.

Funds for major replacement or addition can ordinarily be planned for on the basis of each expenditure individually. Management must know in each case whether surplus working capital will be available or new long-term financing will be required. if new financing is the answer, a source of funds must be found. Plans for a loan or an additional investment by owners should be made well before the time when the funds will be needed.

To have enough working capital at all times without a large oversupply at any time requires detailed planning.

Planning the receipts and expenditures for each month must take into consideration expected levels of activity, expected turnover of accounts receivable and accounts payable, seasonal tendencies, and any other tendencies or circumstances that might affect the situation at any time during the budget period. Relating the inflow plus beginning balance to the outflow for each month shows whether a need for outside funds is likely to arise during the budgeted period.

9.1.6 Income statement and balance sheet projections

From information provided by the four basic budget summaries—sales, production, selling and administrative, and financial—estimated financial statements can be drawn up. These projected statements bring the details together. They serve to check the expected results of operations and the estimated financial position of the business at the end of the budget period.

9.2 Planning for Profit

Projection for income statement and balance sheet figures will indicate the profit, return on investment (ROI), and return on assets (ROA) management can expect. For example, these statements in highly abbreviated form may be as follows:

Net profit before interest and taxes	$ 430,000
Interest cost	80,000
Net profit after interest	350,000
Income taxes (40%)	140,000
Net profit	210,000
Debt	1,000,000
Capital or stock investment	2,000,000
Assets	$3,000,000

The return of investment, i.e., the profit earned in relation to the value of the capital required to produce the profit, is expected to be

$$\$210,000/\$2,000,000 = 10.5\%$$

Further, the return on assets is expected to be:

$$\$258,000/\$3,000,000 = 8.6\%$$

Note. In this last calculation, the return is calculated after tax and before deduction of interest cost. The $258,000 is determined either by adding the after-tax interest cost of 60% of $80,000, or $48,000, to the net profit, $210,000, or by subtracting the 40% tax on the net profit before interest or taxes of $430,000 from this amount. The 8.6%

return on assets is increased to 10.5% return on investment as the result of the leverage provided by borrowed funds at an interest of less than 8.6%.

These ROI and ROA figures may or may not be considered adequate. If they are thought to be inadequate, various parts of the budget should be reviewed to identify possible areas of improvement. In any case, the profit goal should be high enough to provide a reasonable return and at the same time be realistically attainable.

9.2.1 Controlling operations

Budgeting and profit planning are necessary for control of operations, but the forecasts and budgets do not of themselves control operating events. A budget is useless without comparisons with actual operating results, and the comparisons are useless unless the manager takes action on deviations from the budget.

Suppose, for example, that the sales of product 1 in territory A are budgeted at 155 units per month (1850 divided by 12), and actual sales for January and February are 120 and 110 units. (Assume that sales of product 1 in territories B and C are in line with the budget.) On investigating, the manager finds that sales are being lost to a new competitor in territory A.

The actions the manager might take include these:

1. Put more money into the selling effort and try to win back the old customers

2. Lower the price of product 1 to obtain more sales

3. Recognize that the sales are lost, and lower the production budget accordingly

Each of the these actions will have an impact on the budget-planning operations of the company. An increase in the selling effort will affect the selling and administrative budget and the financial budget—more funds will be needed. Lowering the selling price, if the unit sales level originally forecast is just regained, will result in lower cash receipts than in the original financial budget. A decision to reduce production to avoid piling up inventories will affect purchase orders, labor costs, variable overhead costs, income taxes, and other budgeted items. The actions have been signaled by a variation from the sales forecast, but a change in plans affects all the budgets and forecasts.

Any sizable variance from budgeted figures, favorable or unfavorable, should be looked into. Management should find the cause so that when it takes action it will be reasonably sure that it is acting wisely. For example, a favorable variance in direct labor costs might be due to a special incentive program. In such a case, management should put off any plans to use the surplus funds until it is sure that reduction in labor costs will be permanent.

Unfavorable variances from the budget should not be studied in isolation from related budget items. For example, suppose the actual selling expense is much higher than the amount budgeted. It may turn out that the extra expense produced a large volume of new orders. In that case, the added cost may be entirely justifiable. Hasty action to reduce the expense without investigation might result in loss of sales and a net reduction in profits.

Great care must be given to seeking out the causes of variances, and the consequences of possible control action should be analyzed carefully in order to avoid unintended and unwanted results.

9.2.2 Cost and profit analysis

In direct cost, costs are classified as variable (product costs) and fixed (period costs). With these classifications, management can determine which of its products contributes the most and which the least to covering fixed cost and providing a profit. Direct costing provides information for decisions to expand, reduce, or continue production of a given product line or product.

In a standard-cost system, product costs for materials, labor, and overhead are budgeted or planned before the actual production takes place. The preestablished costs then provide a point of reference for analyzing the costs of actual production. By analyzing the variances from these standards, the manufacturer can identify the sources of excessive cost and investigate the causes. This kind of process—the identification of a problem area and the discovery of its cause—can point the way to actions that will reduce costs and improve profit.

9.3 Information Assimilation and Decision Making

Similarly, a business manager should plan ahead for all the activities of the business. That is, the manager should develop a set of budgets for sales, production, selling and administrative expense, and financial requirements. As indicated in the standard-cost system, the actual results of various activities can be compared to the budgeted or planned results. Analysis of the variances found will then help the manager to make reasonable and thoughtful decisions.

These analytical methods—direct cost, standard-cost variance analysis, and budget planning and control—are ways of putting cost-accounting information to good use. The manufacturer can make decisions on the basis of information provided by these analytical tools. There are, however, several other important ways of analyzing costs and profits. Two of these—break-even analysis and incremental analysis—are illustrated below.

9.3.1   Break-even analysis

Sales forecasts are uncertain at best. For this reason, it is important for management to know approximately what cost changes can be expected to go along with volume changes. It must know what levels of production and sales volume are necessary for profitable operation. It must know how much effort and cost are justified to keep volume at a high level. Break-even analysis provides this sort of information.

Break-even analysis is so called because the focal point of the analysis is the break-even point—the level of sales volume at which revenues and costs are just equal. At this level, there is neither profit nor loss.

9.3.2   Break-even chart and formula

A break-even chart is illustrated in Fig. 9.1. A break-even chart can be prepared from budget figures and a knowledge of capacity levels. In Fig. 9.1, it is assumed that fixed costs for production are $20,000 and for selling and administrative activities, $10,000. Variable production costs are $12.50 per unit, or a total of $50,000 at the maximum capacity of 4000 units. Variable selling and administrative costs are $5.00 per unit, or $20,000 at the 4000 level. This gives minimum costs of $30,000 (the total fixed costs) at zero production and maximum total costs of $100,000 at 4000 units. The selling price is $35 per unit.

A total cost line is drawn from the $30,000 cost level for no production to the $100,000 level for maximum production and sales. A total revenue line is drawn from the zero line for no revenues to $140,000 for the maximum sales of 4000 units. The point where these two lines cross is the break-even point—the level of operation at which there will be neither profit nor loss.

The area enclosed by the two lines below the break-even point represents loss, and the enclosed area above the break-even point represents profit.

Figure 9.1 shows that the volume at which the business can be expected to break even is slightly below one-half the maximum point, or between 1500 and 2000 product units. The exact level can be calculated as follows:

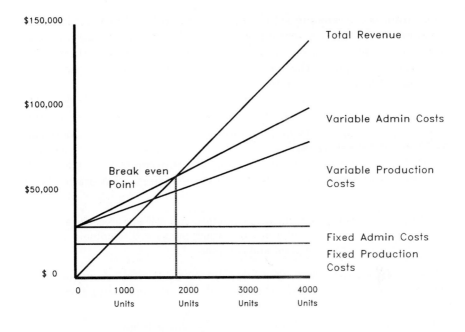

Figure 9.1   Break-even chart.

$$\frac{\text{Fixed costs}}{\text{Selling price} - \text{variable costs/units}} = \frac{\$30,000}{\$17.50} = 1714.29 \text{ units}$$

Verification:

Gross revenue (1714.29 units at $35)	$60,000
Less: variable costs (1714.29 units at $17.5)	30,000
Contribution margin	30,000
Less: fixed costs	30,000
Net profit or loss	$ 0

9.3.3   Utilization of break-even analysis

A break-even chart can give approximate answers to many questions. For example, if the variable production costs are expected to increase 10 percent in the coming year without any change in the selling price of the product, the total cost line will be drawn steeper, and a new, higher break-even point will result. This will show what increase in sales volume is needed to offset the increased costs.

Or suppose a change in selling price of the finished product is being considered. The total revenue line will now be steeper for an increase or less steep for a decrease in selling price. The chart will then show the effect on profit if the sales volume remains the same or if, for example, it drops—as it might if the price is increased.

Suppose the manufacturing firm whose figures are used in Fig. 9.1 is producing and selling 2000 units, thus making a profit of $5000. Now suppose it wants to give its employees a general wage increase. As planned, the increase has the effect of adding $1 to the variable costs of each unit (thereby decreasing the contribution by the same amount) and $3000 to fixed costs. The owner wants answers to these questions:

> How many units must the owner produce and sell at $35 each to realize the same profit—$5000? Can the owner produce this volume without investing any more in plant and equipment?

The required units are calculated by using the following formula. (The calculation can be verified as described in the preceding formula.)

$$\frac{\text{Fixed costs} + \text{decreased profit}}{\text{Selling price} - \text{variable costs/unit}} = \frac{\$33,000 + \$5000}{\$16.50} = 2303 \text{ units}$$

Thus, the manufacturing firm will have to produce and sell 2303 product units instead of 2000 in order to realize the same $5000 profit. Since the plant capacity is 4000 units, management can increase production to 2303 units without any further investment in plant and equipment.

Break-even charts give quick approximate answers. They should not take the place of detailed calculations of the results of anticipated changes, but they do encourage careful consideration of the effects of any increases in either fixed or variable costs. They can also be a constant reminder of the importance of maintaining a high sales volume.

9.3.4 Incremental analysis

Incremental costs and revenues are those that change with increases or decreases in the production level. Fixed costs as well as variable costs may change under certain conditions. Incremental analysis is an examination of the changes in the costs and revenues related to some proposed course of action, some decision that will alter the production level or change production activity. Two kinds of incremental analyses are explained here.

9.3.4.1 The large new production order. Sometimes it happens that a manufacturer has an opportunity to land a new customer and a very large order for goods. Usually, such a large order cannot be landed without adding new production employees and supervisors and increasing overhead costs. The new order thus creates a decision-making problem in which basic information differs somewhat from the current cost data. A study of the incremental costs and revenues—those that arise from the new contract—provides information on which the manufacturer can base the decisions to accept or reject the new order.

Suppose that a manufacturer currently produces 10,000 product units a year. Fixed costs are $20,000 and variable costs, $5 per unit—$2 for materials, $1 for labor, $1 for variable production overhead, and $1 for variable selling expenses. The unit selling price is $10. A new customer wishes to buy 10,000 units a year at $9 per unit. A new night shift would be required to double the plant's production as required by the new customer's order.

Costs at the original level of production (10,000 units) are not necessarily accurate for the new level (20,000 units). Assume that the costs of producing the 10,000 additional units required by the new order are estimated as follows:

Variable costs per unit	
Materials	$2.00
Labor (including night shift premium)	2.00
Variable production overhead	1.25
Variable selling costs	0
Total variable costs	$5.25
Incremental fixed costs	
Additional supervisors	$10,000
Power, light, heat	5,000
Office expenses	3,000
Total incremental fixed costs	$18,000

Note: For accuracy, it is important to use current or projected prices of materials, labor, and overhead in this type of incremental analysis.

Average prices or allocated costs derived from historical cost accounting records (for example, material costs based on first-in, first-out method) may not produce current figures which can be relied on in making a decision.

The financial results of accepting the new contract would therefore be as shown in Table 9.3.

TABLE 9.3   Financial Results

	Original production	New order	Total
Gross revenues	$100,000	$90,000	$190,000
Variable costs	50,000	52,000	102,000
Contribution margin	50,000	37,000	87,000
Fixed costs	20,000	18,000	38,000
Profit before taxes	$30,000	$19,500	$49,500
Return on sales	30.0%	21.7%	26.1%

The incremental analysis shows that the new contract will be profitable, but the rate of return will not be as high as for the original level of production. The computation would have given different results if the original variable costs of $5 per unit and the fixed cost of $20,000 had been used. Incremental analysis makes it possible for a manufacturer to examine the probable results of making major changes in the production level and in the arrangement and scheduling of production activity.

9.3.5   Make-or-buy analysis

Most manufacturing firms buy some parts from outside suppliers for use in assembling the finished products. Often the manufacturing firm could make the part in its own plant instead of buying it from an outside source. The question is whether the cost of making the product itself would be less than the cost of purchasing it from an outside supplier. An analysis of these two alternatives is known as a *make-or-buy analysis*.

Take, for example, a manufacturing firm that now purchases 10,000 small electric motors a year at $4.80 each after deducting quantity discounts. The question it faces is this: can it make the motor itself for $4.80 or less?

Assume that the costs of producing 10,000 motors a year in its own plant would be as follows:

Variable costs	
Materials per unit	$2.00
Overhead costs per unit for the new activity (power, suppliers, etc.)	0.75
Total variable cost per unit	$2.75
Total variable cost per 10,000 units	$27,500

Fixed costs	
Salaries of 3 technicians (capable of producing 15,000 units a year)	$21,000
Depreciation on special machinery ($40,000 spread over a 5-year life)	8,000
Total fixed costs for the new activity	$29,000

Total costs	
Total cost of producing 10,000 motors in plant	$56,500
Total cost per unit	$5.65

Thus, with a production level of 10,000 units a year, the manufacturing firm would be better off buying the motor from a supplier at $4.80 each. However, with the three technicians and the $40,000 in special equipment, it has the capacity to produce 15,000 motors a year. At this level, its total cost would be $70,250 and its unit cost $4.68. It would save at least $0.12 per unit, or $8750 over the 5-year useful life of the special equipment. So the decision to make or buy the motor would depend on the volume needed.

On the other hand, the owner might decide that the $1750 a year would not be enough to offset the extra burden that policing the motor activity would place on key people. This assumes, of course, that the present supplier delivers quality motors on time.

Another analysis is also involved in this decision, however—the determination of whether the $8750 savings is enough to justify investing $40,000 in special equipment, or whether the money could be invested to better advantage elsewhere. This decision is known as a *capital budgeting decision*.

9.3.6   The importance of the basic data

The techniques of cost and profit analysis discussed above are intended as guides. The validity of the method depends largely on the validity of the basic data—the fixed and variable cost classifications, cost estimates, and volume estimates. Errors in any one of these factors will cause mistakes in the computations and may result in erroneous decisions.

Every manufacturing firm should continually review the way its costs are classified and accumulated, and it should be very careful in making cost estimates. Its general accounting system and cost-accounting system should be reviewed periodically to ensure that basic cost data are collected and classified according to an established accounting plan.

9.4   Communications

In a business organization, gathering and using information about the operation of the business may be a fairly simple matter. A purchasing agent, production manager, or sales manager who has only a few employees can observe day-to-day activities and exercise direct supervision and control. When a business becomes larger, however—when more employees and several levels of supervision are necessary—the processes of obtaining information and directing the business become more complex.

The larger business requires a communication system for collecting information about operations and distributing it to managers in the company. Such a system includes four processes:

1. Accumulating

2. Analyzing

3. Reporting

4. Communicating

So far, discussion has emphasized techniques for accumulating, analyzing, and reporting cost-accounting data. Attention must be given to methods of communicating the information to those in the firm who need it in order to accomplish their tasks.

Communication involves both a sender and a receiver of information. A manager sets up an accounting system as a sender. The manager assures that a system produces certain reports and analyses of the basic counting data.

If the information is to be useful, the needs of those who will receive it must be known and considered. Thus, communication depends on a two-way exchange between the senders and the receivers. Figure 9.2 illustrates a communication network in a business organization; the arrows show the two-way flows of information between senders and receivers.

Communication also requires that the information sent and received be useful. To be useful, information must meet the following standards:

1. It must be suited to the needs of the receivers in form and content

2. It must be free from clerical errors, incorrect classification, and unrealistic estimates

3. It must be presented in clear language and not buried under a mass of unnecessary details

4. It must be received soon after the actual events, in time for suitable action to be taken.

9.4.1   The need for timeliness

There is bound to be some lag from the time an activity is completed to the time a report of that activity can be made available. It takes

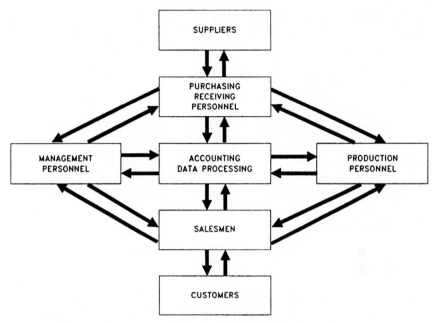

Figure 9.2   Communication network in a business organization.

time for the clerical staff to sort and total materials requisitions and time tickets for posting to job-cost sheets or for use in cost-system calculations. It takes time to prepare data for electronic data processing (EDP), deliver it to a service bureau, and obtain the printed output. If a report is delayed very long, it may be useless by the time it finally reaches the people who need it.

Each manufacturing firm must decide in which areas of operation quick reports are needed. For example, if a machining department is overloaded with work, this should be known promptly so that workers can be shifted to prevent a production bottleneck. Or perhaps work can be routed to another department where machine time is available. For this kind of scheduling, daily or even hourly performance may be needed. Activity reports must be communicated quickly where the situation is likely to change often.

Other situations may require reports on different schedules. Since labor efficiency usually changes slowly, reports on labor hours and labor costs may be prepared weekly or monthly. Material-price variance reports may be prepared at the end of a periodic buying season or after the second week in a month when vendors' invoices are received. Materials usage and efficiency reports may be prepared daily, weekly, or monthly according to the characteristics of the manufacturing process and the needs of the production manager.

In general, the amount of clerical time, effort, and cost put into a communication system is related to the importance of the activities being reported. If the manufacturing process uses a great deal of raw materials and parts, automated equipment, and a minimum of labor time, the accounting system should produce frequent reports on material prices and usage and on equipment maintenance and repair costs. Labor costs can be reported less often because they are not a large part of total production cost.

On the other hand, if labor costs do make up a large proportion of the total costs, the communication system should emphasize labor efficiency and cost. Materials cost can be reported less frequently. The manufacturing firm must decide how much it can afford to spend on its communication system and how valuable the information is to its managers.

9.4.2   Reporting strategy

It takes time to accumulate, calculate, and record cost data. Therefore cost information is generally communicated in monthly, quarterly, and annual reports. The reports usually cannot be delivered in time to be useful on an hourly, daily, or weekly basis. Consequently, the more frequently issued reports contain only physical data—material units, labor units, machine hours, number of employees, machine downtime, work-in-process backlog, material spoilage, etc.

These reports in terms of physical units (without dollar figures) are useful for managers, who can take direct action to affect use of materials, labor hours, and certain types of overhead spending. Cost reports in dollar terms are more useful for managers who can take action that affects prices, wage rates, overhead spending, etc.

The form in which information is communicated can also be important. It varies according to the circumstances and the one who is to receive it. Sometimes an oral report is enough to get the necessary action started, but more often reports are written or printed. Reports issued frequently may be handwritten on standard forms in order to get them out quickly. Other reports may be typed or may be printed out by EDP equipment.

In some cases, visual displays, in the form of large graphs or drawings are used. For instance, displays showing employee output, employee safety performance, and other information of interest to workers can be shown in plant areas. These displays when projected with the name and picture of an employee help to keep the employees informed and interested in their job performance.

All but the smallest of manufacturing businesses are usually organized so that different managerial duties are assigned to different persons. Along with each manager's authority goes a responsibility for managing the segment of business efficiently. Figure 9.3 shows a simple organization chart for a typical manufacturing plant.

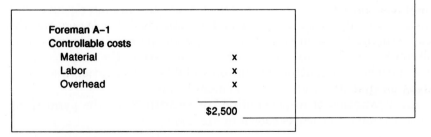

President
Controllable costs
Selling expenses   x
  Accounting and office
    expenses   x
  Production costs $73,000
                              ──────────
                              $101,000

Production manager			
Controllable costs	Prod A	Prod B	Total
Supervisory level	$22,000	x	x
Controllable at manager level:			
Salaries	x	x	x
Depreciation	x	x	x
Overhead	x	x	x
	$35,000	$38,000	$73,000

Production Supervisor A			
Controllable costs	Dept A–1	Dept A–2	Total
Department level	$2,500	x	x
Controllable at supervisory level			
Material	x	x	x
Labor	x	x	x
Overhead	x	x	x
	$10,000	$12,000	$22,000

Foreman A–1
Controllable costs
  Material   x
  Labor   x
  Overhead   x
                              ─────────
                              $2,500

Figure 9.3 Simple organization chart for a manufacturing plant.

In responsibility reporting, information is reported on the basis of areas and responsibility. For example, reports containing information useful to foremen for performing their jobs will be directed to departmental foremen.

This means that the responsibility accounting and reporting system will be only as detailed as the company's organization plan. In a plant with the organization shown in Fig. 9.3, reports could be prepared for the foremen, the supervisors, the purchasing agent, the production manager, the sales manager, the salespeople, and the president. A company that has other organization units must work these into the plan if those units are to be served by responsibility-oriented reports.

A responsibility-oriented accounting system should generate reports that provide information on which each manager can base decisions. The system should also provide reports showing the results of these decisions.

9.5   Responsibility Centers

The responsibility centers may be cost centers, revenue (sales) centers, or profit centers. In Fig. 9.3, each producing department and each of the two service departments (purchasing and accounting) can be considered a cost center if the costs of operations are analyzed along those lines. Each product can be considered a profit center, with the responsibility held jointly by the sales manager (of revenues) and the product supervisors (for product costs). The sales office can be considered a revenue center that produces the gross sales revenues of the company.

9.5.1   Controllable and uncontrollable costs

Costs and revenues are reported to managers according to their controllable and uncontrollable characteristics. A cost item is controllable by a manager if action directly affects the amount of cost incurred.

Assume, for example, that the product managers have sole authority to purchase equipment, thus incurring depreciation costs. The depreciation cost is then controllable at the manager's level but uncontrollable at the department foreman level. Consequently, the department foremen cannot be held responsible for depreciation charges, and that cost is not reported as part of the responsibility. The principle of controllable-uncontrollable classification should be treated consistently at all management levels.

An illustration of responsibility cost reporting and the pyramiding construction of reports is shown in Fig. 9.3.

9.5.2   Construction of the reports

Each report delivered to a manager should be constructed according to the usefulness to the manager of information provided by the accounting system. The actual costs can be compared with the budgeted figures, standard-cost variances can be shown, and unit-cost calculations can be made. Any presentation of the information that promotes for sound decision making is suitable.

The classification of costs as controllable or uncontrollable can be difficult. One of the most troublesome problems is obtaining the agreement and cooperation of the foremen, supervisors, and managers who will be assigned certain responsibilities. It is important to make certain that the reports will be used for making decisions and directing activities. To this end, the supervisory employees involved should take part in planning the reporting system. Preparation of format and content of responsibility reports can serve to involve all the managers in the communication-information system, and the whole process can serve to bring the managers together into a smooth-working team.

9.5.3   Analyzing cost and profit data

The responsibility-center accounting scheme is just one way to organize cost and profit data for analyzing the results of operations. For a firm that produces and sells more than one product in more than one geographical area to customers that have a variety of characteristics, there are many other possible types of analysis. Several are listed below:

1. *Product lines.*   Analysis by product lines can show which products are most profitable. It can point up the need for cost-saving programs or identify products for maximum sales promotion.

2. *Special orders.*   The results of the production of special orders should always be analyzed to determine whether producing to custom-design specifications is really profitable. Such analyses could lead to special pricing formulas, better cost-estimating procedures, and more efficient use of labor and equipment.

3. *Sales analysis.*   The costs of selling products can be analyzed according to various customer characteristics, taking into account the cost of various selling efforts, as follows:

   a. *Customer groups.*   Sales costs and profits can be classified by customer groups. This is especially valuable when a few customers account for a large part of the sales.

b. *Industry.*   Analysis of sales, costs, and profits of sales to certain industries may be useful in planning efforts to sell to growing industries.

c. *Geographical areas.*   The cost of shipping goods to faraway customers may consume a large part of the gross or contribution margin. Analysis by areas can help the sales manager assign salespeople where they can obtain the most profitable results.

d. *Order size.*   Quantity discounts must be watched. Also, classification of sales, costs, and profits by order size may bring out opportunities for new selling categories.

e. *Distribution channels.*   Classification of sales by distribution channels—wholesalers, retailers, and retail customers—may open up profit opportunities and areas for further investigation.

f. *Combinations.*   Combinations of the analyses outlined in items *a, b, c, d,* and *e* may give useful information. For example, sales, costs, and profits classified by both order size and geographical area might lead to the development of a new selling strategy.

These costs and profitability analyses could absorb a great deal of clerical time and perhaps cost more than they would be worth. Management of each firm must decide for itself which ones are worthwhile for its operation. Electronic data processing can play an important part. With proper programming, coding, and processing of the data, a manufacturer can have a vast amount of information and have it more promptly than by manual processing. The task is then that of using the information to plan production and sales strategies.

9.5.4 Communication of business financial status

The primary service of accumulating and analyzing cost-accounting information is to have a sound basis for management decisions. But the information is also important for financial reports and presentation to outsiders.

The financial reports most often issued to outsiders are the balance sheet and the income statement. These statements of financial conditions and the results of operations are of interest to creditors (bankers and suppliers) and to investors and prospective investors.

If the manufacturer takes government contracts, government or agencies will ask to audit the cost of production for various reasons—renegotiation of prices, establishing cost-plus prices, etc.

Because of the close interaction between internal cost accounting and external financial reports, careful attention should be given to the accounting systems and the supporting cost-accounting records.

9.6 Mathematical Methods for Planning and Control

The cost records described in the preceding sections provide the data for more advanced mathematical methods of analyzing and planning business operations. This section outlines briefly some that can be used by many manufacturers:

1. Dealing with uncertainty (probability)

2. Capital budgeting analysis

3. Inventory analysis (economical order quantities)

4. Linear programming

5. Project management

6. Queueing

7. Simulation

9.6.1 Dealing with uncertainty

Business management always involves uncertainty. The manager is never entirely sure what will happen in the future. This uncertainty is of special concern in preparing budgets, establishing standard costs, and analyzing budget variances, as well as other decision situations.

Uncertainty means that the actual events a manager must try to predict or evaluate may take on any value within a reasonable range of estimated values.

The most important concept in dealing with uncertainty is probability. There are two kinds of probability:

9.6.1.1 Objective probability. Objective probability is a measure of the relative frequency of occurrence of some past event. It can be illustrated by the example in Table 9.4.

TABLE 9.4   **Objective Probability and Expected Values**

(1) Hours per unit	(2) Units observed	(3) Expected probability	(4) Expected hours	(5) Expected cost at $5
1	100	0.1	0.1	$ 0.50
2	200	0.2	0.4	2.00
2½	400	0.4	1.0	5.00
3	300	0.3	0.9	4.50
Total	1000	1.00	2.4	$12.00

Assume that a manager observes production, counts the number of the units produced, and tabulates the count according to the number of direct-labor hours used for each unit. This has been accomplished in columns (1) and (2) of Table 9.4. Column (3) indicates the expected probability for each time classification—that is, the odds that each unit will require the number of direct-labor hours. It is computed by dividing the number of observations for each classification by the total number of observations. For example, of the 1000 units whose production was observed, 400 required 2½ direct-labor hours per unit. Thus, 2½ direct-labor hours were needed for 4 out of 10 or 0.40 of the units (400 divided by 1000).

The objective probabilities in column (3) are used to accumulate the expected direct-labor hours in column (4) [column (3) times column (1)]. The expected past figures in column (5) are found by multiplying the labor rate of $5.00 an hour by the expected hours in column (4). The totals in column (4) and (5) are expected values—the average direct-labor hours per unit and direct-labor cost per unit that can be expected on the basis of past production experience.

9.6.1.2 Subjective probability. Subjective probability is not based on observation of past events. It is a manager's estimate of the likelihood that certain events will occur in the future.

Assume that a sales budget is being prepared. The manager estimates that chances are even (0.5) that sales will be the same as last year—1000 units. The probability of selling 1100 units is estimated to be 0.25 and the probability of selling 1200 units, 0.15. There is a 1-in-10 chance, the manager estimates, of selling 1300 units.

The expected sales, based on these estimates of the future, is calculated in Table 9.5. The resulting figure for expected sales, 1085, can be used in the sales budget.

Notice that use of probabilities enables the manager to reduce a range of values to a single value. the manager can still retain for later reference the original data used to make the estimate. The use of probability measures, together with the data a manager must accu-

TABLE 9.5 Expected Sales

(1) Sales in units	(2) Subject probability	(3) Expected sales, (2) × (1)
1000	0.5	500
1100	0.25	275
1200	0.15	180
1300	0.10	130
	1.00	1085

mulate, helps to deal with the uncertainty that is characteristic of every business operation.

9.6.2 Capital budget

Capital is invested in a fixed asset only if the asset is expected to bring in enough profit to (1) recover the cost of the equipment and (2) provide a reasonable return on the investment.

The purpose of a capital budget analysis is to provide a sound basis for deciding whether the asset under consideration will in fact do this if it is purchased.

These elements must be considered in managing a capital budget decision:

1. The cost of the fixed asset
2. The expected net cash flow provided by use of the asset
3. The opportunity of investing funds in capital equipment
4. The present value

9.6.2.1 Cost of the fixed asset. This cost is usually the purchase price, transportation, and installation cost of the asset. However, buying the asset may mean that other investments will have to be made—additional inventories, for example. If so, these too should be included as part of the total investment outlay.

9.6.2.2 Expected net cash flow. The net cash flow may be either (1) the cash cost savings or (2) the increased sales revenue minus the increased cash costs due to using the new asset. No deduction for depreciation is made in calculating cash flow.

9.6.2.3 Opportunity cost. This is the return that the business could realize by investing the money in the best alternative investment. For example, if the best alternative is to place the money in a savings account that pays 5% annually, the opportunity cost is 5% on the amount of the investment.

9.6.2.4 Present value. Scientific capital budgeting is based on the concept of present value, for example, $1.00 put in a savings account at 5 percent interest will be worth $1.05 at the end of one year, $1.1025 at the end of the second year, and $1.1576 at the end of the third year.* Suppose an opportunity to invest some money with the

---

*First year: $1.00 + 0.05\ (\$1.00) = 1.05 \times \$1.00 = \$1.05$
Second year: $1.05\ (1.05 \times \$1.00) = 1.05^2 \times \$1.00 = \$1.1025$
Third: $1.05\ (1.05^2 \times \$1.00) = 1.05^3 \times \$1.00 = \$1.1576$

expectation of receiving $1.1576 (including the amount invested and the interest or other income) at the end of the third year. Investment must not be more than $1.00, because the present value of $1.1576 to be received 3 years hence is $1.00 (based on the best alternative investment at 5%).

This present value is found by discounting the expected future value by the opportunity cost rate. This is accomplished by reversing the process for finding a future value, as follows:

$$\$1.1576 \div 1.05^3 = \$1.00$$

or, in general terms,

Present value = value n years hence $\div (1 + \text{discount rate})^n$

Table 9.6 shows another way of applying the present-value concept to capital expenditure decisions. Consider an investment of $6.50 in an asset that would yield the amounts shown in Table 9.6 if placed in a savings account at 5 percent. At the end of 5 years, the amounts received from the asset investment will total $7.00 (including salvage value of the asset), whereas the savings account will be worth $6.50 × 1.05^3, or $7.52. Therefore, the $6.50 should not be invested in the asset.

The present-value analysis in column (4) of Table 9.6 shows that the present value of the asset investment is $6.30, which is 20 cents less than the amount that would be invested. In other words, the value of the investment opportunity is minus 20 cents. If the amount of the investment were $6.29 or less and brought in the same amount, the present value of the investment opportunity would be positive. The general rule is that investment opportunities with a positive present value should be taken.

Example.   Assume you have an opportunity to buy a piece of equipment for $5000. With this equipment, products can be manufactured with the revenue and cost characteristics described in Table 9.7.

The present value of the cash flows in Table 9.7, computed as in Table 9.6, is $6300. The value of the investment in the opportunity is therefore $6300 −

TABLE 9.6   Application of Present Values to Capital Investment

(1) End of year	(2) Amount to be received	(3) Discount factor	(4) Present value
1	$2.00	1.05	$1.90
2	2.00	$(1.05)^2$	1.81
3	3.00	$(1.05)^3$	2.59
Total	$7.00		$6.30

TABLE 9.7   Revenues and Cost Characteristics

	Year 1	Year 2	Year 3
Sales revenue	$10,000	$11,000	$12,000
Less: variable costs	6,000	7,000	8,000
Less: fixed costs*	2,000	2,000	2,000
Cash flow from production	$ 2,000	$ 2,000	$ 2,000
Sales of equipment	0	0	1,000
Net cash flow	$ 2,000	$ 2,000	$ 3,000

*Not including depreciation charges.

$5000, or $1300. This positive investment opportunity value shows that the equipment investment will have a higher rate of return than the 5 percent available on the next best investment (assumed to be a savings account paying 5 percent). The depreciation is not considered a cash flow item and thus should not be included as a cash cost. The example illustrated does not include depreciation charges. In addition, these figures do not include income tax payments (which are a cash flow item).

However, the cash flow figures in Table 9.8 do take taxes into account. Once taxes are included, depreciation should be considered as well, since depreciation is tax deductible.

This calculation may also be set up as follows:

Net cash flow (ignoring taxes)	$2000	$2000	$3000
Less: Income taxes—as above	80	80	80
Net cash flow	$1920	$1920	$2920

The present value is $6092, rather than the $6300 calculated from the earlier figure, which did not include taxes. The value of the investment opportunity, after income taxes, is $1092, which is still higher than what could be earned from a savings account carrying a 5 percent interest rate.

TABLE 9.8   Cash Flow

	Year 1	Year 2	Year 3
Sales revenue	$10,000	$11,000	$12,000
Less: variable production costs	$ 6,000	$ 7,000	$ 8,000
Less: fixed production costs (including depreciation)	3,833	3,833	3,834
Net profit before taxes	167	167	166
Less: income taxes @ 48%	80	80	80
Net profit after taxes	87	87	86
Cash flow from sales of equipment (add back depreciation)	$ 1,833	$1,833	$1,834
Net cash flow	$ 1,920	$1,920	$1,920

The present-value method of analyzing a capital investment opportunity makes it possible to take into account the timing of the expected cash returns and to compare them with those of other investment opportunities. Note that the cash returns are expected values. The probability method discussed earlier is used in calculating them.

9.6.3 Inventory analysis

Maintaining adequate inventories of raw materials and parts can tie up capital for a manufacturer. Careful management of inventory levels offers possibilities for large cost savings and the release of funds from inventory investment. Two major kinds of costs are associated with maintaining inventories: holding costs and ordering costs.

Holding costs are costs associated with inventory investment. They include insurance, taxes, rent on warehouse space, and the opportunity cost of alternative uses of the funds invested in inventory.

Ordering costs are the costs of processing purchase requisitions and vendors' invoices and of receiving the goods in the warehouse (receiving department salaries, etc.).

The *economic order quantity* (EOQ) is the quantity of goods that should be ordered at one time to ensure the lowest total inventory costs (holding cost plus ordering cost). There is a tradeoff between holding costs and ordering costs. When the order quantity is small and the inventory is kept low, the holding costs are also low. The ordering costs, however, are high, because orders must be placed often and more clerical and receiving department time is needed. When the order quantities are large and inventory is kept high, the holding costs are high (more insurance, taxes, rent), but the ordering cost is low—fewer requisitions have to be processed, and receiving operations are less frequent. This cost tradeoff is shown graphically in Fig. 9.4.

The basic data for the inventory problem are the holding costs, the ordering costs, and the expected demand for raw materials and purchased parts. Other factors that may enter into purchasing decisions are the lead time (the length of time between purchase order and receipt of the materials or parts) and the availability of quantity discounts.

9.6.4 Linear programming

Linear programming is a mathematical method for making the best possible allocation of limited resources (labor hours, machine hours, materials, etc.). It can help management decide how to use its production facilities most profitably. Suppose, for instance, that the firm produces more than one product, each with a different contribution rate. Management needs to know what combination of quantities produced, given the limitations of its facilities, will bring the highest profit.

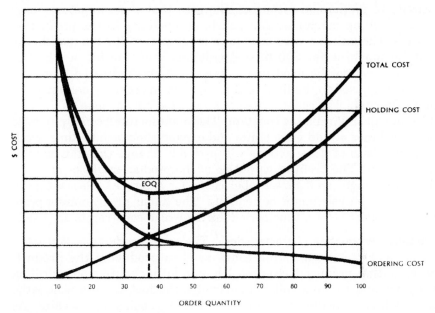

Figure 9.4 Economical order quantity.

Three principal concepts must be considered in seeking a solution to the problem:

1. The profit function
2. The constraints of the problem
3. The production characteristics

9.6.4.1 The profit function. The profit function is a mathematical expression used to show how the profit will vary when different quantities of product 1 and product 2 are produced. It requires knowledge of the contribution each product makes toward overhead and profit.

If the symbols x_1 and x_2 stand for the number of each product to be produced, and if product 1 makes a profit contribution of $2 per unit and product 2's profit contribution is $5 per unit, the profit function is $2x_1 + 5x_2$. The problem is to find the values of x_1 and x_2 that will yield the highest total value (profit), given the constraints of the problem and the production characteristics.

9.6.4.2 The constraints. Sometimes the profit-maximizing solution is obvious—simply produce as many product units as possible with the resources available. However, the manufacturing process may have characteristics that prevent the use of resources in this way.

The constraints of this problem are the limitations imposed by the scarcity of production resources. Suppose that a manufacturer uses lathes, drilling machines, and polishing machines in the production process. The lathes can be used for a maximum of 400 h a month, the drilling machine for 300 h, the polishing machine for 500 h. The machines cannot be used more than the hours indicated, but they need not be used to the limit of this capacity in order to maximize the profit function. The constraints are applicable only as long as the production capacity remains constant. The manufacturer could purchase new machines or add work shifts and expand the machine time available. But both of these actions will probably change the statement of the problem. The profit contribution of the product might be changed (by higher night-shift direct-labor costs, for example). Also, a capital budgeting analysis might be required. The linear programming problem is applicable for a time period during which constraints are fixed.

9.6.4.3 Production characteristics. The term *production characteristics* refers to the machine times used in producing the product. Suppose that product 1 requires 1 h on a lathe, no time on a drilling machine, and 1 h on a polishing machine. Product 2 requires no lathe time, 1 h on a drilling machine, and 1 h on a polishing machine. As stated above, the lathe can be used for a maximum of 400 h a month, the drilling machine for 300 h, and the polishing machine for 500 h. The production characteristics would then be expressed as follows:

$x_1 \leq 400$ — No more than 400 lathe hours are available for product 1, and 1 h is required for each unit. (If 2 h per unit were required, the expression would be $2x_1 \leq 400$.)

$x_2 \leq 300$ — No more than 300 drill hours are available for product 2, and 1 h per unit is required.

$x_1 + x_2 \leq 500$ — No more than 500 polishing hours are available for both products, with 1 hour per unit required in both cases.

Solving the problem. The problem for the values given above is as follows:

Maximize:

$$2x_1 + 5x_2$$

Subject to:

$$x_1 \leq 400$$

$$x_2 \leq 300$$

$$x_1 + x_2 \leq 500$$

There are many solutions that would satisfy all three constraints; there is only one maximizing solution. The possible solutions are shown graphically in Fig. 9.5. The axes represent the number of product units; the remaining solid lines, the production characteristics. Any point in the shaded area will satisfy the constraints, but the maximizing solution will be at one of the four corners of this area— point A, B, C, or D.

Which point is the exact one can be found by drawing a line that represents the profit function $2x_1 + 5x_2$ (the lower dotted line in Fig. 9.5). When this line is shifted upward, always parallel to its original position, it finally reaches the point where it touches the shaded solution space at only one point, B. This point, where $x_1 = 200$ and $x_2 = 300$, is the solution that gives the highest possible total contribution within the limited resource constraints. The solution gives the following results:

Total contribution	$1900 (200 × $2 + 300 × $5)
Lathe time used	200 h (200 × 1 h)
Drill time used	300 h (300 × 1 h)
Polishing time used	500 h (200 × 1 h + 300 × 1 h)

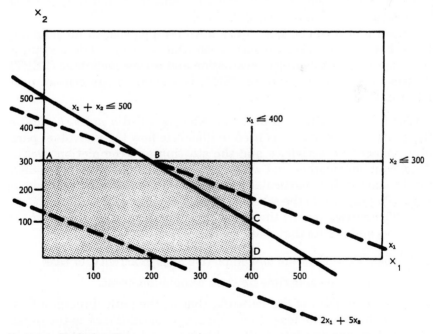

Figure 9.5 Profit function.

There are 200 idle lathe hours, but these cannot be used under existing conditions without reducing the output of product 2. If more lathe time is used for product 1, then more polishing time must also be used; and the polishing time must be taken away from product 2, thus reducing product 2 output. Any reduction in product 2 output represents a loss of $5 per unit, whereas producing additional units of product 1 will add only $2 per unit. The net loss from such a substitution is therefore $3 per unit. On the other hand, the output of product 2 cannot be increased, since there is no more drill time available.

The graphic solution described above is a rough approach to a fairly simple problem. More complex problems can be programmed into a computerized automated system.

9.6.5   Project management

Manufacturing and construction firms may undertake one-time-only jobs, such as the construction of a building or the design and manufacture of a special piece of a large assembly system. In such projects, at any one time much of the company's time, resources, and prospects may be tied to single large job. And whether the job will result in profit or loss for the company may depend on its being completed on time.

If the completion date is to be met, the various segments of the job must be coordinated. Subunits must be scheduled, and they must be finished according to the schedule date.

Project management techniques have been developed for planning the subunits that make up such a job. One of the most important of these methods is the project evaluation and review technique (PERT). Another one, very similar to PERT, is known as the critical path method (CPM).

9.6.5.1   PERT.   PERT is based on what is called a *network plan*. A flowchart—the network—is used to illustrate how the individual parts of a project (the activities and the starting and completion events) depend on one another and which task must be finished before others can be started. It is particularly concerned with dovetailing individual parts of a project into the schedule of the project as a whole.

PERT analysis makes the following estimates from data about the activities and events that make up the project:

1. Earliest expected time for completion of each activity, or task.

2. The latest allowable time for each completion event.

3. The critical path of the network, that is, the path through a flowchart that has the least total amount of slack. (*Slack* is the difference between the latest allowable time and the expected time of completion.)

4. The probabilities that events will occur on a schedule.

A simplified PERT network is shown in Fig. 9.6. The same information is shown as a bar chart in Fig. 9.7. (The bar chart brings out the pattern of activities and slack time clearly, but would be unwieldy for use in a more complex network.)

The circled numbers in Fig. 9.6 represent events—the start or completion of an activity such as excavation for a building or erection of the steelwork. The lines represent work activities extending over time and leading from one event to another. Other symbols used in the figure and discussion have the following meanings:

t_e = expected time
T_E = sum of the expected times up to a given event
T_L = latest allowable time (cumulative)
v = variance

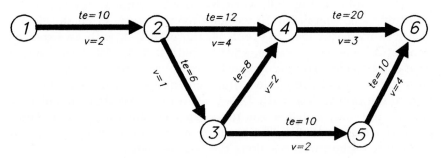

Figure 9.6 Simplified PERT network.

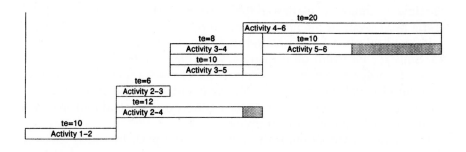

Weeks

Figure 9.7 PERT network as a bar chart.

The values for t_e and v assumed below are those in the illustration.

If o1 represents the beginning of excavation for a building, o2 completion of excavation, and o3 completion of pouring a foundation, then the line from o1 to o2 represents the activity of excavation and the line from o2 to o3 represents the activity of pouring a foundation. The activity of steelwork erection leading to completion event o4 cannot begin until after both the preceding completions, o2 and o3, have occurred. Other intermediate completion events must take place before any entire project is finally completed, event o6.

Time estimates must be made for each of the activities. Since actual activity time is uncertain, probability calculations are used in computing the expected time t_e and the variance v associated with the expected time.

The problem for project management is to determine the expected time for completion of the entire project and to identify the activities that could cause bottlenecks. PERT enables the management to accomplish this and thus schedule work activities for the earliest possible completion.

For example, in Fig. 9.6, work leading to completion event o2 is expected to be completed 10 weeks after the start. From this point, two network paths lead to event o4: the path 2–4, with an expected completion time of 12 weeks, and the path 2–3–4, with a completion time of 14 weeks. The total expected completion time for event o4 is larger of the total completion times on the two network paths that lead to the events. The path 1–2–4 has a total expected time of $T_E =$ 22, and the path 1–2–3–4 has total expected time of $T_E = 24$. Thus, the T_E for event o4 is 24. The variance associated with $T_E = 24$ is $v = 5$, the sum of the variances along the 1–2–3–4 path.

Thus, the manager can expect completion event o2 in 10 weeks, event o4 in 24 weeks, and event o6 in 44 weeks after work is started. One other time variable must be taken into account, however: the latest time T_L that can be allowed for an event without disturbing the T_E of the final event in the network (event o6).

If event o6 must be completed in 44 weeks, then T_L for event o6 is 44. Since T_E for event o6 is also 44, the allowable lag time is zero ($T_L - T_E$). In order to complete event o6 in 44 weeks, event o4 must be completed in 24 weeks. Thus, T_L for event o4 is 24 and allowable slack time is also zero. T_L and slack time can be calculated for each event by working backward from event o6.

The critical path through the network—the path in which there is the least amount of slack—is identified by noting the events where allowable slack time is zero. In Fig. 9.6, the critical path lies through events 1–2–4–6. Any construction delay in these events will increase the total completion time for the project. Event o5, however, is not on the critical path. The tasks on the path segment 3–5–6 could be delayed for as much as 8 weeks in all without affecting the final completion time for event o6. The activities represented by the path segment 2–4 could be delayed 2 weeks without delaying the final completion of the project. Resources could, if necessary, be shifted from these activities to other tasks to ensure completion of the entire project on schedule.

Note that T_E and T_L are expected values. There are statistical variances associated with each of these measures. The expected time estimates and variance estimates are used in calculating the probabilities that events will occur on schedule.

9.6.5.2 CPM. The critical path method of project management is closely related to PERT. Both PERT and CPM involve determining expected times of completion for individual events and for the entire project. PERT goes further to include variance analysis; CPM does not. CPM, on the other hand goes beyond PERT in another direction. It uses cost data to assess the financial effects of setting up crash programs in the network's critical path segments to ensure completion on schedule.

9.6.5.3 PERT/cost. PERT can be augmented with cost data to facilitate financial planning for large projects. The cost data are budgeted according to work activity and completion-event classifications. Actual costs should be accumulated according to the same classifications so that they can be compared with the budget figures.

9.6.6 Queueing or waiting line analysis

A queueing or waiting-line situation exists when there is a service center of some kind that receives customers, processes their work, and sends them on their way. Three types of waiting lines often operate in business. One is seen in a doctor's office where arriving patients form one line and are served through only one station—the doctor. Another is seen in a barber shop where people form one line and are served through several stations—any of the barbers. A third type of queueing is in a supermarket where customers form many lines and are served through many stations—the checkout counters.

A waiting line can occur in a manufacturing plant when product units—"customers"—arrive at a processing department or work area for completion and transfer to the next department. A manufacturing queueing situation is illustrated in Fig. 9.8.

The arrival of raw materials at processing department 1 can be controlled by the timing of materials requisitions, so it is unlikely that a waiting line of materials units will form at department 1. However, the processing or service rate in department 1 determines the timing

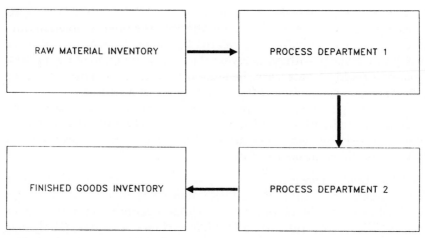

Figure 9.8 Manufacturing queueing situation.

of the release of the semifinished units to department 2.

As long as department 2 can accept the units for processing, no waiting line will form. Units will be processed as they arrive at the work area. However, if department 2 is busy, a bottleneck will occur, and semifinished units will form a backlog. The time spent in line carries a cost to the manufacturer similar to the holding cost of inventory. (Note that the backlog of semifinished products is a part of the work-in-process inventory.) The queue, and therefore, the waiting cost, can be reduced by speeding up the processing or service rate in department 2.

The processing or servicing of additional units, however, adds to the cost of the manufacturing process. A move to accelerate processing in department 2 and shorten the queue may add costs beyond normal production costs for materials, labor, and overhead. This would be the case, for instance, if it were necessary to add a night shift or new equipment. Thus, there is a tradeoff between waiting costs and some processing costs similar to the EOQ tradeoff between holding and ordering costs. Queueing analysis takes into consideration the following factors:

1. Arrival rate

2. Service rate

3. Queue lengths

4. Utilization of the service facility

5. Total service time for produced units

6. The cost of waiting

7. The cost of servicing

These variables are combined in such a way as to provide data on the efficiency of the present service and the economic facility of adding service units.

9.6.7 Simulation

Simulation is a method used to represent a real situation in an analytical framework. Verbal descriptions, diagrams, flowcharts, and systems of equations are all forms of simulation. A budget is the result of simulating the financial results of operations, based on assumptions about volume, cost relationships, and planned production activities.

Simulation allows management to try out various actions without risking its resources in the real marketplace. With simulation, management can calculate events on the basis of alternative assumptions on paper and then attempt to select the best course of action for the real application.

In the terminology of modern management science, *simulation* refers to a computerized model of a company or some segment of it. The model includes sets of mathematical equations that quantify volume-cost-profit relationships. The computer can accept changes in the relationships and calculate the effects of the changes on operations. Suppose, for example, that management is contemplating substituting a new raw material that will reduce materials requirements but increase labor. Management can program the changes into a computerized simulation and learn what the overall effect will be on manufacturing operations.

Simulation is a powerful management tool, but it is complicated by the fact that the operating characteristics have to be specified in great detail. Nevertheless, it can be used to great advantage to test and examine alternative courses of action without having to put them into actual practice. The volume, cost, profit relationships, and physical production characteristics described in this chapter can provide much of the data required for simulation models.

9.7 Where Do Sensors and Control Systems Take Us?

Currently, most industries employ equipment with limited intelligence, but in the near future, the advancing level of adaptive sensors and control systems and of computer technology will rapidly increase the use of "smart" equipment. On the other hand, as computer integrated manufacturing strategies are introduced on a wide scale and as technology moves forward, the size of the work force involved in engineering will be reduced dramatically, while the pattern of working seems destined to change completely (Fig. 9.9).

Already sensors and control systems play a significant role in manufacturing. An example is tool wear and breakage monitored by sensors using the torque or power at the spindle. Macotech Machine Tool Co. Inc., Seattle, Washington, has taken that concept a step further to

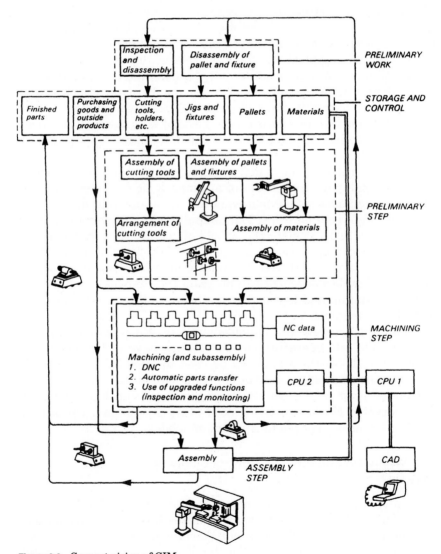

Figure 9.9 Current vision of CIM.

adjust the feed rate and torque to obtain the optimum cutting rate whatever the conditions. As a result, cutting times have in many cases been reduced by 50 percent, while the need to program the tool in detail is eliminated; sensors and control systems no longer need to be told the material, cutting speed, and feed rate.

In one case, the concept of employing sensors and control systems in manufacturing has been applied to deep drilling, where a constant load is applied. The feed rate changes automatically to maintain the constant load on the cutting tool, but if as a result of the buildup of swarf the feed rate is reduced to preset level, the drill is withdrawn to clear the swarf, and then starts again. Not only is the speed optimized through sensors and control systems, but as a result, drill wear is more consistent, and tools do not break.

The aim in increasing the level of intelligence of machine or assembly controls is to reduce the need for attention and programming. The extra "senses" needed are vision sensors, force sensors, optical sensors, and metrological sensors. With a vision system, the controller of the robot or machine will be able to recognize the workpiece, determines its precise position, and instruct the machine to carry out the necessary operations. Thus, it will not be necessary to load a workpiece precisely on a pallet or install limit switches and other sensors to identify the workpiece and check its position. Since location need not be precise, sensors in robot loading will be easier to arrange. The force sensors will optimize the cutting speed whatever the material, and for a robot, will ensure that a component is installed with the correct amount of force. The use of such sensors and control systems will allow planning of manufacturing operations to be automated, with such programming as is needed done off-line.

A lathe or milling machine would be equipped with adaptive sensors and control systems on the spindle, to adjust the speed for turning, drilling, tapping, and milling. In that case, the vision sensors would be used to identify the workpiece for the robot. Lathes and milling machines can already be programmed off-line, and with vision sensors, the robot could be programmed in the same way, with the sensors adjusting the position and size of the gripper or changing the gripper to suit the workpiece. Also, the controller could assess whether a new set of chuck jaws is needed, and if so, instruct the robot to change it. Thus, without the need for in-line programming and adaptive control the lathe or milling machine would be able to machine a wide variety of components completely unmanned.

Just as the productivity of lathes and milling machines has improved, quantitatively and qualitatively, so other machines can be improved through sensors and control systems. For example, several machining centers have an extra attachment to allow machining of the fifth face, whereas some machines made by Mandelli Machine Tool, Italy, can machine the fifth face without the need for an attachment. This is a worthwhile development, since it reduces downtime. With the introduction of modular universal machining centers for turning, milling, boring, and grinding operations, manufacturers will encourage more moves in this direction. In some cases, handling equipment to permit the fifth face to be machined without the need for a full resetting operation is also likely to come into use, as are the laser welding and laser sensor measurement systems on the machine tool itself.

But the major developments will continue to come in sensors and control systems, computers, and software. In the next stage, the computer at each flexible manufacturing system might have access to an "expert" system, which is in effect a huge database. Included in the database are data concerning all the components to be produced—input from the CAD system—such as machining methods for different speeds, surface finishes, and tolerances. In fact, the database will be a complete encyclopedia of machining. This will be updated continually and directly from the CAD system and from a remote database on machining (Fig. 9.10).

Once sensors and control systems are perfected, this will allow machining to be performed without programming. When the workpiece is fed into the system, a vision sensor will identify it in detail. Then the data will be fed back to the database and compared with data for the workpieces that are to be machined, and full dimensional data on the workpiece will be accessed. The vision sensor will then scan the workpiece, and its dimensions will be compared with those of the finished part in the parts library.

The data will then be fed back to the controller, which will instruct the machine how to start machining, by reference to the database and not to a specific program. When necessary, the workpiece will be transferred to other places for automatic heat treatment or grinding, and later to assembly. In this way, machining without programming and without manual intervention will be practical. Lathes are unlikely to need such a control system, and instead, the database would contain a vast number of programs, the best one being selected automatically for the shape of the component to be produced. Of course, these advanced systems in controls and sensors would not be needed to produce all components, nor would they be installed in all flexible manufacturing systems. Lower-level machines would still continue in use for more routine, medium-volume parts. However, in planning for flexible manufacturing systems with computer-integrated manufacturing strategies, managers need to take this sort of development into account.

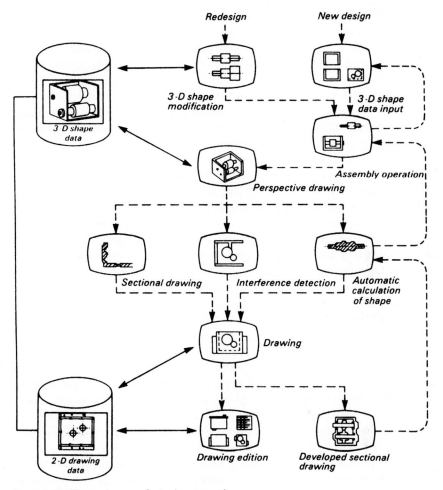

Figure 9.10 Dynamic manufacturing operation.

In assembly, a similar sensor and control system could come into use, except that there will be more stages, because of the need to provide the right group of components in suitable positions to be assembled. One handling robot, on a trolley, might be used to supply components to the assembly cell, where some robots and standard units, such as presses and CNC nut runners, would carry out assembly. With the combination of these expert systems and sensors and control systems, flexible manufacturing systems will be able to machine or assemble in batches of one with the efficiency expected in high-volume production runs. Thus, the real aims of flexible manufacturing systems—high productivity in small batches, produced within a very short time after the order is placed—will be achieved.

The other major development is in computer systems for functions from purchasing through delivery. All the systems will be connected together, and will perhaps be incorporated in one computer (which will have voice recognition and voice response), so that for most jobs the keyboard will be unnecessary. Indeed, by the end of this decade, computers with conventional languages and vocal human/machine interfaces should be available. Anyone will be able to program and make the most complex transactions using normal conversation, so the problem of computer illiteracy will disappear. People will also have access to databases of products, materials, and stock and to expert systems on costing and other relevant subjects.

In the case of production control and order processing, this means that, once an order is accepted, the data will be fed to the CAD system, where the designer will produce a new design, modify a design, or merely call up an existing design. Planning will be automatic; from the data generated, the schedule will be produced and components and materials ordered. On arrival, they will be fed through the system to be packed and delivered with hardly any manual intervention. Needless to say, the materials will not wait around in a storeroom, but will come straight off the delivery truck into the system to be processed.

Overall, therefore, we are at the beginning of an era which will lead to the end of the factory as we know it. Some machines will be able to carry out simple routine maintenance themselves through advanced sensory and control systems. Very complex robots with a number of sensors will be available to carry out repairs, so that, in theory, operators will be needed only to work with the software—in designing and updating the databases and expert systems. Both the machines and robots will be extremely complex, so the operators who are responsible for their installation, let alone maintenance, will probably not be able to understand the complete system; that is already the situation with large-scale computers.

Some engineers expect to continue to rely on people to carry out some jobs. They argue that there is little point in taking every one out of the plants just for the sake of doing so, and in any case, the devices that can do these jobs are likely to be very expensive. In addition, even with all these elaborate sensors, there is possibility of some catastrophic failure, and it will be a very long time before managers are prepared to go home at night and leave a plant completely unattended. In any case, working conditions will be very good, and working hours will be short, so there is every reason to employ some people in the plants.

However, other experts say that there is no point in thinking in terms of staffing the plant at all, except in an external supervisory role. They say that, because the machines will be quite complex, maintenance fitters will not know how to rectify faults. Then, because the sensors and control systems will be so advanced, the machine will in any case be better able to trace faults and put them right, so rectification by humans may not be helpful at all. Clearly, that stage is still a long way away, but managers need to be ready for it.

But just how many people will be needed in manufacturing by the year 2000? Some experts estimate that the work force in manufacturing will be cut by 60 to 75 percent in the next 10 years, while the number of people actually employed on the shop floor is likely to drop by about 90 percent. The effect of a 70 percent reduction in the work forces of the major industrialized nations is shown in Table 9.9.

Although those figures appear alarming, they reflect similar trends that have taken place in agriculture, process industries, automotive industries, and computer industries. However, those changes took place far more slowly, and the rate of change is bound to cause considerable problems to politicians and industrialists alike. With the prospect of 20,000,000 jobs being lost in the United States, 13,000,000 in Japan, 5,000,000 each in France and Italy, and 3,500,000 in the United Kingdom over a 10-year period, people are bound to be alarmed. Figure 9.11 shows the expected relationship between manufacturing technology advances and manufacturing employment.

In this period, though, many new products and systems will go into production, and, in theory, with the reduction in prices that the increased productivity will bring, demand worldwide should increase. Equally, it is clear that the development and maintenance of all this software will provide a large number of skilled jobs. In fact, there is likely to be a shortage of skills in many areas, which will have the effect of retarding development of new systems.

Clearly, the nations that train people in the relevant skills will have a head start. In any case, the changes in the number of employees are based on the assumption that each nation maintains its current share of the market for manufactured goods. In practice, this is unlikely, since some countries are more competitive than others,

TABLE 9.9   Worldwide Work Force Reduction

Country	Current industrial work force	Industrial work force reduced by 70%
United States	30,000,000	9,000,000
Japan	19,500,000	5,800,000
West Germany	11,500,000	3,400,000
Italy	7,700,000	2,300,000
France	7,500,000	2,250,000
United Kingdom	5,300,000	1,600,000
Sweden	1,350,000	400,000

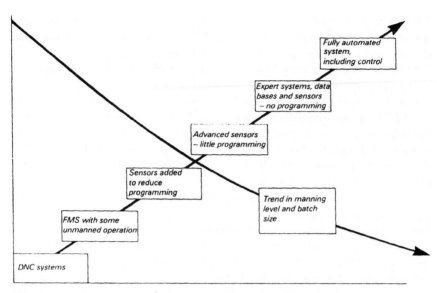

Figure 9.11 Role of sensors in moving manufacturing technology forward to the year 2000.

while the newly industrialized nations will also take some of that business.

Although it is imperative that people are educated in the right skills, that is not enough; the whole way in which people are educated and think of work needs to be changed. The introduction of computer-integrated manufacturing strategies and computerized sensors and control systems in manufacturing, transport, and perhaps some other services will lead to the need for people to work fewer hours. It is likely that a three-day week will appear quite soon, both to spread work around and to ensure that people are prepared to work the unsocial hours that are occasionally needed.

But what will people do when they are not working? Another worry is that, as we move from active to supervisory roles, we may tend to lose skills in dexterity and become less active. The education level needs to be raised, so that people can cope with the jobs available in this new society, while education also needs to be expanded so that people are better equipped to face life with more time on their hands and perhaps do more for themselves in maintaining their homes and their possessions.

Perhaps many people will have more than one job, one in a company for 3 days a week, and another outside for 2 days. Of course, others will not want much work at all, and they will probably be able to fulfill that wish. But none of those futuristic notions will be possible unless a great deal of value is being added in manufacturing or business.

We are entering an exciting era of significant opportunities, opened up by sensors, microelectronics control systems, and computer-integrated manufacturing strategies. But the competition will be intense, and only those that take big chances now will reap the benefits. As a psalmist once said, "They that sow in tears shall reap in joy. He that goeth forth and weepeth, bearing precious seed, shall doubtless come again rejoicing, bringing his sheaves with him." (Psalms 126. verses 5–6, *The Bible.*)

Further Reading

Ansoff, H. I., "Planning as a Practical Management Tool," *Financial Executive,* **32:** 34–37 (June 1984).

Bacon, J., "Management of the Budget Function," Studies in Business Policy, report no. 131, National Industrial Conference Board, New York, 1990.

Chambers, J. C., S. K. Mullick, and D. D. Smith, "How to Choose the Right Forecasting Technique," *Harvard Business Review,* **49:** 45–74 (July–August 1991).

Gentry, J. A., and S. A. Pyherr, "Stimulating an EPS Growth Model," *Financial Management,* **2:** 68–75 (Summer 1983).

Gershefski, G. W., "Building a Corporate Financial Model," *Harvard Business Review,* **47:** 61–72 (July–August 1989).

Gordon, M. J., and G. Shillinglaw, *Accounting: A Management Approach,* 4th ed., Chap. 16, Homewood, Ill., 1969.

Hamermesh, R. G., "Responding to Divisional Profit Crises," *Harvard Business Review,* **55:** 124–130 (March–April 1977).

Meyer, S. C., and G. A. Pogue, "A Programming Approach to Corporate Financial Management," *Journal of Finance,* **29:** 579–599 (May 1974).

Pappas, J. L., and G. P. Huber, "Probabilistic Short-Term Financial Planning," *Financial Management,* **2:** 36–44 (Autumn 1987).

Parker, G. C., and E. L. Segura, "How to Set a Better Forecast," *Harvard Business Review,* **49:** 99–109 (March–April 1991).

Smith, G., and W. Brainard, "The Value of a priori Information in Estimating a Financial Model," *Journal of Finance,* **31:** 1299–1322 (December 1976).

Soloman, S., *Modern Welding Technology,* TAB Books, Blue Ridge, Pa., 1982.

Wagle, B., "The Use of Models for Environmental Forecasting and Corporate Planning," *Operational Research Quarterly,* **22** (3):327–336.

Warren, J. M., and J. P. Shelton, "A Simultaneous Equation Approach to Financial Planning," *Journal of Finance,* **26:** 1123–1142 (December 1991).

Weston, J. F., "Forecasting Financial Requirements," *Accounting Review,* **33:** 427–440 (July 1958).

Index

ABOUT THE AUTHOR

Sabrie Soloman is manager of advanced manufacturing technology at Pfizer Pharmaceutical Inc., as well as professor of advanced manufacturing technology at Columbia University, where he lectures on Sensors and Control Systems in Manufacturing, Affordable Automation, Computer-Integrated Manufacturing, Flexible Manufacturing Systems, and Product Design for Manufacturability. Dr. Soloman holds numerous patents and is also the author of *An Introduction to Electromechanical Engineering* and *Modern Welding Technology*. He is a fellow of the Society of Manufacturing Engineers, the Royal Society of Mechanical Engineers (England), and L'Ordre Des Ingenieurs Du Quebec (Canada). He received several awards from the American Society for Mechanical Engineers (ASME), the Society for Manufacturing Engineers (SME), and the American Management Association (AMA). Dr. Soloman is considered an international authority on advanced manufacturing technology and automation in the microelectronic and automotive industries. He has been and continues to be instrumental in developing and implementing several industrial modernization programs through the United Nations to various European and African countries. He is the first to introduce and implement unmanned flexible synchronous/asynchronous manufacturing systems to the microelectronic industry and the first to incorporate advanced vision technology in a wide array of robot manipulators. Dr. Soloman was selected to deliver the closing address, "Innovative Sensors," at the Universal Design Conference, New York, May 14, 1992.